NAVAL RADAR

Norman Friedman

NAVAL RADAR

Drawings by John Roberts

Conway Maritime Press
Greenwich

First published in Great Britain 1981
by Conway Maritime Press Ltd,
2 Nelson Road, Greenwich,
London SE10 9JB

ISBN 0 85177 238 2

Designed by Richard Johnson

Printed in Great Britain by
R J Acford, Chichester

Contents

Acknowledgements

This book reflects the assistance of many friends, such as A D Baker III; David Brown of the Naval Historical Branch (London); David Lyon of the National Maritime Museum, Greenwich; Giorgio Giorgerini; Chuck Haberlein; John Lewis; Norman Polmar (who disclaims responsibility for the result); Antony Preston; Alan Raven; Robert Scheina; Larry Sowinski; and Robert Sumrall. I am extremely grateful to Jim Glander of the US Naval Sea Systems Command for his assistance in identifying many obscure systems; he also supplied data on many declassified ones. The lists of US sets could not be nearly as complete as they are without his assistance. Robert Hill of NAVSEA commented on the theoretical portion of the text and clarified many points of great importance in current radar practice; I hope that I have not disappointed him by the errors which must inevitably remain. Preston E Law of NAVSEA and Tom Walkowiak of The Floating Drydock provided many unique (and labeled) photos of the antenna arrays of modern US warships; I am also indebted to that great photographer, Giorgio Arra. Dr Dean Allard and the staff of the Classified Operational Archives supplied considerable information on wartime US naval radars, supplemented by the draft US Radar Survey of 1945 supplied by the Polar Archives Branch of the US National Archives; I am also grateful to the Navy and Old Army Branch of the National Archives, for access to wartime attaché material on British radars. Tom Greene of RCA and Don Hall of Raytheon were extremely helpful, and I would like to thank public relations departments of Cardion Electronics, Grumman, Lockheed Electronics, ITT-Gilfillan, Hughes, Raytheon, RCA, Sperry and Thomson-CSF for photographs and other material; uncredited photos are US Navy Official.

I hope that the finished product justifies the labors of all who helped me; I am of course responsible for whatever errors remain in it. My wife Rhea deserves special thanks for her patient assistance during the lengthy preparation of this manuscript, particularly including assistance in obtaining some of the photographs.

Norman Friedman

In memory of my father, Leo Friedman

PART 1
Theory, Function, Performance

1. The Basis of Radar Systems

Perhaps the greatest *visual* difference between the warships of 1940 and those of 1945 was the growth of a forest of electronics, most notably radar – and the forest has continued to grow ever since. Although the basis of radar is well known, the very rich variety of antennas and systems in service is due to a series of complications less widely understood. Yet an appreciation of these complexities provides a valuable link between the many designs visible on warships and the very considerable technical ingenuity underlying them.

The present volume attempts to describe in some detail the array of naval radars from their inception in 1937 to the present, and to examine some of the ideas behind their designs and also behind the way in which they have been mounted aboard warships. Some major pieces of non-radar electronics will also be described and illustrated.

Nearly all radars operate by sending out a stream of pulses made up of short radio waves; the pulses reflect off objects, and the radar detects them on their return. Since radio waves travel through the atmosphere at a constant and very high speed, the short time interval between pulse and reception measures the distance to the radar target. The radar system, then, consists of a pulsing sources of electrical energy coupled to an antenna which emits corresponding radio pulses in a more or less well-defined beam; and a receiver to detect the returning echo and use it to operate some kind of display device. Each of these simple descriptions hides enormous technical problems, only some of the solutions to which will be touched on here.

FUNDAMENTAL REQUIREMENTS

All of the major radar functions can be reduced to the determination of the *range* and *bearing* of an unknown target. The latter is very much a matter of how sharply the sending and receiving antennas shape radar beams. At least one of the two must be highly directional, otherwise the radar cannot know from which direction an echo originates. In fact, of course, modern radars use the same antenna for both sending and receiving. This was no mean achievement, as we shall see; as late as 1943, for example, British naval radars required separate antennas.

The measurement of the *range* to a target is a matter of precisely measuring time. Radio waves travel at 328yds every microsecond, *ie* every millionth of a second. Since the time in question is the time between transmission and reception, the radar pulse must make the journey to and from its target, which means that every microsecond represents 164yds of range. It seems remarkable in retrospect that the accurate measurement of such short times was one of the least difficult problems in radar development.

Indeed, in most radars the time is not measured explicitly. Instead, the outgoing and incoming pulses are displayed on a cathode-ray tube, very much like a television tube. In the simplest radar displays ('A-scopes'), the beam of electrons which 'writes' on the face of the tube sweeps across at a constant (very high) rate, and is controlled by the radar receiver, so that a 'blip' appears both at transmission and, a short distance later on the face of the tube, on reception. The face of the tube lights up where the beam touches it, and it remains bright long enough for an observer to watch. Moreover, the radar produces pulses at a very high rate, so that what an observer really sees is the result of many hundreds or thousands of pulses superimposed. The accuracy of measurement then depends of how well the radar can reproduce the short time interval between pulses, which is a very different matter from making a clock accurate to a millionth of a second. As long as the pulse repetition rate is known, the time interval between pulses – the interval during which the electron beam sweeps across the A-scope – equates to a known distance, which can then be marked on the face of the scope.

We dwell at length on the A-scope only because it provides some particularly simple and apt visual examples of how radar works. Later we will describe the great variety of other displays which have appeared. However, the reader should note that the very simple means of measuring distance here described makes sense only if the output from the scope is *seen*. Translating it into numbers, for example to feed ranges automatically into a computer, is a far subtler task.

INTERFERENCE

Matters would be fairly simple, then, were it not for the fact that radar signals returning to a receiver are mixed with noise – static, extraneous returns such as sea clutter, and, once radar was a going concern, jamming. It then became important to pick a real signal out of the background of false ones, and in many systems the important limit on range is the one imposed by the ratio of signal strength (which decreases with range) to noise level. Considerable ingenuity has been expended to improve the signal-to-noise (S/N) ratio by sophisticated processing of the signal inside the radar. For example, for some years it has been known that a pulse with a

Since World War II, radars have come to determine, to a considerable extent, the effectiveness of other naval weapons. This is the new battleship New Jersey, *in October 1943. The spotting telescope normally mounted above her conning tower has been replaced by a long Mk 3 main-battery fire control radar: atop her secondary-battery directors are Mk 4s of similar design. Just visible on her tower foremast is an SG surface search antenna, placed there (rather than higher up) to reduce losses between antenna and receiver.*

complicated (coded) shape can, with complex processing, be picked out of a sea of noise far more easily than can an ordinary blip. A simpler antinoise measure is a sharp increase in the rate of pulsing. The location of the returning pulse on the A-scope is fixed by the range to the target, which varies very slowly in comparison to the rate at which pulses are generated; but the noise is random, and for successive pulse cycles it may well cancel out. Then if the radar or its display unit can 'remember' successive cycles of pulses and add them up, a signal may be extracted from the noise more efficiently. However, a very sharp increase in the pulse repetition frequency (PRF) will limit a radar's maximum range.

That is, the radar 'sees' unambiguously only those signals which return to it between successive pulses. Otherwise it will not be able to gauge distance: did the echo return from pulse 1 or from pulse 2? To see how this works let us imagine a radar with a PRF of 1000 pulses per second; each pulse is roughly 1000 microseconds (328,000yds) from the next. For a pulse to go out *and* return before the next pulse is generated, it must strike a target closer than 164,000yds, about 82 nautical miles. Imagine that S/N is unsatisfactory. Doubling the PRF to 2000 halves maximum range. On the other hand, the laws of signal propagation are such that an echo returning from half the previous range will have 16 times the power (S) of one from full range; an increase in power may buy more than PRF. By the same token, to achieve at 80nm the S achieved at 40 requires 16 times the power – indeed, to maintain a constant S at 50 per cent more range requires *five* times the power, since the signal falls as the fourth power of the range.

Power is generally thought of as the peak power per pulse; but in fact what counts is the *average* power a radar delivers over its operating cycle. For example, a 100kW, 0.5-microsecond pulse repeated 500 times per second is roughly equivalent to a 25kW, 1-microsecond pulse repeated 1000 times per second – or to a 100kW, 2-microsecond pulse repeated 125 times per second. The longer pulses tend to overcome noise because on the average pulses of noise do not last very long; on the other hand, they reduce the radar's ability to discriminate between nearby objects. Long pulses are useful as a means of achieving long range, *ie* as a means of providing good S/N performance without using a high

GIORGIO ARRA

Modern surface warships depend largely on their radars for their combat efficiency. The missile cruiser California, *shown in 1977, displays a combination of systems illustrating most of the alternative modes of operation currently employed. From left to right (fore to aft) she shows an SPG-60 pulse-doppler fire control radar, a pair of SPG-51D missile control radars (pulse doppler tracking and CW Illumination) for her Standard Missiles, then an SPS-48 frequency-scanning three-dimensional radar and an SPS-10 surface search radar. On her mainmast she shows a TACAN air navigational beacon (secondary radar), essential for effective aircraft control; an SPS-40 long-range two-dimensional air search radar; the enclosed antenna of the SPQ-9 Track-While-Scan (TWS) surface search and gunnery radar; and one of another pair of SPG-51Ds. The small square antenna alongside her forward superstructure is a satellite communications antenna (AS-3018/WSC-1), normally mounted in pairs aboard modern US warships. At the after end of the forward superstructure is an array of ECM antennas.*

(range-limiting) PRF or an excessive peak power. The latter might, in a multi-megawatt radar, cause 'arcing', in other words literally short-circuit the antenna. As we shall see, radars show a variety of balances between PRF, peak power and signal length; in addition there are means of signal processing, most notably pulse compression.

None of these measures shows in the structure of the radar antenna, which is all an observer sees. In a sense many of the basic developments in radar have been attempts to improve S/N: higher power, which raises S without affecting the random and externally generated noise (except for the noise the radar *itself* generates), and more directional antennas, which see less of the surrounding noisy environment in comparison to the limited angle from which a signal can return. Both provide limited relief. Sharp increases in power increase effective range only slightly, because the power which returns falls off very rapidly with range. More directional antennas are sometimes very useful, but in at least one important application, search, they are definitely harmful.

TRACKING

Generally the goal of a naval radar system goes beyond the simple determination of the range and bearing of a static target; instead, it is important to be able to *track* the target over time, to predict its future position. The known target track history is in itself an important means of target identification. For example, in a modern highly automated combat system radar track data is used to rank incoming radar targets in order of the threat they pose to the ship, and to guide weapons against them on this basis. In an ideal system it should be possible simultaneously to detect all nearby targets (search) while at the same time keeping track of all which have already been detected. In practice, the need for a combination of long range and rapid search (to detect fast-moving targets) limits the accuracy of most search radars, and tracking often requires a dedicated radar locked onto a single target. Much of the impetus for the development of computer combat systems such as the US Navy's NTDS and the Royal Navy's ADAWS has been the desire to improve the track-keeping efficiency of search radar systems by entering indivdual plots (range and bearing of individual targets detected) in computers, which could associate successive plots to form target tracks. Such systems can also integrate data from a range of radars with varying accuracy and range characteristics, to form the best available target tracks. The new US AN/SYS-1 computer system will provide an automated detection and tracking system for most US missile-equipped warships.

VARIANTS

Several variations on the usual radar determination of range and bearing deserve mention:

Moving Target Indication (MTI). The rejection of clutter, or unwanted echoes, is a major consideration in many radar applications. In many cases the sorting can be done by speed: echoes of objects moving in the appropriate range of speeds are worthy of notice, whereas other echoes can be rejected. MTI is a general term for such filtering, often on a pulse-to-pulse basis. It implies some form of memory within the radar system, so that successive echoes or the results of successive radar scans can be compared. MTI is often considered an anti-decoy (*eg* anti-chaff) or anti-ECM measure.

Doppler radar measures speed by means of the *doppler shift*, a shift in the frequency of a returned echo proportional to the speed of the echoing object. Pulse radars use (in effect) a mixture of frequencies and hence tend to swamp out at least some of the doppler shift, which in any case is small: the shift as a fraction of the basic frequency equals the speed as a fraction of the speed of light. Thus 600mph gives a frequency shift of only 0.00008 per cent, 160 cycles in the 200 million of a P-band radar. Matters would be simpler at higher frequencies; for example, on the 30,000mc/s of a K-band (1cm) set, the shift would be 24,000 cycles. The doppler effect is used in two principal forms: without pulsing at all (CW, or continuous wave mode), and in a specialized *pulse doppler* form. The latter is quite common in airborne radars, which must overcome the clutter effect of the ground, but there are also several important naval pulse doppler radars in service, the most important being the Tartar guidance radar SPG-51.

Track-While-Scan (TWS). Like MTI, this is a radar with some form of 'memory', so that it can correlate the differing positions of particular targets on subsequent sweeps. The first TWS systems were fire control radars which could present multiple targets while maintaining track on the one under fire: the US Mk 8 surface gunnery set, which could observe both splashes and target, was an early example. Unlike a conventional tracker, a modern TWS radar can maintain track data simultaneously on several targets. Generally that requires a very rapid data rate which precludes long range; the current SPQ-9 fire control system, for example, employs an antenna spinning once a second. The Aegis radar (SPY-1) is able to achieve TWS performance by electronic scanning, without the limitation of very rapid scan rates and consequent limitations on range.

Automatic Target Detection (ATD). Human detection of individual radar contacts can be extremely effective, but a human operator cannot remain permanently alert against a continuing threat. It is possible, however, to design a radar which can itself decide whether a given signal should be considered a real target. Typically that means that the radar measures a mean noise level and then attempts to generate a constant (acceptable) false alarm rate, or CFAR. As the noise level rises (due, for example, to jamming), more and more real signals are disregarded. By way of contrast, a human operator can often identify signals buried in noise – but he cannot retain his alertness continuously, and he will therefore miss a percentage of signals which the ATD radar may catch. ATD technology also permits a radar to feed its data directly into a computer via a radar video processor (RVP).

Automatic Detection and Tracking (ADT). Air defense systems tend to be concerned with target *tracks* rather than with particular target detections. ADT combines an automated decision that a blip *is* a target (ATD) with automatic tracking: in effect, ADT systems turn search radars into TWS systems. Target-tracking is often enhanced by fuller exploitation of the details of the returning target echo, which may include the extraction of doppler shifts to estimate and project target motion. ADT is generally an attempt to counter saturation raids by fast aircraft and antiship missiles. Generally the human operator serves as a monitor of system failures; in an ADT-based combat system, he can abort an attack but usually does not have enough time to decide whether to fire in the first place. The new US SYS-1 system forms tracks from the ATD outputs of several shipboard radars, but such current radars as the SPS-48E and SPS-49 are designed to operate in an ADT mode when they are not connected to a larger track-forming system.

Continuous Wave (CW) Radars. In a CW system the radar transmits continuously; there is no pulsing, and operation is generally at a pure frequency. Since there is no pulsing, separate transmitting and receiving antennas are generally required. Such a set cannot measure range, but it can measure speed very accurately by means of the doppler shift – as in SPN-12. CW radars are commonly used to illuminate targets for missiles, one example being the illuminating component of SPG-51. Another is the two-antenna radar used to control the Nato Sea Sparrow (Mk 91). Very highly coherent radars such as the Hughes Target Acquisition System approach CW operation, using, in some cases, a modulation of the frequency for ranging. In effect, there are no pulses, but the signal changes enough over time for different elements of it to be distinguished.

Transponders, or 'secondary radar'. A transponder reacts to a radar pulse by sending out a fresh signal of its own, in effect a delayed and transformed echo of the original signal. Such a signal can greatly exceed the strength of a reflected signal, even though the transponder may have a relatively weak transmitter, Important applications include IFF (Identification Friend or Foe), radar beacons, and LORAN (Long-Range Navigation). Many ECM devices are essentially transponders.

2. Radar Signals

A radar set generates a stream of pulses, bursts of radio energy. Usually it is charactersized by the peak power it puts into the pulses, by the pulse repetition frequency (PRF), and by the radio frequency at which it operates. The latter is equivalent to a wavelength and very largely determines the character of the radar. Thus radars are usually referred to as 'X-band', 'L-band', etc, which mean a radar operating in the 'X' frequency band, 'L' frequency band, etc. Bands, or ranges, of frequency were originally substituted for characterizations of precise frequency during World War II as a means of assuring security, for example against potential jammers. Moreover, radars were often designed so that they could operate over a range of frequencies, so that several could be operated nearby without mutual interference. The band designations remain useful because radars operating within a band have many important characteristics in common (the accompanying tables show in approximate form the radar bands and their frequencies and wavelengths).

The reader may see a contradiction between the idea of a frequency and the characterization of a radar output as a series of sharp pulses, usually presented pictorially as little square bumps. Actually the bumps are short periods of oscillation at frequencies very close to the radar's nominal frequency; in fact it is very important that the PRF be small compared to this fundamental frequency. This will generally be true: PRFs are of the order of a kilocycle* or two, radar frequencies in the thousands of kilocycles. Such operation is analogous to that of a radio, in which speech,

Radar Bands (World War II)

	Frequency (mc/s)	Wavelength	Notes
P	200–500	1.5–0.6m	Original US radar ('pulse' equipment) band
L	500–1500	60–20cm	'Long' waves; often used for modern air search sets
S	1500–5000	20–6cm	'Short' waves; early surface search sets such as SG
C	*See note*		
X	5000–15,000	6–2cm	Exotic, or very short, microwaves
K	15,000–40,000	2–0.75cm	Lower limit of most radars. The designation may derive from the use of a klystron generating tube in early experimental systems

Band boundaries are approximate. The notes are the author's explanations for designations, and appear plausible in view of other evidence.

Special sub-bands were devised for IFF systems: 'A' in the P-band and 'B' in the L-band. 'C' was a postwar designation for the S/X boundary (4000–6700mc/s, 7.5–4.5cm). There were also extensive subdivisions of existing bands, most of which are rarely used; Ku and Ka are exceptions.

In 1945 efforts began to define radar bands for military use, as the civilian economy began to exploit wartime UHF technology. Some assignments were made, but two bands, 600mc/s and Xb (about 6400mc/s), had to be abandoned under a new international agreement of 1959, which was implemented in 1960–61. This agreement had serious consequences for several radars, for example SG-6, SPS-5, SPS-38 and SPS-45.

Radar Bands (current)

	Frequency (mc/s)	Wavelength (cm)	Radar bands (mc/s)
VHF	30–300	300–100	133–144, 216–225
UHF	300–1000	100–30	420–450, 890–942
L	1000–2000	30–15	1215–1400
S	2000–4000	15–7.5	2300–2500, 2700–3700
C	4000–8000	7.5–3.75	5250–5925
X	8000–12,000	3.75–2.5	8500–10,680
Ku	12,000–18,000	2.5–1.67	13,400–14,000, 15,700–17,700
K	18,000–27,000	1.67–1.11	24,050–24,250
Ka	27,000–40,000	1.11–0.75	33,400–36,000

Note that the band designations given here do not quite agree with World War II practice, but they are close enough not to cause undue difficulties. The wartime designations are generally used in this book.

Recently US ECM experts (but not the radar engineers) have adopted a new series of band designations in a logical alphabetical order. These designations sometimes appear in print, most notably with reference to Soviet systems. In this scheme K approximates K/Ka, and J straddles upper-frequency K. These movements back and forth mean very little in an absolute sense. A is VHF, B and C are UHF, D corresponds to L; E and F to S-band; and G and H to the former C.

*In the past few years the term 'cycle' has been replaced by 'Herz', or Hz, after the discoverer of radio waves. Hence 200mc/s (megacycles per second) would more properly be written 200MHz. A billion Hz is now called a Giga-Herz, so that 2000MHz = 2GHz. It may be useful to bear in mind that 1mc/s = 1000kc/s (kilocycles). Wavelength is given by the speed of light divided by the frequency: in centimeters, about 30,000 divided by the frequency in megacycles.

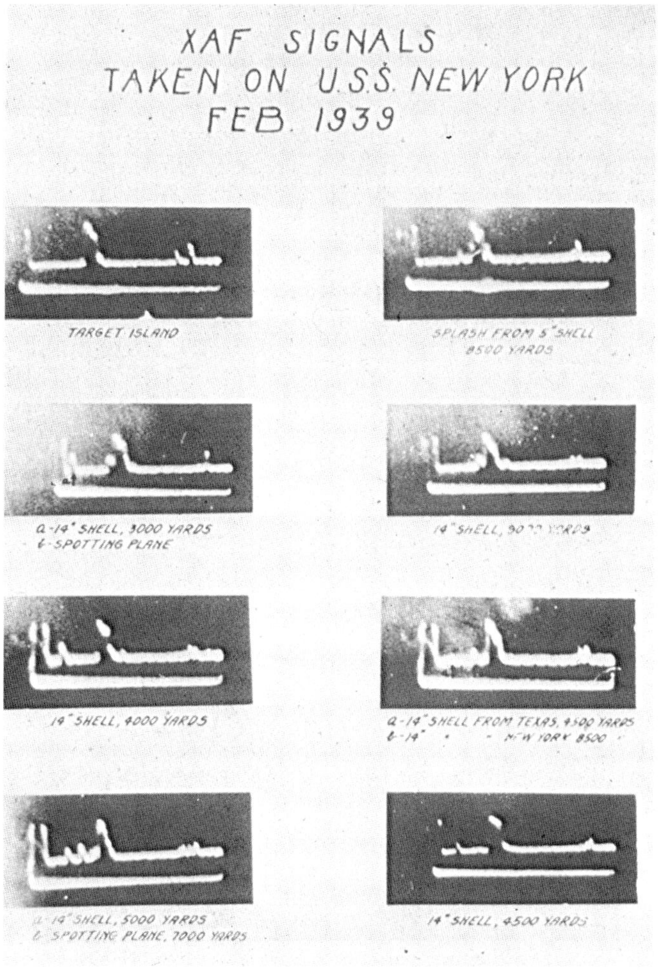

These A-scope traces were photographed aboard the battleship New York *during Fleet Exercises in the Caribbean in February 1939. The image is actually a folded A-scope display, with range increasing from left to right on the upper line, and from right to left on the lower; the total is 50 miles. Aircraft, ships, and land masses were all visible; in the trace at the lower right hand corner, a signal has been reflected from a target beyond 50nm, and it appears instead as a very weak signal at very short range — a 'Second-Time-Around' (STA) echo. This was a very early example of the relation between PRF and unambiguous range,* ie *range free of STA signals. The radar, XAF, was the prototype of all later US Navy long-range air search sets. Note that these displays showed weak signals clearly (as would not be the case in a PPI), but that they did not provide a clear image of radar targets around the ship. Again, that is the classic trade-off between a A-scope, showing the intensity of the signals, and a PPI, showing their location in bearing as well as in range.*

with kilocycle frequency, is superimposed on a 'carrier wave' with a frequency of hundreds or thousands of kilocycles.

BANDWIDTHS

A carrier with some signal imposed on it is no longer a pure wave of a fixed frequency. The train of radar pulses is itself a mixture of pure waves spread over a

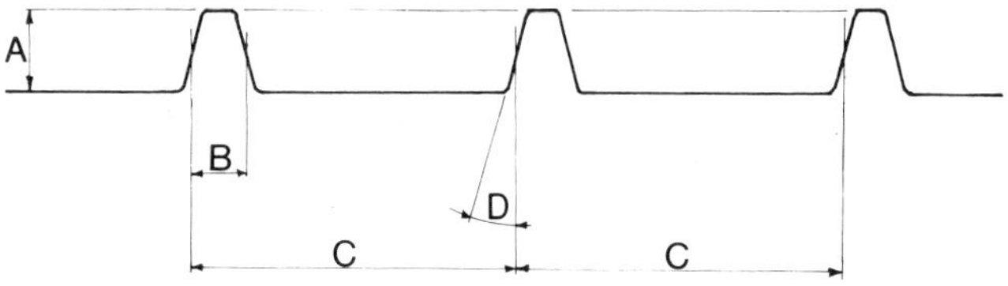

Radar signals, in time (not to scale) and strength. **A** *Peak power, usually in kilowatts (kW).* **B** *Pulsewidth (length/duration), usually in microseconds.* **C** *Period between pulses, usually in milliseconds (and close to 1/PRF, since the pulses are far shorter than the resting period).* **D** *Rise time or angle of leading edges (ideally this should be vertical for perfect accuracy, but angle is governed by transmitter modulator circuits and receiver bandwidth).*

very wide range of frequencies, but concentrated over a spread given roughly by the inverse of the pulse-length: 1 megacycle spread to accommodate 1 microsecond pulses. That is, waves generated by a 200mc/s radar using 1-microsecond pulses would be spread largely between 199.5 and 200.5mc/s, *ie* they would have a 1-megacycle bandwidth. Such a phenomenon limits the number of alternative 200mc/s radar 'channels', as it were; a similar effect limits the number of television channels. In general what limits the number of channels is the ability of a receiver to discriminate between nearby frequencies. At low frequencies this theoretical limit is washed out by the bandwidth required; but as the carrier frequency rises, bandwidths shrink *as a fraction of the carrier frequency*. For example, the same 1mc/s would involve, at 2000mc, a band extending between 1999.5 and 2000.5mc, in other words would be 0.05 per cent of the carrier frequency, not 0.5 per cent as at 200mc. Thus higher frequencies (or shorter wavelengths) permit either greater frequency diversity to counter possible interference between neighboring ships, or else frequency agility as an anti-ECM measure.

Not all frequencies within the radar bandwidth are represented. Rather, the radar operates at a series of discrete frequencies whose separation is determined by the pulse repetition rate. This separation between adjacent frequencies makes doppler radar operation possible, as frequency shifts within the gap between frequencies can be measured. A familiar analog of this effect is the difference between the number of VHF (conventional) and UHF television channels. Each channel can be visualized as a complex signal with some given bandwidth centered on a particular frequency. Bandwidth in this case measures the information content of a TV picture, and therefore does not vary appreciably between VHF and UHF television. But the range of carrier frequencies is multiplied by ten, so that roughly ten possible VHF channels become about one hundred UHF channels.

Frequency and wavelength are inversely proportional: thus a 200mc wave (one which repeats 200 million times per second) has a wavelength of about 1.5m. However, 200mc is an extremely high frequency for radio and a relatively low one for radar, as we shall see. Classically, high-frequency (HF) radio has operated on wavelengths of hundreds of meters, equivalent to tens or hundreds of kilocycles. Short-wave radio, or VHF (very high frequency), operates at 30–200mc, *ie* from about 10 down to 1.5m.

Above that are the traditional radar bands, although there is nothing holy about the boundary at 1.5m. (The main radio bands are given, very schematically, in the accompanying table.) At still higher frequencies the radiation becomes visible as light: in many ways radar waves behave like light. One important difference is that radar waves are bent, or refracted, more sharply by

Frequency Bands

	Frequency	Wave-length	Notes
Low (LF)	30–300kc/s	1000–10,000m	Follows the curve of the earth. Used for LORAN. Too low for voice transmission
Medium (MF)	300–3000kc/s	100–1000m	Shows a sky bounce effect, hence can be used for long-range radio
High (HF)	3–20mc/s	10–100m	Sky wave reflection from the ionosphere ends beyond 30mc/s. The first British CH radars operated at 50m and later at 12
Very High (VHF)	30–300mc/s	1–10m	Line-of-sight transmission: voice radio, television, some radars
Ultra High (UHF)	300–3000mc/s	0.1–1m	UHF television. Waves can scatter troposphere to achieve range greater than line-of-sight
Super (SHF)	3000–30,000mc/s	1–10cm	Microwaves, *eg* microwave radar
Extremely (EHF)	30,000–300,000mc/s	1–10mm	

Note that light has wavelengths of the order of 0.005mm, *ie* frequencies of about 600,000,000 megacycles, or 600,000,000,000,000 cycles per second.

The frequency and wavelength ranges shown are suggestive rather than precise; no one really cares, for example, whether 2950mc/s is in the UHF or SHF range. However, it is often useful to be able to categorize a device as, say, VHF rather than LF.

the atmosphere, so that the radar 'sees' a horizon more distant than the visual one.

WAVELENGTHS

The bands matter because much of a radar's performance depends on its *wavelength*: an S-band radar really does differ fundamentally from one operating at P-band wavelengths about ten times longer. Radio waves of a given length will bend around an object much smaller than that length; a radar cannot 'see' objects smaller than half its wavelength. A corresponding wavelength effect is the extent to which a given antenna can focus its radar beam. The larger the antenna in *wavelengths*, the sharper the beam. Even very large antennas, however, cannot approach the performance of optical systems, simply because light waves have such short wavelengths – typically of the order of 0.5 *millionths* of a meter, *several hundred-thousandths* of the length of the shortest waves used in radars. It would take an antenna 200m across to equal the effect of a 1cm lens using light waves.

It would seem, then, that the aim of radar developers should always have been shorter waves, *ie* higher frequencies. Shorter-wave radars would have superior definition for given antenna sizes and indeed would make more compact antennas practical. Historically, land-based radars followed just this pattern. The earliest British radars, the fixed early-warning sets of the Home Chain, operated at first on a standard wavelength, 50m, or about 60mc. This frequency had been chosen deliberately on the theory that bombers with wingspans of about 25m would be the primary targets. As soon as possible, however, the British went to a higher frequency and hence the shorter wavelength of 10 to 15m. Early US naval radars operated at about 200mc, or 1.5m (P-band). In each case, improvement depended on the development of special transmitters capable of putting out high power at higher and higher frequencies: the great wartime development was *microwave* radar, *ie* radar operating at about 10cm or less (S-, X- and K-band). This was entirely a matter of the perfection of specialized tubes: magnetrons and klystrons. Later attempts to work at millimeters failed because the waves produced were easily absorbed by water vapor in the air; however, centimetric radars saw very extensive application from 1942 on.

The reader will note that in naval service long-range air search radars operate at metric rather than centimetric wavelengths. Such operation clearly involves considerable penalties, but it is necessary, given the character of the naval environment and of the signals themselves. Each pulse carries not merely a duration but also a phase, a timing of the oscillations within it. Two signals arriving together in the same phase add together, but if one is shifted precisely out of phase with respect to the other they cancel out. At intermediate phase shifts the two signals partially reinforce or cancel

CONWAY PICTURE LIBRARY

each other. Phase is measured in degrees, so that zero phase difference is equivalent to 360°. Full phase cancellation occurs at 180° difference and at odd multiples of 180°. The phase of a signal varies along its path according to the ratio of distance to wavelength. Thus when echoes return to a radar along two alternative paths, the length of those paths determines the extent to which the two series of echoes cancel out.

That is the case in a radar just above the surface of the sea. The sea, at least when it is smooth, is a reflector, and the echoes can return either directly or after reflection off it. The two path-lengths depend upon the angle at which the radar is looking and upon its height above

The British Type 267 submarine radar system illustrated the values of different wavelengths with particular clarity, combining a P-band air search set based on Type 291 (at left) with an S-band 'cheese' for surface search (right). It is shown aboard the submarine Tally-Ho, *June 1954.*

For the US Navy, one of the nastier postwar surprises was the failure of shorter-wavelength (L-band) air search radars such as SPS-6 in the face of jet aircraft. Carriers therefore retained their wartime P-band radars far longer than might otherwise have been expected. Here the carrier Coral Sea *(CVA-43) still shows her old SK-2, aft, in April 1957. Forward she carries postwar radars, the SPS-8 height-finder, SPS-6, and a surface/zenith search set, as well as TACAN.* Essex *class carriers retained SC-series sets for the same reason.*

OURTESY OF A D BAKER III

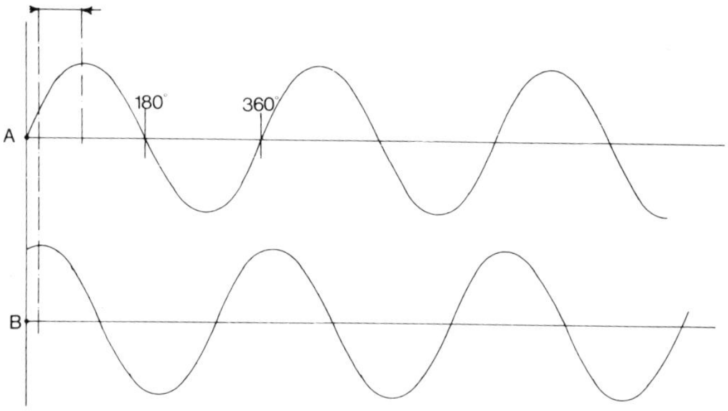

The concept of 'phase'. Radar signals are a mixture of many different pure frequencies, added up. The distance shown (between the arrows) in the diagram is a measure of how far **A** *is* out of phase *with respect to* **B**. *When two signals are 180° out of phase, the cancel completely. However, every 360° they add (reinforce). All electronically scanned radars are based on the idea that signals add when they are in phase, and cancel when they are out.*

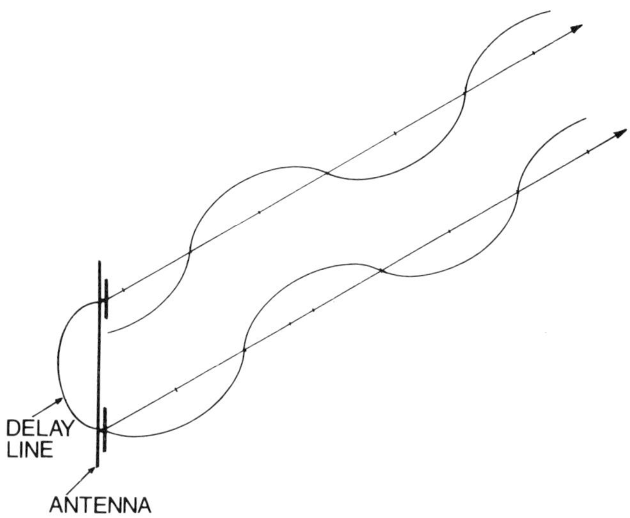

In a FRESCAN radar, the delay line adds just 360° to the phase of the signal in the appropriate direction. *Note the change in phase to match direction of wave; in other directions the two waves are out of phase and largely cancel out.*

the sea in wavelengths. At some angles the radar will see far better than it would without the reflection effect, but at others the signals will cancel out; they are said to 'fade'. The shorter the wavelength, the lower the minimum 'seeing' angle – but the more numerous the fades. At longer wavelengths, then, coverage at low altitude will be poorer but medium and high altitude coverage will be far better. The intermediate L-band of many postwar air search radars is thus an attempt to compromise between the great directionality of short-wave radars and the long range achievable through constructive interference (addition) at longer wavelengths – hence the use of S- and X-band radars for surface and low-altitude air search (*eg* SG, SPS-10), and P- and L-bands for long-range high- and medium-altitude air search. The shorter-wave radars might indeed detect a high-altitude target, but the large number of 'fades' in their coverage would tend to reduce too sharply the probability of detection.

Since the whole issue of fading depends upon the reflection of radar signals off the surface of the sea, short wavelengths present the possibility of radar beams so finely focussed that they may not touch the sea, except at great distances. Such radars have, in effect, no fade at all, except at very low elevations; but the price they pay is very narrow beam unsuited to search. SM and SP fall into this latter category.

LOBES AND NULLS

In a broader sense, the reflective effects that produce lobes and nulls (*ie* regions of constructive and destructive interference) depend upon the *smoothness* of the sea. A rough sea breaks up radar beams striking it, so that only a small fraction of radar energy has a chance to interfere destructively – or constructively – with the direct path signals. In such a situation, radar performance reduces very much to its 'free space' condition; wavelength matters much less, although it still enters in that the atmosphere absorbs more or less of the signal depending on wavelength. Very crudely, the complex pattern of lobes and nulls disintegrates at Sea State 3 and above. Short wavelengths still make sense in terms of low-angle coverage and high directivity, but the sacrifices associated with metric wavelengths no longer seem so worthwhile. Such reasoning has never really appealed to the US Navy, which has to operate in all seas, including very calm ones such as the South Pacific. However, the reader may note that a major Soviet air search radar, the 'Head Net' series, is generally described as S-band – and Soviet home waters, in the North, tend to roughness.

On the other hand, a radar rolling in a heavy sea will pick up so much sea return that surface targets will be obscured. Hence the attempts at stabilization in the SU series of late-war surface search sets. A more exotic effect of rolling was experienced with S-band versions of the TDY jamming transmitter. Originally it had been feared that nulls analogous to radar fades would limit its effectiveness, but the rolling of a ship washed out fades and lobes quite throughly.

The land is far less conductive, and hence less reflective, than the sea; it also tends to roughness, which further reduces lobing. It is therefore no great surprise that *land*-based early-warning (air search) radars were operating in the microwave band as early as 1945.

BLIND SPEEDS

The phase structure of the radar pulse can b

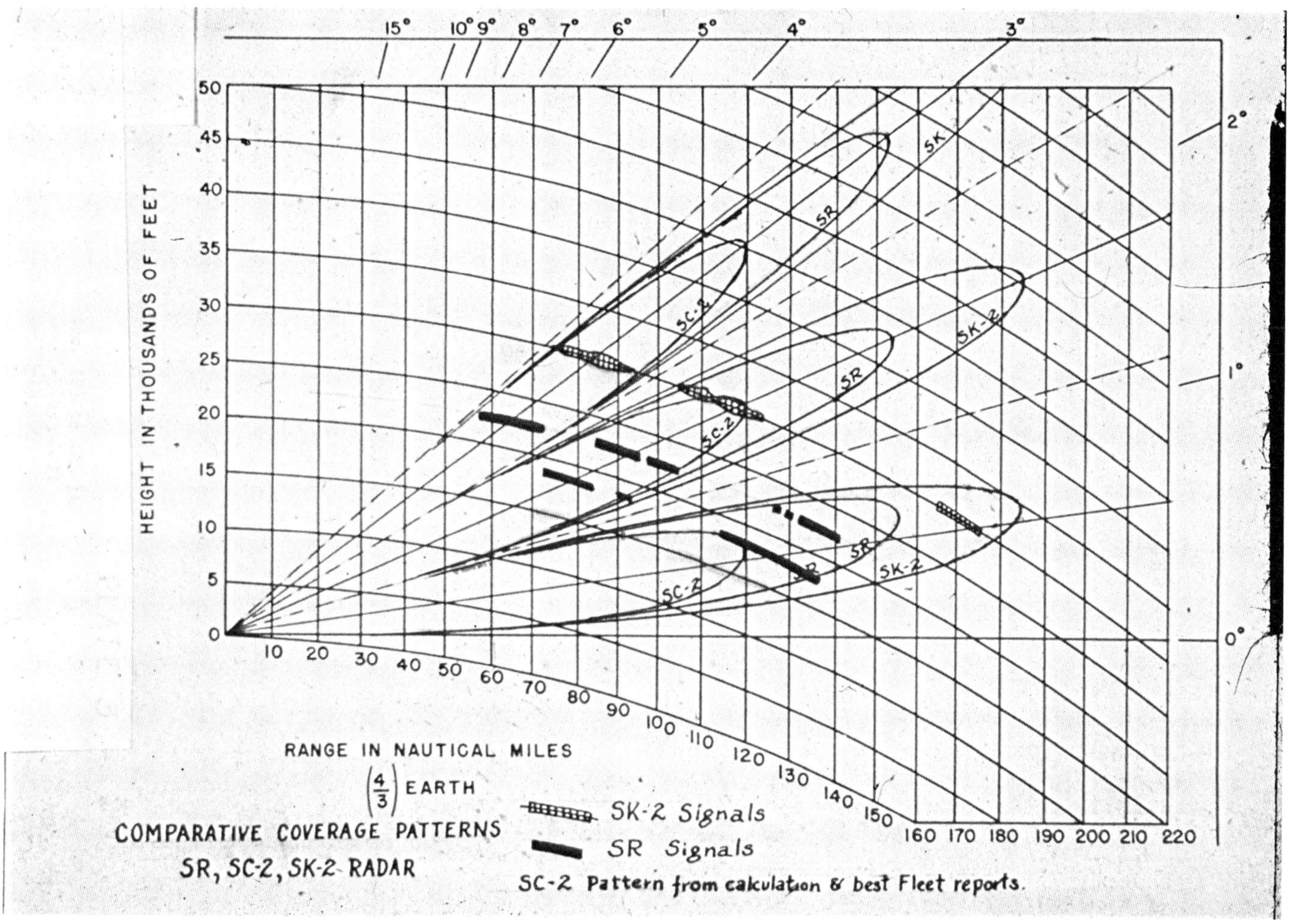

Every two-dimensional naval radar shows breaks in its coverage of air targets due to reflections from the water; the shape of these 'fading' regions varies with the shape of the antenna, with its height, and with its frequency. Moreover, their distance from the radar is a measure of the height of an airplane the radar detects. Before the wide availability of pencil-beam (three-dimensional) radars, 'fade charts' like this one were often used to estimate target altitude; the radar would track an incoming airplane or raid, and points at which the signal faded out would be noted and compared to the chart. This one shows the fade zones for three of the most important US sets of the war: SC-2 for destroyer air search and for carrier secondary air search; SR, its replacement aboard destroyers; and SK-2, whose dish antenna produced a relatively narrow beam, and which was used for carrier primary air search at the end of the war.

extract information: for example, MTI systems commonly compare successive echoes to detect target motion. If the target moves, successive echoes arrive at slightly different intervals, *ie* with slightly different phases relative to the radar transmitter proper. Adding them up in the receiver therefore allows motion to be detected. There is one drawback however: at some particular speeds the change in phase is exactly 360°, and the radar detects no motion at all. These speeds, proportional to the PRF and to the wavelength, are called blind speeds. The lowest is approximately

$$0.29 \frac{\text{PRF}}{\text{f}} \text{ kts},$$

where PRF is pulses per second and the frequency is expressed in thousands of megacycles. For example, for a typical radar with PRF of 900 operating in the S-band (3000mc) the lowest blind speed is about 87kts, and higher ones are multiples of this figure. The higher the PRF, the higher the lowest blind speed, but in that case the radar loses its ability to detect range unambiguously, as returning pulses from long ranges become mixed with those from much shorter ones.

This MTI system is called *coherent*; it requires some means within the radar of maintaining phase relationships ('coherence') between successive pulses. However, the magnetrons commonly used at S-band, for example, do not maintain any fixed phase relationships between successive pulses, and the radar must incorporate some means of maintaining a phase standard. Early MTI systems, by contrast, were *non-coherent*: they compared the *strength* of successive returning pulses. Background fixed clutter was needed to detect a change in echoes from a moving target; thus non-coherent MTI is independent of ship motion, but requires a fixed land

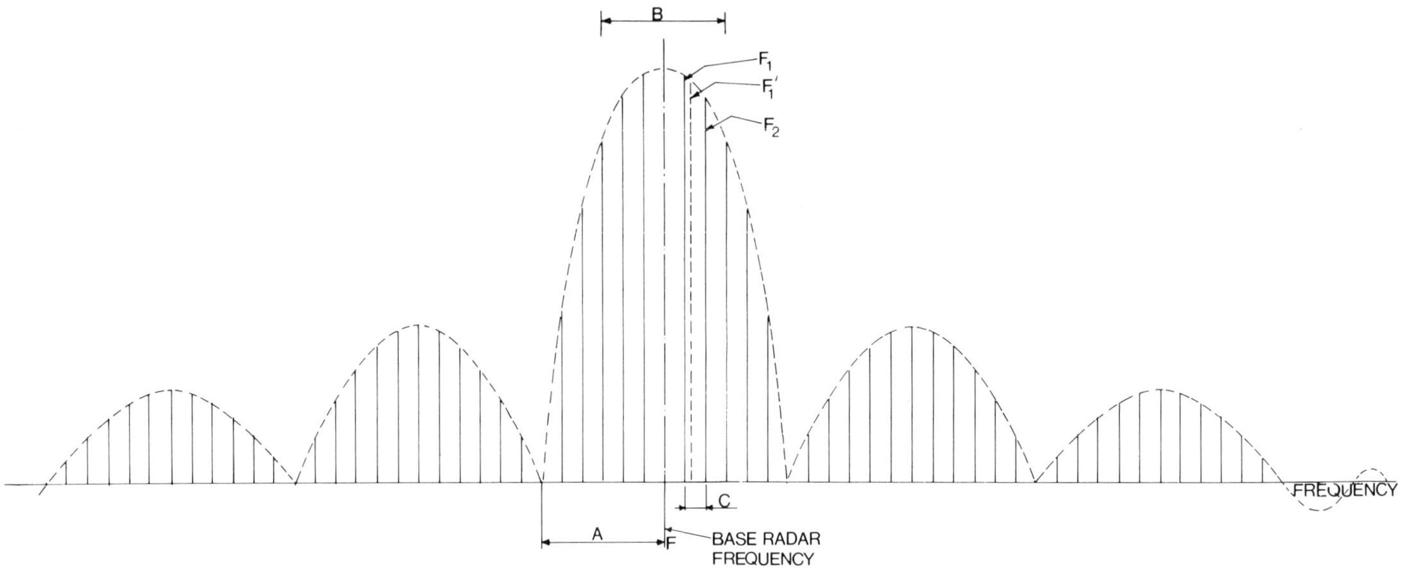

Pulse-doppler, showing frequencies represented in a radar signal at (base) frequency **F**, *with pulse duration* **D** *and a given PRF. The most prominent frequency is F itself, as indicated, but the spectrum shown extends infinitely far out. Here* **A** *= 1/D; for 1-microsecond pulses, A is 1 megacycle. Most radars are designed to receive only frequencies within the bandwidth* **B**, *which is usually set at 1/D, half the distance between the first zeroes.* **C** *is the* spacing *between frequencies, determined by the PRF. It in turn determines doppler blind speeds. For example, the frequency line* $\mathbf{F_1}$ *will be shifted to* $\mathbf{F_1'}$, *by doppler effect, the amount of shift proportional to the speed. The further apart the frequencies, the further* $\mathbf{F_1}$ *can shift before it becomes confused with* $\mathbf{F_2}$. *C is equal to the PRF.*

background against which to work. Coherent MTI works at sea, but requires compensation for ship motion (*eg* roll and pitch).

There is also Area MTI, in which the radar cancels out returns which have not moved sufficiently since the last complete scan. Scan rate and beamwidth determine the minimum velocity needed to escape cancellation; in the SPS-50 prototype, for example, it was 150kts in a radial direction. At that time, about 1965, AMTI was considered 'the processing mode which should result in the greatest capability of clutter cancellation. The AMTI is basically an electro-mechanical velocity filter designed to remove residual point clutter and chaff echoes which pass through the MTR and auto-gate circuits.' Here the MTR was a moving target resolver which separated targets moving at different velocities (using doppler shifts) and the auto-gate eliminated such spread-out targets as chaff clouds in favor of point targets. It should be noted that AMTI is more commonly applied to Airborne MTI systems, as in airborne early warning aircraft.

PULSE DOPPLER RADARS

In effect a pulse radar emits a spread of discrete frequencies, spread around the radar frequency and spaced in multiples of the PRF. Most of this spread is concentrated within bounds set by the pulse-width: the longer the pulse, the narrower the net spread. As long as the doppler shift of a target is within the spacing determined by the PRF, it is possible for the radar to measure that shift by breaking a returning echo into its pure frequency components. This is the basis of *pulse-doppler* radars. Very high PRFs are chosen, and they introduce range ambiguities; the radar receiver concentrates on only one pure frequency of the received signal, and measures its doppler shift, which has only to be significantly less than the PRF.

Range ambiguity is generally resolved by establishing some pulse-to-pulse relationship which can be used to distinguish returning pulses. For example, the pulses themselves can be coded. Alternatively, the PRF can be varied systematically, so that the same targets produce a sequence of different range ambiguities related to the varying PRF. Then these measurements can be compared. The latter technique gives the best range accuracy, but introduces considerable complexity – and weight. Thus airborne pulse-doppler systems usually employ pulse coding techniques, a single coherently varying signal being 'chopped' to form the train of pulses.

Pulse-doppler has advantages in overcoming clutter, such as sea returns competing with signals from a low-flying airplane or missile. Its ability to detect target velocity improves tracking accuracy, as in the SPG-51 and the new Target Acquisition System Mk 23.

3. Factors in Radar Performance

The world's naval radars present an immense variety of characteristics, which can be understood in terms of (a) problems of installation (not covered here); (b) theoretical performance factors (pulse-length, repetition frequency, power level; beam size and antenna gain; and fade patterns); (c) the character of noise and clutter the radar may encounter; (d) atmospheric effects (transmission losses, refraction and weather losses, all of which are consequences of a choice of radar frequency); and (e) target characteristics at different frequencies. So far we have referred mainly to the theoretical factors. For example, PRF determines the maximum *unambiguous* range (although modern pulse-doppler airborne radars get around this problem); but the same high PRF which limits range also improves radar performance in the face of noise and clutter the radar must face. Pulse *length* determines how well the radar can resolve distance. As we have noted, a 1-microsecond pulse is 328yds long. If it strikes a pair of targets 100yds apart, the reflection of the second target will be swamped by that of the first. On the other hand, only so much power can be pumped into a pulse. A longer pulse carries more power and hence may permit operation at longer range: several of the radars described in Part 2 of this book show the effect of longer *vs* shorter pulses in this regard.

Another consideration is bandwidth. The shorter the pulses, the more bandwidth they require: very roughly, bandwidth is inversely proportional to pulse-length, *eg* 1mc/s for 1-microsecond pulses, or 10mc/s for 0.1-microsecond pulses. Thus very short pulses are practical only at high frequencies.

Similarly, beam shape (angular dimension) determines how well a radar can distinguish objects separated by a small angle. A narrow beam means better resolution, except that it presents problems in scanning. Radar detects targets on a *statistical* basis – so many per cent chance of detection per pulse. The wider the beam, the more pulses fall onto a target for a given scanning (rotational) speed. That is, a radar 'sees' as equivalent all bearings within a beamwidth. All pulses sent while the radar is pointing *within* a beamwidth at a target have a chance of registering detections. The narrower the beam, the fewer the pulses per scan.

We have only touched upon the problem of *generating* radar power, which depends very much on frequency. Historically, the first radars operated at long wavelengths precisely because high power sources were relatively easy to build. Relatively high powers later became available at microwave frequencies; but the reader must keep in mind that wavelengths approximate to the physical dimensions of transmitters. One centimeter is very little space in which to confine kilowatt power pulses!

C&S TAYLOR

There is a great divide between the theoretical performance of a radar and its effective capabilities; one problem is blockage by other parts of a ship. Here the Soviet 'Kashin' class missile destroyer Obratsovy *shows two identical 'Head Net' S-band air search radars during a visit to Portsmouth in 1976. They may operate together to scan the area around the ship more rapidly, but that seems unlikely, since the forward of the two has a clear after arc past the after radar. More probably the after radar is intended as a reserve in the event of the failure of the primary system, in which case its forward arc will be blocked. Clearly visible here are the wind-balancing vanes characteristic of Soviet naval radar practice, as well as the massive stabilized mountings. Blockage is almost certainly a major factor in the actual (*vs *theoretical) performance of Soviet naval combat systems.*

NOISE AND CLUTTER

A real radar receives both the signal it seeks and considerable irrelevant signals: noise and clutter. *Noise* in the random pulsing which fills space, as well as the random signal a radar receiver inevitably produces. In many cases jammers produce noise uniform over some frequency range. *Clutter*, on the other hand, is an echo from an object considered irrelevant by the radar operator. For example, a radar seeking low-flying aircraft will show on its display returns from waves and perhaps also from sea birds. Both obscure the target and hence are 'clutter', even though they are by no means spurious radar targets. In a sense, then, clutter corresponds to noise, radar decoys playing a role analogous to jamming. The choice of radar frequency is important in both cases.

External noise falls off as frequency rises: P-band is significantly better than HF, L better than P, etc. However, external noise rises once more at K-band. Moreover, against the improvement in noise performance must be set weather effects (which begin to be significant in S-band) and transmission losses. Clouds are effective enough blocks of radar transmission at X-band that such radars are often used for 'weather avoidance'. In fact C-band was adopted for many postwar naval radars as a compromise between the good weather penetration of S-band and the sharp directionality possible at X-band. K-band suffers from low power and, in many cases, from crippling absorption losses. Within any band, noise can be imagined as constant across a range of frequencies. Then a restricted bandwidth reduces external noise – but is equivalent to a larger signal, *ie* to a less accurate radar.

The other great antinoise measure is 'post-detection integration', the addition of echo after echo so that the systematic blip remains while the random noise cancels out. The greater the number of pulses added up, the better the *effective* signal-to-noise ratio; but that means higher PRFs and hence shorter maximum ranges. Coding pulses, *ie* making their shapes complex, corresponds in a sense to integration. In theory a receiver should find it relatively easy to pick a long coded pulse out of noise.

Probably the best-known form of coding is *pulse compression*. The pulses are 'chirps', in which the frequency changes during the pulse. In some sense a perfect receiver would be able to distinguish instants *during* each pulse. Two partly superimposed pulses could be distinguished, so that the radar would not lose range resolution. Upon its return, the pulse is compressed by a dispersive filter, *ie* one in which the delay varies with frequency. For example, a 100-microsecond chirp might be compressed into a 1-microsecond signal; the 'compression ratio' of 100 to 1 is theoretically equivalent to the (idealized) integration of 100 pulses, which means an improvement in signal-to-noise ratio by a factor of 100. By way of comparison, an operator watching 100 pulses add up on a radar display would probably achieve an improvement closer to 30.

In effect, a chirped signal substitutes bandwidth for short signal length. That is, high resolution requires a very short signal, corresponding to a considerable bandwidth. A succession of signals at different frequencies supplies the same bandwidth, and signal-processing can make the two approaches equivalent, in that both supply the radar with the same information content.

Unfortunately the chirped signal requires great bandwidth – a bandwidth increased by the compression factor. Noise increases similarly, and much or all of the *direct* improvement vanishes. In fact the complex signal processing makes matters worse. However, the saving grace of pulse compression is that it is far easier to package high power in long than in short pulses. For example, 1-microsecond 10MW pulses cause problems like electrical breakdown; but such a pulse is equivalent to one of 100 microseconds and 100kW. The remaining problem is minimum range. The radar receiver must be turned off during transmission, so the minimum radar range somewhat exceeds the length of the chirped pulse, *eg* 8nm for a 100-microsecond pulse.

Chirping is relatively complex conceptually and electronically, and it requires a massive structure of delay lines and filters. In the past decade digital signal-processing has become more and more compact. In a radar signal, coding can be achieved by phase changes within the pulse. When the pulse is received, it can be decoded by matching against a series of coded filters, the phase changes being equivalent to a code of 0s and 1s, and, like the frequency shifts, serving to differentiate between portions of the same pulse. The new Aegis radar, SPY-1, is probably the first to employ digital signal synthesis and processing on a very large scale, although many of the ideas involved were previously tested on the SPG-55 missile fire control radar. Given the compactness of digital electronics, it would appear that phase-coding will become more and more popular in pulse-compression radars of the future.

Just what a radar sees depends very much on its wavelength; in effect the smallest object a radar can see is about half a wavelength in size. That is why raindrops have a devastating effect on K-band radars but are invisible at P-band. Clutter is generally a matter of relatively small, randomly located reflectors – ocean waves, birds, chaff, etc. Compared to noise, clutter fluctuates more slowly, since it results from the motion of sizable physical objects; it shows a relatively narrow bandwidth. Moreover, clutter does *not* vary perfectly randomly with time: the average bird, for example, moves slowly from one moment to the next, does not jump 500ft instantaneously, and does not cancel out his neighboring bird. Hence integration does little good; anticlutter measures tend, rather, to exploit distinctive features of clutter. For example, doppler-filtering or

CONWAY PICTURE LIBRARY

At sea, radar performance depends strongly on radar wavelength, due to 'lobing'. The Royal Navy found it valuable to combine radars with different wavelengths, and therefore with different lobe structures, to achieve, together, fuller coverage. For example, the carrier Victorious *(shown here in July 1949) had Type 281B aft and Type 279 at her masthead, the small dipoles visible above the 281 antenna being the associated IFF (Type 243). Visible on her foremast are the twin cones of Type 252, and the array below appears to be the Type 243 associated with the Type 279 air search radar above, its drive motor visible below it. A Type 242 IFF, typically associated with radars such as Type 293, is just visible above the after edge of the funnel. Forward, she shows a pompom control set (Type 282) and a medium-caliber antiaircraft fire control radar (Type 284), all of her sets being of wartime type. The British height-finder, Type 277, surmounts her bridge, and is seen edge-on, and what appears to be the 'cheese' of Type 293 is visible aft. This radar configuration was typical of Royal Navy ships at the end of World War II.*

MTI will pick a fast-moving target out of a flock of birds.

Many naval radars are limited by sea clutter, by echoes from the randomly moving waves. Several considerations compete here. First of all, the longer the radar wave, the more it tends to average out over the patch of sea it strikes. On the other hand, a narrow beam limits the area of clutter (sea) as compared to the target, which suggests the virtues of highly directional, short-wavelength radars. Similarly, short pulses tend to 'freeze' the sea motion and hence to avoid clutter. However, a pulse-compression signal is as effective and it permits the use of higher power.

High PRF does little good: the sea changes rapidly, but not rapidly enough for its effects to cancel out between one pulse and the next. Experience suggests, rather, that what counts at sea is the correlation of signals from successive scans; this is sometimes expressed in terms of a 'decorrelation' or randomization time of 0.01 second. For example, in one experiment a rapidly rotating antenna (600rpm, PRF 5000) was used. The postwar periscope detection radars (SPS-19 and -20) operated on this principle.

ATMOSPHERIC EFFECTS

Similar considerations apply to other forms of clutter, except that in some cases clutter is literally invisible at low enough frequencies. That is particularly the case in rain, the drops of which are important only below 10cm. Although 10cm is far beyond the size of a single raindrop, it is comparable to a correlation length in rain, *ie* a distance over which raindrops fall in some coordinated manner. Matters worsen considerably, of course, as the radar wavelength approaches raindrop dimensions, as at K-band. Even at X-band the very small droplets of clouds reflect substantial quantities of radiation: X-band radars are commonly mounted aboard airliners as storm detectors. Larger particles, such as snowflakes, affect the longer-wave radars. As in the case of sea clutter, the proper choice of polarization can reduce these problems.

Very high resolution is a defense against such spread-out clutter. That is, a radar capable of differentiating objects over very short distances can, in effect, divide the space it searches into small cells. The smaller the cell, the better the chance that a target in a cell will produce a significantly larger echo than the clutter surrounding it, if the clutter is at a roughly constant level in the entire area being searched. Thus, the finer the resolution (to a point), the better the chance the radar can overcome weather clutter and even chaff. In this respect an inherently high-resolution (pulse-compression) radar such as SPY-1 is at a considerable advantage compared to an inherently lower-resolution set such as the frequency-scanned SPS-48, the resolution of which is limited by the bandwidth available per step in elevation.

As for chaff (strips of reflecting foil which create artificial targets), one solution is to adopt a wavelength so long that the radar will not see the strips. Generally the ideal length of a strip of chaff is half the radar wavelength. During World War II the German night-fighters adopted 200 rather than 400mc/s radars as a countermeasure after the Allies deployed chaff over Germany. Similar considerations make frequency diversity among the radars of a fleet attractive. Alternatively, measures such as Moving Target Indication and high resolution may overcome the weather-like clutter associated with chaff.

The atmosphere has two other major effects on radar performance. First of all, it absorbs energy as the beam passes through, and this absorption rises with the radar frequency. Typical transmission losses for air search radars pointed at aircraft 200nm away might be as shown in the accompanying table. From the table we can see that at X-band radar range in air for a given power output might be reduced by 37 per cent as compared to free space (the fourth power of range effect), whereas at L-band the reduction would be only about 2 per cent. These are merely suggestive; in actual cases a great deal depends on beam elevation and atmospheric density. However, the high losses at L-band and above imply the virtue of P-band long-range air search radar, although against the lower transmission loss must be set higher noise losses at lower frequency.

Typical Transmission Losses

Band	Loss factor (divide signal strength by)
P-band (1.5m)	1.13
L-band (30cm)	1.91
S-band (10cm)	2.18
X-band (3cm)	3.55

Really spectacular losses occur at K-band. At such short wavelengths, radar signals begin to match the resonant frequencies at which individual molecules in the atmosphere can absorb energy. In particular, the original K-band radar (1.24cm, 24,000mc/s) was so close to an absorption frequency of water vapor (22,200mc/s) that range proved negligible. The Ku- and Ka-bands were chosen to avoid this problem.

DUCTING

The other great atmospheric effect is refraction: the layers of the atmosphere bend radar waves. Under normal conditions, refraction allows radars to look slightly beyond the geometric horizon, but under some conditions a very strongly refracting layer (a 'duct') is formed. Signals directed up into the duct are trapped, as in a gigantic horizontal waveguide, and aircraft above or below the duct may approach undetected.

Perhaps the most spectacular ducting effect is caused by evaporation from the sea surface. Signals trapped in a 'surface duct' so formed can achieve incredible ranges, ranges which might at first glance seem absurd in view of the low height of the radar. For example, a submarine operating under duct conditions could generally detect destroyers at ranges far beyond the latter's since the destroyer's masthead search radar would lie above the duct.

The classic cause of near-surface ducting is evaporation into a dry atmosphere. The refractive powers of the atmosphere depend to some extent on its water content. A very dry atmosphere over water shows considerable moisture close to the water, but the level of moisture falls rapidly. The high rate at which refractive powers change is responsible for the duct, which is typically 30–90ft thick. During a wet season, more of the atmosphere is damp, refraction changes more slowly, and the duct vanishes. For example, the record range for a surface radar is held by a P-band set at Bombay, India, which once detected points on the Arabian peninsula, 1500 miles away – during the dry season. During the monsoon, the same radar, which was unaffected *directly* by the rain, reverted to a more understandable 20-mile range. Surface ducts are so common that it is tempting to make regular use of them, but even though the phenomenon was well known during World War II it is still unpredictable. Indeed, that may be said of most naval radar effects due to atmospheric conditions.

Higher-altitude ducts are also caused by rapid variations of temperature with height, for example by 'inversions' in the meterological sense. Often temperature inversions occur together with abrupt changes in humidity – in fair weather, such bad weather tends to smear out the layer structure of the atmosphere, *eg* via rain.

Ducting depends for its effect on the *rate* at which the index differs from one frequency to another; and the measure of rapidity is really wavelength. Generally longer wavelengths tend to avoid ducting, which is one reason the US Navy reverted to P-band for long-range air search.

Other related atmospheric effects are responsible for spurious radar echoes. In 1943 a US battle group in the Aleutians fought a lengthy engagement (the 'Battle of the Pips') against just such a false target, and even at present there is no general prescription for avoiding them.

DETECTION AND RANGING

There remains the question of just what the radar 'sees' at the target – which is also a question of wavelength. In most cases, the smaller the wavelength, the larger the effective cross-section. However, in practice other factors enter. For example, longer-wave radars tend to average over the details of their targets. At shorter wavelengths target design features, such as streamlining, may reduce the effective radar return. This type of effect was experienced with L-band air search radars in 1949–50, and was partly responsible for the revival of the P-band.

It was later found that the wavelength effect was strongest for very streamlined targets, such as jet aircraft carrying no bombs or fuel tanks; for such reflectors, the radar cross-section is proportional to the square of the wavelength. In practice attacking aircraft would be quite 'dirty' – no modern attack aircraft carries its load internally. On the other hand, attacking *missiles* are generally of very 'clean' design, and so the choice of P-band was fortunate for a US fleet faced with what was largely a cruise missile threat. Such an analysis suggests that intrinsically lighter L-band radars might well prove satisfactory for a fleet faced with the main offensive arm of the US Navy, *ie* with carrier aircraft.

The actual performance of a radar depends, in the end, on what happens to its signal at the receiver. Noise submerges most radar echoes, but there is a probability of detection among the echoes of each scan, typically expressed as a 'blip–scan ratio'. The net probability that a target will be detected before it reaches some minimum range, then, is determined by the probability of its detection, scan by scan, as it approaches. This in turn depends on operator ability in most radars. In automatic detection systems, the internal electronics of the system is adjusted to accept some Constant False Alarm Rate (CFAR) based on a measurement of the noise in which signals are submerged. The higher the level of noise (or jamming), the greater the probability that real signals will be missed, or that too many false signals will be accepted as real ones.

How, then, is radar range to be defined? Typically a range for 50 or 90 per cent probability of detection on a given target is used. For example, figures of 300nm and 165nm have both been used for the big SPS-2 radar of the early 1950s. The former is a 50 per cent figure, the latter 90 per cent. However, 50 per cent is not an unreasonable measure of range in a task group, where there will be several similar radars. Two 50 per cent radars have a joint probability of detection of 75 per cent; three, of 87.5 per cent; four, of over 93 per cent. Enough independent radars, linked by a system like NTDS, make much lower detection probabilities worthwhile: five 20 per cent radars add up to a joint probability of detection of better than 67 per cent. In fact detection made by only one radar of the five might not be believed, but even so the combination of relatively ineffective systems can produce a relatively impressive result.

Radar range capability is sometimes estimated from PRF, *ie* from the maximum range at which a target can be detected without ambiguity in range due to second-time-around (STA) signals falling among more normal echoes. This measure is most often applied to foreign systems for which signal-processing and normal performance data are unobtainable. However, given all of the factors which detract from effective radar performance (particularly performance against difficult targets) it can be misleading. Probably its chief justification is the belief that no intelligent radar designer would knowingly choose a PRF lower than necessary, as that in itself would reduce average power and average detection range.

A related issue is *data rate*, literally the rate at which the radar renews its store of data concerning the area around it. For example, a radar rotating at 10rpm 'looks' in each direction ten times per minute, or once every six seconds, for a *data interval* of six seconds.* The faster the target, the shorter the data interval must be if they are to be tracked, *ie* if target location on one scan is to have any relation to target location on the next. On the other hand, the probability of detection on any one scan is in part a matter of the number of pulses available, in other words of the PRF *per scan*: the higher the PRF, the shorter the maximum range of the radar. This dilemma arose very forcefully just after World War II, as aircraft speeds rose dramatically at the same time that their effective cross-sections fell due to streamlining. Radar scanning rates were driven up at once (compare SR-3 to SPS-6: 2.5 or 5rpm *vs* 5 to 15), but that only aggravated the PRF problem. One solution, adopted in some long-range radars (see the radar characteristics) has been a high rotation rate coupled with long pulses and a moderate PRF.

For a fixed-antenna (electronically scanned) radar, the appropriate measure is *frame time*, the time it takes the radar to carry out a complete operating cycle. This may include not merely a complete scan of some volume of space, but also the generation of special tracking beams in the direction of particular targets of interest, to detect target motion since the previous frame. The term is adopted from television usage, in which the electron beam which actually makes the picture on the screen requires a fixed time to fill out the frame, or image, being transmitted: it is only the generation of successive frames which produces the moving picture, as in a movie

*Some writers prefer a data rate to a radar rotational rate, using *seconds per rotation* (spr) as a measure.

4. Displays and Antennas

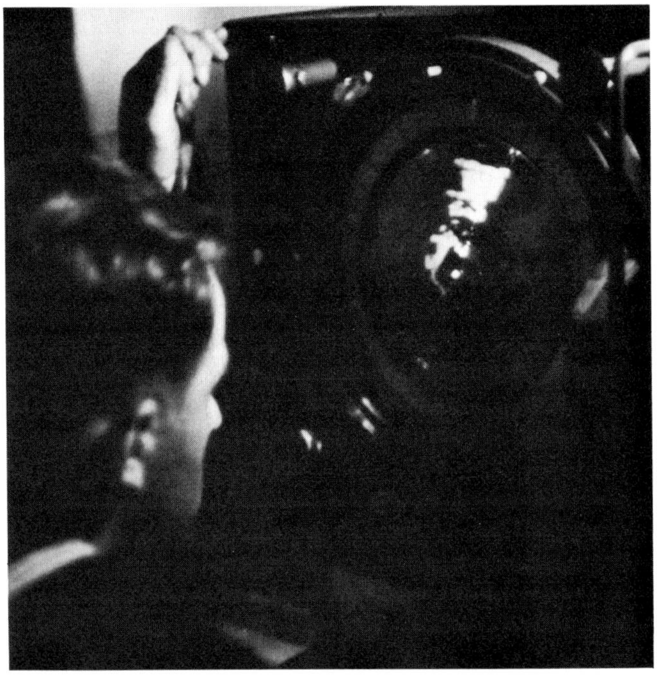

CONWAY PICTURE LIBRARY

The earliest radar display, the A-scope, was a simple indicator of range but not of bearing. A great variety of other displays were invented, most of them during World War II; below we list the most important of them. The A-scope was deficient in that it provided no direct read-out of the complex variety of targets around the radar; the radar operator would have to form a kind of radar map in his mind as he scanned. Even for gunnery, a pure A-scope would provide no indication of relative angular error. Hence the development of alternatives:

Plan Position Indicator, or PPI. Probably the best-known display of all, the PPI produces, in effect, a

◀ *A typical Plan Position Indicator (PPI), this one aboard HMS* Vanguard *in a photo taken in 1952.*

▼ *A summary display, at the Test Control Center at the US Navy's Aegis Combat System Engineering Development site at Moorestown, New Jersey, contrasts dramatically with the equipment shown in the previous photograph. The human element is now 'coordination', rather that 'operation' as in the early days of radar development.*

RCA

radar-made map, the center of which is the location of the radar itself. Echoes appear as bright spots on the circular scope, the radar direction as a bright line sweeping around the scope as the antenna itseft sweeps in bearing. A PPI is in effect a rotating A-scope in which the spikes of the former are replace by brightening on the face of a cathode ray tube (CRT). PPI was the key to effective search radar; it was invented independently in Britain and the United States almost simultaneously. Later forms sometimes incorporated an off-center origin, so that part of the radar 'map' could be emphasized.

A major defect of PPI, and indeed of all the spot-brightening scopes, is reduced efficacy against noise and jamming: an A-scope is the best means for blip, as opposed to noise, recognition. Hence many sets equipped with PPI or similar displays also had A-scopes.

B-Scope, or Range–Bearing Display. Unlike PPI, this display is distorted as compared to the scene it describes: it is a plot of range against target bearing angle represented in rectangular coordinates. As such it proved useful for fire controls: B-scopes were used in Mk 3 and Mk 13 main battery fire control sets.

Range–Height Indicator, or RHI. In effect a PPI displaying range and altitude instead of range and bearing angle, RHI was the natural display for height-finding radars such as SPS-8 and forms part of the display for such three-dimensional search sets as SPS-39.

E-Scope. A distorted RHI in which elevation *angle* rather than elevation forms one coordinate. As in the case of the B-scope, this is a useful presentation for error indication, for example in fire control.

C-Scope. An elevation/azimuth (bearing) display. If one thinks of an RHI as the side view of a scene, this is the end view; it is the type of display a CCA (Carrier Controlled Approach) radar such as SPN-35 requires.

Comparison Scopes (K- or L-Scopes). These were developed for lobing radars, to permit comparison between blips from alternative lobes. K provided side-by-side, L back-to-back comparison. Both were variations on the A-scope.

In fact just about every conceivable type was tried, in some cases to match unusual types of scanning (*eg* spiral); there was even a J-scope, in effect an A-scope wrapped around the circular circumference of a CRT.

EARLY DISPLAYS

Until well after World War II, radar detection was exclusively a matter of an operator *seeing* something on a display and then measuring its position on some appended scale. There were no means by which radars could automatically detect targets. Even so simple a fire control system as Mk 57 required that an operator see the appropriate return and mark it electronically. The precise character of radar displays, then, became a very important factor in radar efficiency and accuracy. For example, by 1945 it had been discovered that actual detection of a target on any given scan of a rotating radar was a matter of probability – of the probability that a given return would stand out above the noise, and of the probability that a given operator would notice. Operator performance depended on training, on the character of the radar display itself, and quite strongly on the extent to which the operator might expect a target to be present, *ie* to which he had been *alerted*.

Similarly, a great part of the anti-ECM armory was a combination of radar-scope modification and operator training. One might say that correct interpretation of a noisy (jammed) radar display is a problem of pattern recognition, and that a human is far more adept at it than is a machine. On the other hand, a machine can maintain a high level of alertness for much longer than can a man – hence current interest in ADT, or Automatic Detecting and Tracking. In essence an ADT system accepts some percentage of false alarms to achieve a very high probability of catching real targets consistently and even in high densities. A human operator may do better at times, or on an individual target, but he may not consistently bet on contacts apparently lost in noise.

INTEGRATION

Much of radar display development has, naturally, been devoted to improving operator performance by attacking the effective signal-to-noise ratio. The usual method is *integration* over successive pulses to allow the fixed echo to dominate the random noise. In particular it was discovered that radar scopes allowing traces to persist for some noticeable time permitted effective 'post detection' integration. More recent radars use signal processing (*eg* range correlators and video processors) to achieve some measure of integration before the signal is displayed.

Of all the types tried, the A-scope was most effective for post-detection integration: although a spot on a PPI could become only so bright, the height of an A-scope echo could grow and grow as returns added up. In many cases A-scopes were provided as antijam aids. The radar could be stopped in a likely direction, and pulses added on an A-scope until a target became clear.

Another problem has been the *apparent* size of radar 'blips', which imposed an effective limit on accuracy. Many radar displays incorporated variable range scales which permitted high accuracy at short range. For example, the wartime SL surface search radar had a PPI 7in in diameter, *ie* with a 3.5in range scale – which could be made equivalent to 4, 20 or 60 miles. The minimum dot of light was about 0.03in in diameter, about half a mile in the 60-mile scale. One attempt to resolve this problem was a *delayed* display, which deleted all returns *inside* a set range. For example, a 40-mile delay would permit display of targets at 40–60 miles on the 20-mile scale, with far better effective resolution (0.03in is, then, about 300yds). Of course,

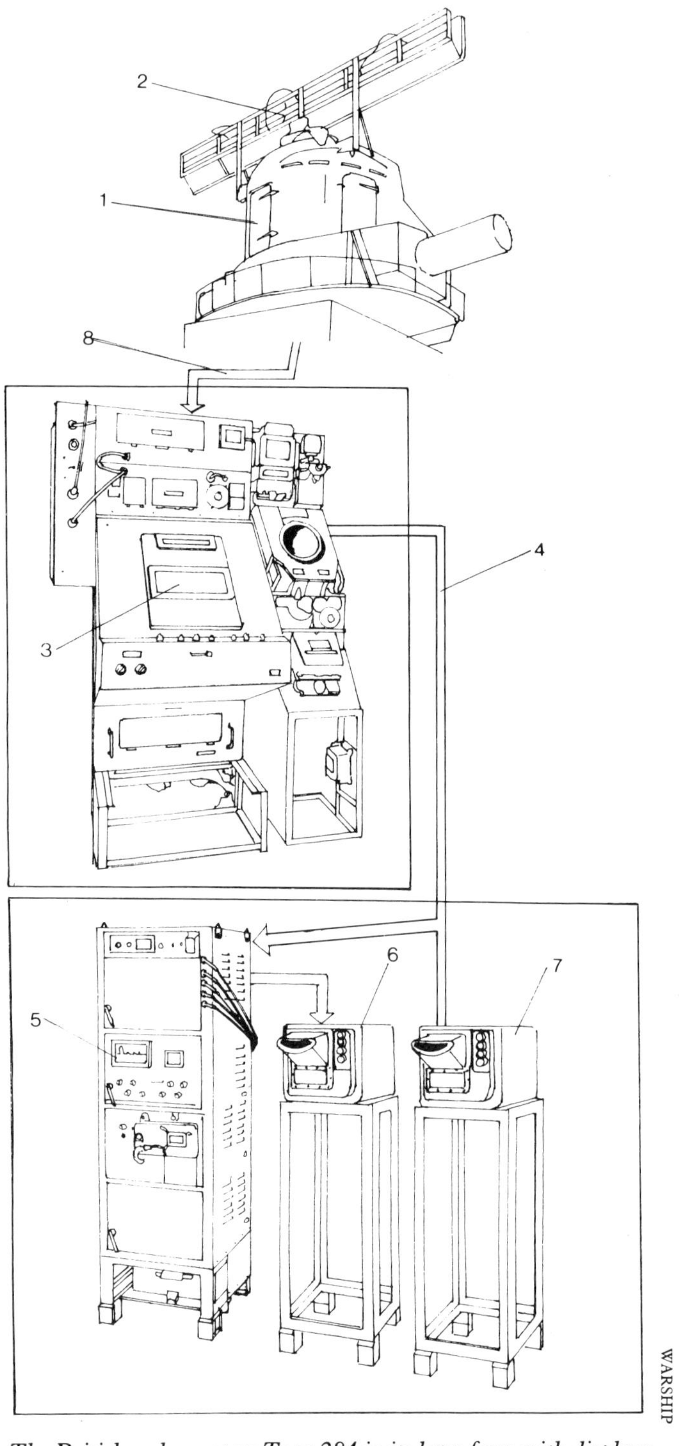

The British radar system Type 284 in its later form with displays in the transmitting station (TS). **1** *Director Control Tower (DCT).* **2** *Radar antenna.* **3** *Ranging display Panel L12 (normally removed when display panels fitted in TS).* **4** *Radar signals to displays in TS.* **5** *Ranging display Panel L24.* **6** *Spotting tube.* **7** *Training tube.* **8** *Radar signals to radar office.*

In addition to the equipment shown, the TS was fitted with a second ranging display (Panel L18 — part of the Type 273QR) for use as a standby or with the after gunnery radar set.

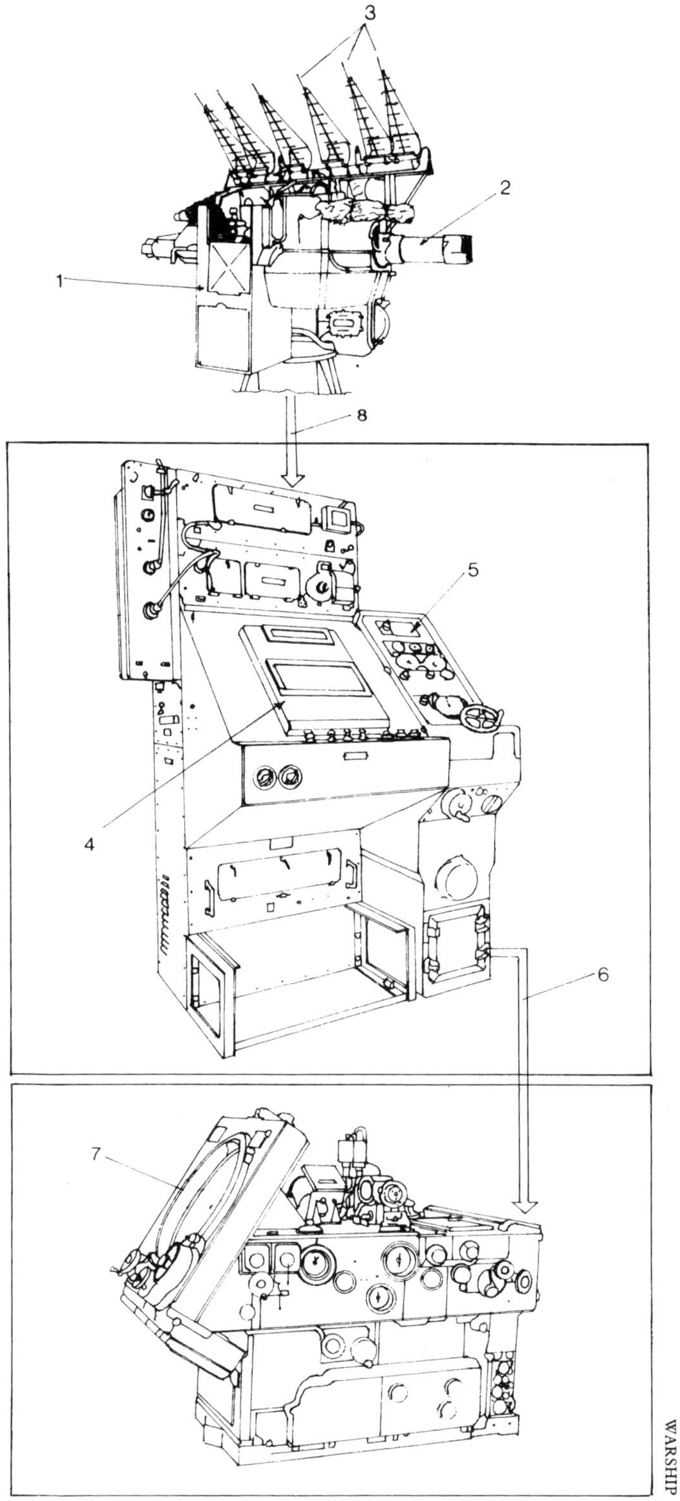

The British radar system Type 285 as applied to the High Angle Control System (HACS) Mk III. **1** *HACS director.* **2** *Rangefinder.* **3** *Yagi antennas.* **4** *Ranging display Panel L12.* **5** *Range Transmission Unit (RTU).* **6** *Range information transmitted to HACS in High Angle Control Position (HACP).* **7** *HACS Mk III (mechanical computer for processing fire control information for AA weapons).* **8** *Radar signals to radar office.*

In its later development, two ranging displays (Panels L24 and L27), together with a spotting tube, were fitted in the HACP, and the ranging panel in the radar set was removed.

The DASH antisubmarine drone helicopter was tracked by ship radars to attack a target visible only on sonar; a combined radar and sonar display in CIC permitted coordination, but the system suffered because sonar and radar ranges often did not quite match. DASH suffered, too, from the absence of feed-back from helicopter to operator: when the operator could not see the drone, it often maneuvered in unexpected, even dangerous, ways. For example, it could return to its ship upside down, with unfortunate consequences. DASH was abandoned before solutions to the coordination problem could be found. One proposal called for the helicopter to drop a sonar buoy which could serve as a reference point, while the helicopter itself flew high enough for ship radars to track it.

the delayed display introduced new problems of distortion, for example in the relative bearings of different targets. Expanded sweeps were also employed on A-scopes ('R', or Range-scopes) to achieve high-precision ranging; in many cases such sweeps were calibrated by the electronic 'clock' which times the sweep itself, for very high precision.

A variety of display-improvement devices for PPI were devised during World War II. Sensitivity-Time-Control (STC) reduced gain at very short ranges to reduce the blind spot at the center of the PPI generated by the initial (outgoing) pulse; it also reduced sea return and helped reduce side lobe echoes from very close targets, *eg* other ships in formation. In at least some cases a *smaller* PPI tube was used to sharpen images. Other mechanisms, IAVC and FTC, are discussed elsewhere in this book under counter-counter-measures.

A major wartime development was the use of standardized remote displays, which could be concentrated in a Combat Information Center. Since 1945 most naval radars have been connected to a series of standardized displays (AN/SPA series). New features included display-generated lines (cursors for bearing, strobes for range in PPIs) which could be used to make readings more precise; continuous range-scale variation; and off-centering so that, in one system, any target within 250 miles could be made the center of the PPI display (cursors did not have to have the same center, so that relative data could be read off). Off-centering was particularly useful in displays used as shipboard terminals for airborne early warning (AEW) radar.

In some cases postwar displays could be used to show the outputs of several sensors simultaneously. DASH, the Drone Anti-Submarine Helicopter, presented perhaps the most interesting example. DASH was tracked on a destroyer air search radar while its submarine quarry was tracked on sonar; inputs from both were combined in the CIC, from which DASH was controlled. In fact this system was not entirely successful, partly because sonar range was far more ambiguous than radar range. Combined displays are also necessary in weapon systems requiring hand-off from one radar to another, *ie* in the missile ships.

THE FUNCTION OF THE ANTENNA

To a considerable extent – there will be important exceptions – the appearance of a conventional radar antenna betrays the range of wavelengths on which it works. There are two tell-tales: the structure of the antenna and the means by which the radar waves are projected.

Conventional antennas consist of *radiation sources* and *reflectors*. Each source of radiation – microwave horn or dipole – creates a relatively undirected shower of radio waves. Several sources, fixed in relative phase, can combine to form a beam which a reflector sharpens or directs; in some cases it is sufficiently effective to permit the use of a single source. Typically one speaks in terms of an *array* of radiating elements.

It should be remembered that, at least in all US naval radars, the same antenna *receives* as well as transmits. During the receiving portion of the radar cycle, the radiating elements receive radar signals gathered by the reflector, in much the way the mirror of a reflecting telescope gathers light. The great problem in using the same elements to transmit and then to receive was always the vast difference between transmitted and received power. Radars commonly put out pulses of a megawatt or more; what comes back may be a millionth of a watt, *ie* a millionth of a millionth of the power sent. A receiver sensitive enough to pick up so little power would be burnt out by the pulse leaving the radar a short time earlier. The difficulty, then, has been to concoct a fast-acting switch for transmission and reception, a 'TR box'. How fast it acted would determine the 'dead time' between the emission of a pulse and the earliest time the radar could pick up an echo, in other words a minimum range.

Although most of the discussion below will be orientated to radar *transmission*, the reader should keep in mind that reception is in many ways a precisely parallel process. For example, an antenna generally has identi-

The missile cruiser Albany, *shown after a refit in 1974, illustrates the complexity of modern warship topside arrangement, given the great number of radars and radio antennas required. Forward, she has a discone antenna (***12***) for the data link to her NTDS; it could not be accommodated elsewhere in the ship. Similar installations in gun-armed ships limit effective fields of fire, but missiles are not generally fired dead ahead in view of the blast they create. Originally* Albany *had SPS-30 long-range height-finders fore and aft, presumably in large part to make her more useful as a Task Force air control ship; ultimately one was removed, the other remaining (just under the number* **36***). Missile control required a combination of long-range air search (SPS-43A;* **58***) and high-data-rate search (SPS-48, not numbered, on her forward 'mack'). In this class centerline space was so scarce that masts and funnels had to be combined, to form 'macks'; the later US missile frigates were similarly arranged.*

cal directional patterns for transmission and reception, even though one might imagine considerable differences at first blush.* In particular, the importance of sheer antenna size is clearest for reception, since a larger antenna will sop up more of the returning echo than will a smaller one. But it will also be more directional, which will mean that more of the energy sent out will hit the target. In effect, then, an increase in antenna size would have a double effect on radar range, possibly far more of an effect than greater power. The first US naval air search sets provide a good example. The earliest set, CXAM, was a big 'mattress' too large for anything short of a heavy cruiser. A new, smaller antenna was then developed, but the resulting SC had poor range performance: power had to be *doubled* before performance even approached that of the earlier set. However, merely providing a much larger antenna (SK, three times the height of SC), roughly doubled effective range.

*There are important exceptions. Some antennas, such as those of monopulse radars, employ different feeds for reception and transmission. Phased arrays, such as SPY-1, have very different reception and transmission characteristics.

Radar beam size depends upon antenna size *in wavelengths*: the larger the antenna, the narrower the beam. If antenna size is fixed by other considerations, the shorter the wavelength, the narrower the beam and the more directional the system. Directionality, for example, makes very short wavelengths attractive in aircraft and missile radar systems. Thus the illuminating radars of semiactive homing missile systems tend to operate in X-band, the shortest wavelength which still provides for high power and long range. The radar receiver in the missile, which is necessarily small,

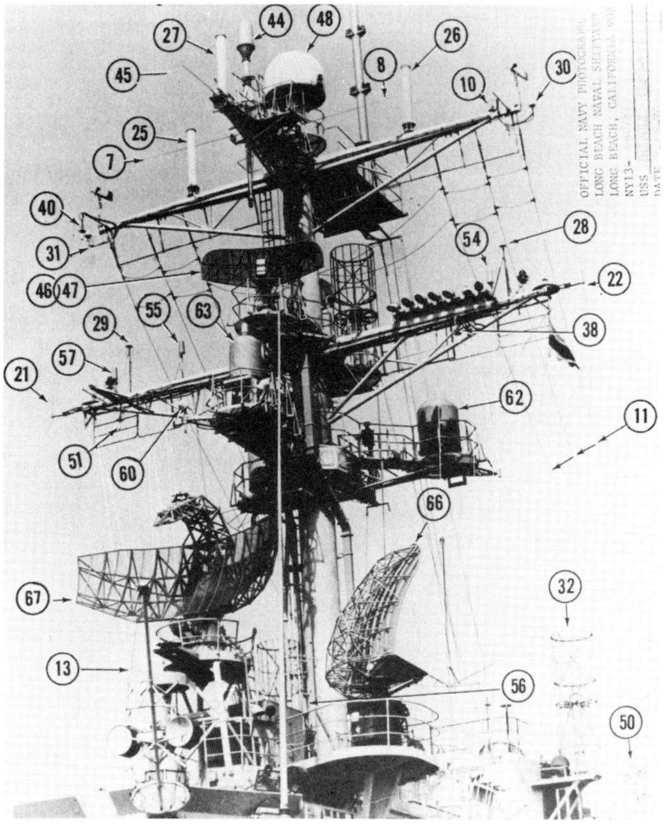

Even helicopter carriers require extensive electronic installations, as USS Tripoli shows in this January 1976 photograph. Most of the antennas are broad- or narrow-band radio or ECM; the only radars visible are a Sea Sparrow illuminator/director (not numbered, at the lower left hand corner of the photograph; note its separate receiving and transmitting antennas for CW signals); an SPN-6 CCA marshaling radar (**66**; the large radome-enclosed SPN-35 is not visible); an SPS-40 (**67**); an SPS-10F surface search radar fitted to operate also in the SPS-58 low-flyer detection mode (**46/47**); and a commercial LN-66 navigational radar (not numbered, in the small circular radome forward of and below the main air search set, SPS-40). As in all US carriers, TACAN (**48**; in this case, URN-20) surmounts the mast, closely surrounded by the broadband dipole-like antennas of AS-1018/SRC (**25**, **26** and **27**; for SRA-33 radios). The other standard antenna type present is AS-390/SRC, a small radome surmounting angled, splayed-out dipoles (**28**, **29**, **30**, **31** and four spares, **38**, **40**, **51** and one not visible here). Surely the most unusual antenna present is the cage, AS-2231, just abaft the mast (not labeled); the ship had a total of five. She also had the usual whips, such as a 16ft sleeve antenna, the top of which is visible as (**13**). As seen here, the ECM installation matched those of earlier ships, as the active jammers are not visible; Tripoli shows ECM intercept antennas (paired port and starboard, and feeding a WLR-1C: only one of the four, an NT-6613, **60**, is clearly visible here) as well as AS-571/SLR (**62**), AS-899/SLR (**63**) and AS-616A/SLR (a radome hidden by the mast, symmetrical with **62**). Also standard were the URD-4B direction-finder (**44**) and four IFF antennas (**54**, **55**, **56** and **57**; AS-177/UPX). Not visible is a series of satellite communications antennas.

Carrier reconstruction in the early and mid-1950s emphasized the need for clear antenna arcs: Bennington, photographed in that period, is a case in point, with a single massive radar mast, topped by a TACAN beacon. The two small mattress-like antennas near it are the AS-571 and AS-616 antennas of her SLR-2 radar direction-finding system, in effect replacing the wartime DBM. In later ships, such as FRAM destroyers, they were generally enclosed in radomes, but early installations were not covered. The narrower radome is a URD-4 direction-finder. Wind-balancing vanes on the SPS-6B and SPS-8 antennas below are particularly noticeable, as is the wartime-type SC-series set (with a postwar IFF replacing its Mk II system) abaft the heavy radar mast.

achieves far better directional performance at X- than at, say, L-band. On the other hand, an antiradar missile behaves much like a semiactive homer, seeking the beam of the radar it is intended to destroy. It, too, is limited in diameter and it, too, will perform better its chances against antiradar missiles. In this sense a 400mc/s radar such as SPS-40 may be a better compromise between precision and immunity to antiradar attack than a narrower-beam, higher-frequency system such as SPS-49.*

The first point worth noting is that the shorter the wavelength the more solid the reflector must be: to

* As a useful approximation, beamwidth in *radians* (*ie* in units of about 57.3°) is equal to one divided by antenna size in wavelengths. That is, an antenna 100 wavelengths long will produce a beam about 0.1 radians, or 0.6°, wide.

metric waves chicken wire looks perfectly solid and reflective, but to microwaves it is full of holes. Thus the P-band SK search radar is an open meshwork 'mattress', whereas the S-band SG has a solid antenna. Moreover, the reader can be quite certain that, wherever possible, mesh *has* been substituted for solid pieces of metal in the interest of saving weight; indeed, a few microwave radars have had smaller-mesh antennas (*eg* SPS-10).

The other characteristic is the *type* of radiator. Metric radars typically employ dipoles, but microwave radars use horns; the former terminate coaxial cables, the latter waveguides. This distinction leads us to consider very briefly radar 'plumbing', the means by which the product of a radar wave generator proceeds from generator to antenna.

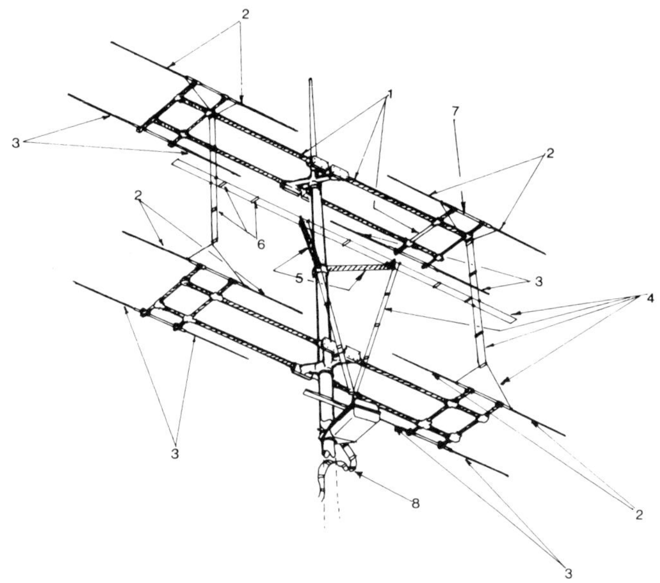

WARSHIP

Radar antenna for Type 281 (virtually a doubled version of Type 279 with the two large horizontal 'H' antennas replaced by four of smaller dimensions). **1** *Supporting framework.* **2** *Transmitting dipoles.* **3** *Reflecting dipoles.* **4** *Open wire feed carrying radar signals to transmitting dipoles.* **5** *Support arms for feed wires.* **6** *Insulators between feed wires.* **7** *Insulator separating dipole halves.* **8** *Feed wires arranged in spiral form around mast to allow antenna to rotate through a reasonable arc.*

POWER TRANSFERENCE

The generator does no more than cause oscillations in an electric current – very rapid oscillations, to be sure, but in principle comparable to the cycles on a home electric outlet. The currents (signals, if one likes) on lines attached to such an outlet are quite happy to travel along the interior of thin wires. However, as the frequency rises, the current comes to occupy only the outer skin of the wire, and indeed as the volume of wire available to it falls off, resistance to its passage rises – and energy begins to dissipate in heat.

Conventional HF radio operates at about the upper limit of frequency for efficient use of conventional wiring. Above that some effort must be made to use or contain the radiation (radio) field carried along with the current. One method, the coaxial, was simply to surround a signal-carrying wire with a hollow conductor. In effect the latter would confine the radio waves associated with a signal carried by the former; the name coaxial derives from the fact that both conductors share the same direction-axis. The radiator associated with a coaxial would consist of a pair of wires, one connected to the inner conductor, the other to the outer: a *dipole*.

In all cases of dipole feed, the dipole dimensions show the radar wavelength: dipoles are generally very nearly half the radar wavelength, because they are by far more efficient at such a size. In fact the few microwave radars fed by dipoles show their short wavelengths in the very short length of those dipoles. Postwar US 'mattress' radars such as SPS-29 show a new type of dipole radiator, folded to form a closed loop. Like their simpler forebears, those dipoles are by far more effective at a total length of half a wave; but they permit a somewhat simpler electrical arrangement. Yet another variation is a 'printed dipole', embedded in a dielectric plastic. Such radiators figured in the SPS-32/33 planar arrays, and in the 'broadside arrays' of IFF, *eg* the IFF antennas of air traffic control radars.

Coaxials could not be used with microwaves. The latter come to resemble light, and do not travel well along wires. However, they can be directed down a pipe

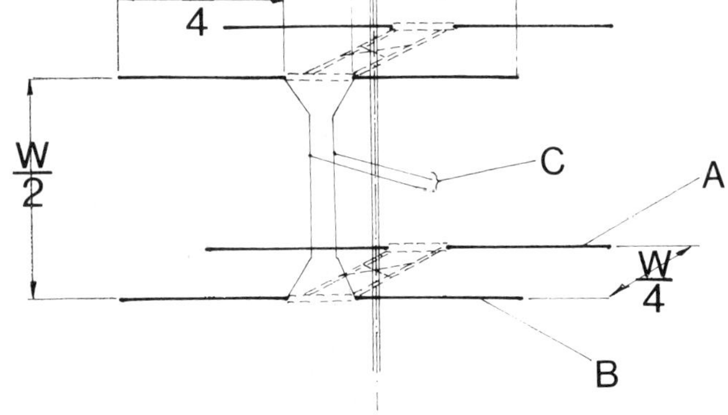

WARSHIP

Radar dipoles were normally half-wavelength, ie *the total length (excluding the insulated gap between the two poles) was W/2, where W = wavelength; each half was, therefore, W/4 in length. In parallel dipoles (*eg *British Types 279 and 281) the upper and lower were a half-wavelength, and the reflecting (***A***) and transmitting (***B***) dipoles were a quarter-wavelength apart. In the diagram,* **C** *represents the twin open-wire feed.*

BY COURTESY OF A D BAKER III

The missile cruiser Galveston *was equipped with the long-range Talos missile, but its SPS-39 radar did not provide corresponding long-range data. An enlarged planar antenna was tested in 1962 as SPS-39B, but it did not provide sufficient improvement. The smaller planar array antenna tested at the same time became standard; its sinuous-feed waveguide was contained in the fin on one side of the antenna.*

made out of conducting material – a *waveguide*. A waveguide open at the far end pours out its energy, and if the open end is properly shaped into a horn, the energy is somewhat directional. Some microwave horns ('cheeses') are so directional that they are effective antennas. A 'cheese' consists of two flat parallel plates surrounded by what would, in a more conventional antenna, be the narrow reflector; the feed can be from inside or, more conventionally, via a horn in front. In either case the pattern is broad parallel to the plates, narrow perpendicular to them. 'Cheeses' were particularly favored by the Royal Navy. On the other hand, the radiation pattern of a dipole is symmetrical about its axis: a dipole radiates about equally in all directions *at right angles* to its axis. In this it is something like the filament of a light bulb, which needs a reflector to produce any kind of well-defined beam at all.

Both coaxials and waveguides look very much like storm drains running down the masts and funnels of warships. The waveguides in particular require careful design to avoid sharp turns which might reflect microwaves and thus cut down the energy delivered to the antenna. In both types a major problem is dampness, which can cause short circuits, for example between the central conductor and the outer skin of a coaxial. The usual solution is to pressurize the air in the waveguide or coaxial to keep out damp sea air; without continuous air pressure, electrical breakdown soon occurs. Moreover, both are points of vulnerability in a warship – a radar with its waveguide cut is a dead radar.

The waveguide or coaxial transmits radar signals out from the wave generator and echoes back from antenna to display. A serious problem in early microwave systems was waveguide losses, which limited the practical distance between antenna and radar generative/display. That is why early US installations of the microwave surface search set (SG) placed the antenna near the bridge, even when such placement severely limited the arc the radar could cover. There was no such problem with the coaxials of metric radars; many battleships and at least one cruiser showed, in 1942–43, big metric air search sets on the mainmast and a diminutive SG forward near the primary conning position. The solution to the waveguide problem shows in the widespread revision of this arrangement in 1944: SGs were moved aloft to topmasts to clear arcs for the big SK air search sets.

SINGLE-FEED ANTENNAS

The reader will note that quite a few radars do not quite fit these prescriptions. For example, some postwar air search radars (*eg* SPS-6, -12, and -40) share the open mesh of metric bands and the horn feeds of microwaves; they occupy that intermediate band, the L-band. On the other hand, some microwave antennas are fed by dipoles.

In some arrays there are no substantial reflectors, only radiators. The radiators are moderately directional, since their emissions add up to form a well-defined beam. In all such systems, what matters is the relative phase (in effect the timing) of the elements of the array. In most arrays, such as the large dipole-fed air search sets (*eg* SK, SPS-43), relative phases are fixed to achieve a directional beam in a fixed direction. A more sophisticated concept is an array whose beam is *steered* by varying the signals sent to different array elements. For example, the beam of a *phased array* is steered by varying the timing (phase) relation between elements. The idea is not new: the first US phased array radar was the Mk 8 L-band fire control set, in which a series of moderately directional radiators was mechanically scanned in phase to produce a fairly narrow moving beam. Mk 8 was mechanically complicated, and it was soon displaced by a more conventional X-band set (Mk 13); for quite some time no more was heard of phased arrays.

Indeed, one of the great appeals of such single-feed antennas as SPS-6 must have been the relaxation of any need to control relative phases among radiators. For shorter wavelengths, precise phase relations imply closer dimensional tolerances – for example, relative phase in a slotted waveguide is very much a matter of the distance between slots. Thus microwave (slotted) arrays are a relatively recent development, and require a high constructional standard. They repay such attention by producing very sharp beams; moreover, an array permits electronic beam steering. Hence the use of a planar array in the new AWACS long-range airborne radar.

FRESCAN

During the 1950s it began to be appreciated that an *electronically* scanned array could steer its beam much more rapidly than could any mechanically steered radar; and phasing could be used to produce a very narrow beam indeed. Moreover, if the antenna did not have to move it could be made very large, so that even long waves could be used to form a narrow beam. Indeed, the same antenna could be used both for wide (search) and narrow (fire control) beams, or for that matter it could produce several beams at once. The most impressive developments of this type were the gigantic fixed SPS-32 and SPS-33 radars of USS *Enterprise* and USS *Long Beach*. In a series of much smaller sets, such as SPS-39, the radar scans mechanically in azimuth but the beam elevates electronically, to give a very rapid three-dimensional search with a narrow beam.

Two means of scanning were used: frequency-scanning (FRESCAN), and phase-switching. In the former, an indirect form of beam steering by phase variation, the beam angle varies as the frequency varies. In FRESCAN radars such as SPS-39 the frequency varies from pulse to pulse, so that subsequent pulses sweep out elevation angles. Such a design is far simpler than a directly phased array; on the other hand a wide frequency band is required and the radar cannot employ frequency diversity as an anti-ECM measure. It is, however, possible to design a FRESCAN radar to employ more than one frequency per angle of elevation, since what matters is the phase shift due to the delay between slots. At a higher frequency suitably chosen, the phase shift may be 360° greater, *ie* it may be equivalent to the original one. As of 1981, Marconi was offering a FRESCAN radar with six alternative frequencies per elevation.

FRESCAN radars suffer from inherent limits in their range resolution. That is, each beamwidth in elevation may be equivalent to a frequency shift of 1 or 2 megacycles. Any greater bandwidth, which would be required for shorter pulses (or for pulse compression) smears out the beam in elevation and is therefore self-defeating. Electronic scanning in elevation permits programming of the elevation pattern of the radar, *eg* to compensate for the ship's motion. Thus US FRESCAN radars can achieve considerable savings in weight compared to more conventional height-finders, which were mechanically stabilized.

The means employed to achieve phase shifts shows clearly in FRESCAN radars such as SPS-39 and SPS-48. What determines the angle of the beam is the timing delay between the emissions of neighboring radiators *in wavelengths*. The adjacent radiators in these radars are

Most modern radar types were developed for the first time during World War II, including even frequency-scanning. The US Mk 8 gunnery control radar, seen here atop the main-battery director of the cruiser Portland *(1944) employed a mechanical form of frequency scanning to achieve very rapid movement of its beam without loss of accuracy: it was described as the prototype of modern Track-While-Scan (TWS) radars. Above it is visible the large mattress of an SK air search radar, the small rectangular array above the mattress being an IFF. The large dipoles at the base of the fire control tower are AS-56/SPR-1 ECM antennas, searching for enemy emissions. Just visible to the left of the secondary-battery director (Mk 33) is the dish of a Mk 28 fire control radar; these directors were too light to carry the later 12/22 combination.*

cuts in a long waveguide folded intestine-like: the radar signal travels up the waveguide, leaking out of the slots as it goes. The folds are needed to achieve appreciable delays and hence large beam angles. It follows that FRESCAN is easiest to achieve at relatively short wavelengths, since otherwise the delay lines would not be long enough *in wavelengths*. In SPS-39 this microwave 'intestine' appears as the prominent spine running up the antenna; in SPS-48 and SPS-52 it forms an end plate. FRESCAN is relatively simple and has therefore

The large phased-array antenna was an attempt to escape the overcrowding of electronics aboard conventionally designed warships. It appeared, at least at first, peculiarly well suited to nuclear-powered ships, and the carrier Enterprise, *shown here in 1962, carried two of the first, in big 'billboard' arrays: SPS-32 on the right and SPS-33 on the left, the latter a pencil-beam set. As back-up, she was fitted with a conventional SPS-12, whose antenna is visible to the right; this photograph looks forward, as is evident from the position of the small radome for an SPN-10. The SPN-6 marshal radar also looks aft. The complex array of antennas surmounting the island was a series of ESM collectors, in principle far superior to rotating antennas such as AS-616/SLR. One peculiarity of the two massive electronically-scanned radars was that they had no integral IFF; indeed, one reason for the SPS-12 was that it could supply IFF identifications. The 'cheese' visible on the yardarm, then, was an independent directional IFF antenna, AS-1065/UPX. No later carriers had the SPS-32/33 array, although modified versions were often proposed through the 1950s and 1970s: it was far too costly, and indeed in 1981 the* Enterprise *is having hers removed in favor of quite conventional radars.*

been used widely in US service, whereas the alternative electronic scanning technology, the phased array, was for many years represented only by the huge 'billboard' of SPS-33.

The relative simplicity of FRESCAN may also explain its adoption in the 'Top Sail' and subsequent Soviet Navy three-dimensional radars; the Soviets never did field an equivalent of SPS-33 or of the later SPY-1. It appears that the fire control radar of the new *Kirov* will be their first true electronic-scanner; until its advent the only large fixed arrays in Soviet service were land-based antimissile radars. Some recent British (Plessey) land-based (airport traffic control) FRESCANs employ a long pulse through which the frequency is varied. In effect the returning signal is coded by frequency; the pulse is spread out in elevation angle.

Probably the greatest defect of a single-beam FRESCAN is the limited number of pulses it can project at any given target. Generally the radar scans rapidly in bearing, and it elevates quickly to cover the entire hemisphere on each scan. However, with only a very few pulses returned from any particular target, detection range suffers, and Moving Target Indication (MTI) becomes difficult at best; MTI is, after all, a pulse-to-pulse comparison process. Even angular accuracy can suffer. The problem cannot be resolved for any simple system, because it is inherent in the radar design that each portion of the hemisphere (technically, each resolution cell in elevation and bearing) be visited only briefly. One solution is to employ several simultaneous beams, scanning as a group in elevation, to increase the time each can spend on any given cell. This is the system employed in SPS-48.

PHASED ARRAYS

In true (directly) phased array, the phases of radiators are directly controlled electronically. SPS-33 used ferrite phase-shifters, electronic components which could be manipulated to impose delays on radar signals passing through them, *ie* to change relative phases in a programmed sequence so as to shape a beam. More recently it has become possible to manufacture small solid-state microwave generators which can *themselves* be used as the elements of a phased array, switching on and off in sequence, to form a beam. Such a system may be far simpler that the massive array of SPS-33, consisting of FRESCAN pipes in one direction – all fed from one radiation source – feeding into banks of ferrite phase-shifters in the other.

However, the new SPY-1 phased array radar of the Aegis system still uses a single powerful radio-frequency energy generator feeding a network of phase shifters. It therefore generates a single beam, and has a single signal processor. Final amplification is by a network of 32 amplifiers, and identical systems are placed in the pair of arrays at each end of the ship, to cut losses. Solid-state technology has drastically reduced the complexity (and failure rate) of the radar receiver involved, but the era of individual solid-state receiver-transmitter array elements is as yet far in the future, if indeed it is ever to arrive.

Planar arrays of elementary radiators are already very common: the radiators are adjusted in phase to produce a well-defined beam. The simplest form is a series of slots along a waveguide, an early example of which was the postwar SR series. Several recent airborne radars consist of two-dimensional planar arrays of radiators, which present opportunities for lobing by electronic steering plus mechanical steering for scanning.

RCA

Phased array radar antennas look quite simple externally, but their insides are a very different matter. This is the back of one face of a SPY-1A under assembly by RCA, showing waveguides leading to the face of the antenna.

ANTENNA SHAPE

The *shape* of an antenna is intimately related to the shape of the beam it produces. Very generally, a large antenna dimension is associated with a narrow beam dimension, so that a rectangular antenna produces a beam with roughly the opposite rectangular shape. Many antennas are curved to focus beams in much the way the mirror in a car's headlight focusses its beam: SK-2 corresponds (in shape) almost exactly to such a mirror. In some cases a short dimension is made more effective by curvature: the US Mk 3 and Mk 4 fire control radars show considerable curvature vertically but they are much longer that they are high. A more recent example is SPS-39, whose small dimensions and solid reflector proclaim it to be a microwave radar. It is sharply curved to produce a beam narrow in bearing, *ie* in its horizontal dimension; *vertical* focussing is assured by the height of the antenna and of the radiating elements which occupy a vertical spine along its curved face.

The postwar SPS-8 height-finder is another good example. Far taller than it is wide, it produces a horizontal fan beam, *ie* a beam much wider than it is high,

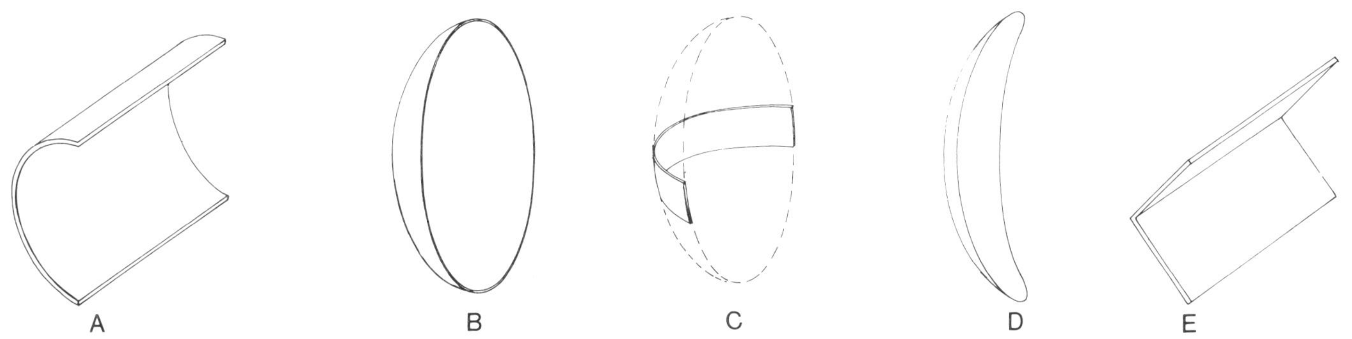

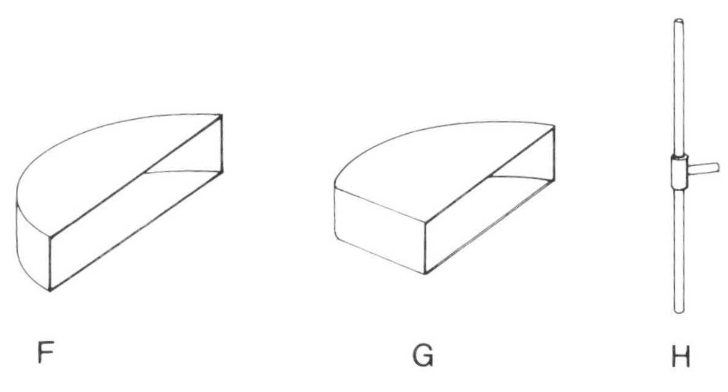

Types of antenna. **A** *Parabolic cylinder.* **B** *Paraboloid.* **C** *Truncated paraboloid.* **D** *'Orange-peel' paraboloid.* **E** *Corner reflector.* **F** *'Cheese'.* **G** *'Quarter-cheese'.* **H** *Dipole reflector (passive dipole).*

for good determination of an angle of elevation. On the other hand, the SPS-30 which replaced it is a circular dish, producing a roughly symmetrical pencil-like beam for very precise height *and* bearing determination.

Generally the reflectors of naval radars are either flat 'mattresses' of wire mesh (*eg* SPS-29), pieces of parabolas (*eg* SPS-6) or dishes (*eg* SK-2). There are, however, interesting variations. Thus the 'W' shape of SPS-43A can be regarded as a pair of vertically stacked corner reflectors. Such reflectors tend to reflect radiation back in exactly the direction from which it originates, and hence provide very strong reflection. In radar operation, corners in a target provide particularly bright reflections; In a big mattress, they provide a particularly directional beam.

However, at least at metric wavelengths, the simplest reflector is an appropriately placed dipole – in fact the reflector of the first US naval air search set, CXAM, was a dipole array. Later US radars employed mesh or

A typical example of the 'mattress' type antenna is the SPS-29 seen in this photo of the radar-picket destroyer Turner, *shown here in February 1962. Destroyer (and, for that matter, cruiser) radar outfits of the period after World War II were relatively simple;* Turner *had only a single surface search radar, and a single air search radar (SPS-29); her 5in battery was controlled by the Mk 25 dish atop her director. However, her main battery was the Combat Air Patrol she controlled, requiring the TACAN and the SPS-8 visible aft, as well as improved communications. The squeeze in topside space is evident in the placement of an AS-1018 (here shown without its usual covering) atop her No 1 gunhouse, as well as the proliferation of whips. Surface-to-air radio antennas surround her URD-4 direction-finder, at the fore masthead. At this time she had not yet been subjected to FRAM modernisation.*

BY COURTESY OF A D BAKER III

SPS-6 featured a parabolic antenna. The major US early postwar air search radar, it was in fact produced with three alternative antennas, with alternative beam forms. The first generation of US radar-picket destroyer escorts received two radars to achieve better total coverage. Here the newly converted Harveson *(March 1951) shows SPS-6B forward, and SPS-6 aft; there was also an SPS-6A. The empty platform atop her bridge was to have taken an SPS-8 height-finder, delayed (as of this time) by production problems. In the end, production problems with SPS-6 series radars made it impossible to provide two even for such important units as Continental Air Defense radar pickets, and the ineffective SPS-6s were landed.*

USS Hornet, *late in the war, shows the large 'dish' of an SK-2 air search radar, and the smaller one of an SP fighter-control set, both atop a platform for maximum cover. Carrier operations always required a massive array of radars merely to achieve effective control of aircraft. During World War II both the United States and Britain evolved a wide variety of long-range search (warning) and aircraft control systems, including beacons permitting carrier aicraft to home reliably on their carriers. The long dipole in the SK-2 dish and the one in SP are IFF antennas, without which the carrier controllers would be unable to distinguish friendly from enemy aircraft. Between them is DBM, a dome-enclosed radar direction-finder to facilitate the jamming of enemy radars; the antenna sponsoned out from the funnel is a TDY jammer with two dipoles to allow it to cover two frequency ranges. The prominent antenna topping the mast is a YE aircraft homing beacon, with a surface search radar (SG) below it. Finally, the carrier's two Mk 37 directors are topped by Mk 12 and Mk 22 ('orange peel') fire control radar antennas, the latter particularly important against low-flying aircraft. Not visible, but certainly present, is a secondary air search radar (SC series), a back-up against the failure of the SK-2. There were also back-ups to YE, and numerous ship-to-ship, ship-to-air, and ship-to-shore radio antennas, all of which greatly complicated ship design (and, incidentally, caused very considerable mutual interference).*

The 'cheese' is, if anything, the characteristic Royal Navy radar antenna, as demonstrated aboard the cruiser *Superb*, with a Type 293 target designation radar aloft and a Type 274 main battery gunnery set, a 'double-cheese' (here covered but visible), below it. Such antennas produce a very narrow horizontal beam, for high bearing precision, although their beams are wide in the vertical. In the case of Type 293, that increased the probability that aerial targets at unknown altitudes could be detected. Given an accurate bearing, search in elevation could be swift. It was generally carried out by the pencil beam of Type 277, the dish on the platform just abaft the main battery director. The small double-cone antennas are IFFs: the vertical one is Type 252, that at an angle to the vertical Type 253.

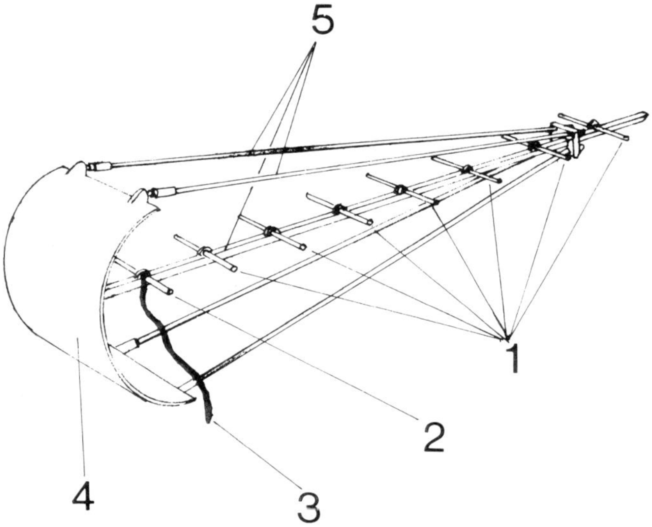

Yagi antennas are both simple and directional, and therefore were particularly popular early in the development of radar. They were employed for radar Types 282, 283 and 285. **1** *'Passive' director dipoles.* **2** *Active transmitting dipole.* **3** *Feed to active dipole.* **4** *Parabolic reflector.* **5** *Nonmetallic supports.*

solid reflectors, but the British long-wave air search radars (such as Types 279, 281 and 960) did use dipoles, one for each dipole radiator. The British also used a more sophisticated dipole-reflector antenna, the *yagi*. Such an antenna uses reflectance as well as the principle of *parasitic radiation*: an antenna struck by a radio wave absorbs some and re-radiates some. How much is re-radiated, and with what phase relative to the original signal, depends on the antenna characteristics. In fact reflection can be visualized as re-radiation with a phase such that the original and re-radiated waves just cancel beyond the reflector, and reinforce on the reflecting side. A yagi is a stack of dipoles: a *reflector*, a *radiator*, and several *directors*, all perpendicular to the direction of transmission.

Yagis are most effective at metric wavelengths. They were used in several British naval fire control radars and in many airborne installation. The yagi directs radiation down its length, and exemplifies *end-fire* systems. An array of such radiators produces a more directional beam; such arrays were used in the British fire control sets. The US Mk 8 employed another type of end-fire radiator, the 'polyrod', to achieve similar results. Yagis are probably most familiar as elements of outdoor television antennas.

MICROWAVE LENSES

The analogy between radar signals and light extends to lenses. They are not, of course, glass, but rather are generally fabricated of metal plates or of plastics. Their great advantage is that the radar can employ a complex radiating assembly, which would block off the surface of a conventional reflector. For example, early monopulse tracking assemblies were massive: the early US monopulse missile trackers, such as SPG-49 and SPQ-5, employed microwave lenses. As monopulse 'plumbing' became smaller, this expedient was dropped, and newer US missile tracking systems employ more conventional microwave reflectors. The other major application of a microwave lens was the huge British Type 984 pencil-beam ('3-D') set mounted in several carriers during the 1950s and 1960s. Several radiating elements were mounted behind the big lens, close enough together for it to collimate all, though at different elevations. Switching among the elements produced an elevating beam. A higher-powered low-elevation beam was provided for long-range search; trunnions at the sides of the 'searchlight' were used to stabilize it against ship motion. Once more, the radiating elements were far too long to be suitable for mounting in front of a reflector.

Lens technology also has important applications to phased-array systems. For example, it is possible to focus a single beam through a lens whose properties are

The Type 984 microwave lens was the British equivalent to the US SPS-2: an ultra-long-range S-band radar capable of height-finding as well as detection. It appears, however, to have been rather more successful, and is shown here aboard HMS Hermes *in November 1959. In effect, Type 984 replaced the former trio of Types 960, 982 and 983, and finally solved the problem of limited island space for radars. Its very high cost appears, however, to have led ot is demise in favor of the much simpler Type 965, even before the Royal Navy had to abandon fixed-wing full-deck carriers. The small radome at the peak of the lattice mast is a British equivalent of the US TACAN.*

controlled electronically, scanning it. The Sperry Dome Radar uses a passive lens (*ie* one with fixed properties) above a flat, electronically scanned array. It focuses appropriately-formed signals from the flat panel over more than a hemisphere. Because there is only a single scanned face, the Dome Radar is only about half as expensive as a comparable multiface electronically scanned system covering the same area, such as SPY-1. On the other hand, compromises inherent in the use of the special lens limit beam-forming quality to some extent, *eg* in the generation of side lobes.

Cassegrain antennas are an alternative approach to the same problem of blockage. As in a lens, the radiator is set at the rear of the antenna, emitting in the direction ultimately desired. However,it fires directly at a small auxiliary reflector at the front of the antenna, which in turn reflects back onto the main reflector. SPG-55 is a current example of this type of construction.

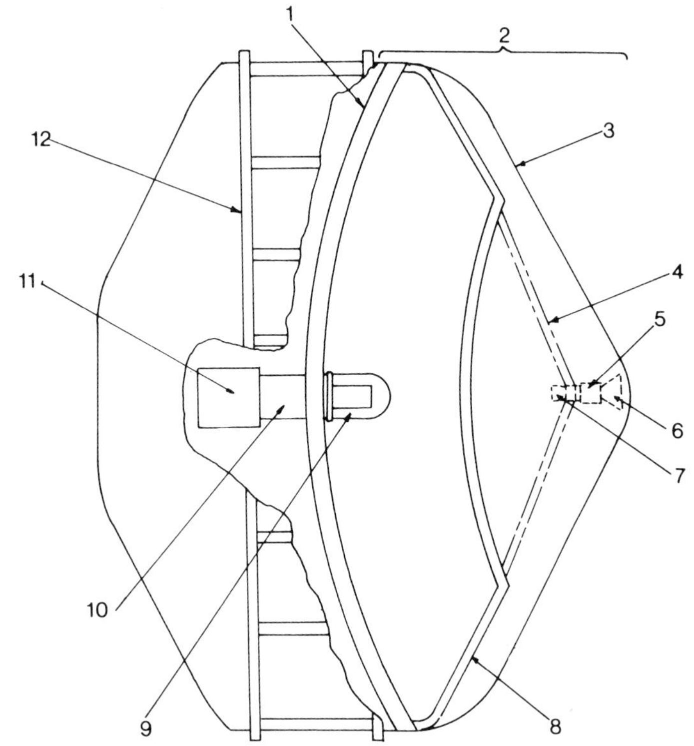

Cassegrain operation can simplify the design of a high-powered radar. The transmitter can be mounted directly at the rear of the antenna, illuminating a small secondary reflector, which in turn illuminates the main antenna to form the radar beam. The result is extremely compact, and the secondary reflector can be polarized to reduce losses as the beam passes through it on the way out of the antenna proper. The US SPG-55B Terrier guidance radar, shown here in cross-section, is a current example. **1** *Main dish (parabolic reflector).* **2** *Antenna subassembly.* **3** *Radome.* **4** *Support structure.* **5** *Continuous-wave (CW) feed horn assembly.* **6** *X-band secondary feed horn.* **7** *X-band primary feed horn.* **8** *Subdish (hyperbolic reflector).* **9** *Feed horn assembly F(50-15)1 (C-band, 4-horn cluster).* **10** *Comparator and feed horn assembly UD50-15.* **11** *Comparator UD50-02.* **12** *Antenna support structure.*

LOBES AND SIDE LOBES

Throughout we have spoken in terms of well-defined radar beams much like the searchlight beams of a car; in fact, matters are never nearly so simple. A radar beam is not very well defined. Commonly, expressions like 'ten degrees wide' really mean that power in the beam falls off by half within five degrees to either side of beam center. Moreover, because radar wavelengths – even for microwaves – are so large compared to reasonable antenna dimensions, antennas generally radiate to some extent in directions other than the desired one. Their reception patterns duplicate this behavior; in general an 'antenna pattern' consists of a primary 'lobe' as well as undesirable 'side lobes'. Reduction in unwanted side lobes was a major element in World War II radar development, for example in the change from the flat SK mattresses to the parabolic SK-2 dish. Similarly Mk 12 differs from Mk 4 largely in the addition of minor ('parasitic') antennas to reduce side lobes.

Originally side lobes were undesirable because they drew power out of the main lobe and thus reduced the effective range of a radar, but they have other unfortunate effects as well. In particular, they mean that a radar pointed in one direction can receive energy from another without giving any indication that the latter signal comes from any other direction. Such an opening makes deceptive ECM very practical: an enemy who knows the antenna pattern of a radar can concoct a false target, complete with false target motion, merely by pouring strong signals into side lobes when the radar is pointing in the right direction. Moreover, 'leakage' of radar signals through a side lobe can be used by anti-radar homing missiles fired by platforms other than the one being tracked by the radar in question. For example, in the air campaign over North Vietnam, US aircraft were equipped with such a missile – Shrike. For their part, the North Vietnamese required radar to guide their missiles, fighters, and AA guns; but the mere existence of Shrike inhibited them from very full use of radars. On the other hand, were the US strike aircraft to carry only Shrikes, they would have little

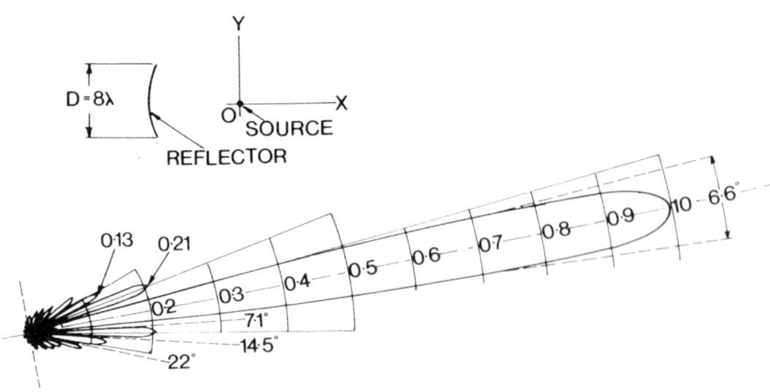

Radiation pattern of a parabolic cylinder, showing typical beam shape and side lobes; the pattern is that in the vertical XY plane of the antenna. The width of the cylinder (D) is 8 wavelengths (8λ).

capacity left for bombs. To some extent the details of the radar operated by the North Vietnamese determined the nature of the campaign – how easily they could be switched on and off, how easily data from different radars could be correlated, how quickly Shrikes could lock-on or, alternatively, what US attack tactics could be developed to force the North Vietnamese to keep their radars switched on long enough to allow Shrikes to operate.

We shall return later to other ECM considerations but it is worth noting here that one of the great creators of side lobes is the abrupt edge separating the illuminated reflector of the radar antenna from empty air space. To some extent radar designers have tried to smooth out this transition by reducing illumination at the antenna edges and even by shaping the antenna edge, hence the shapes of such nominally parabolic antennas as SPS-6 and SPS-40. On the other hand, it would appear that 'mattresses' like SPS-28 and SPS-43 work sufficiently well without smoothing. An important factor here may be the difference between horn feed and an array of dipoles.

There is a limit to side-lobe reduction. Antenna theory suggests that for a single radiator with a reflector there is a trade-off between the sharpness of the main beam and the size of the side lobes. A common recent effective lobe-reduction measure is an auxiliary 'guard antenna', with a relatively undirectional (fat) pattern. A signal received strongly by both guard and main antennas is probably coming from a side lobe, and hence can be cancelled out. However, such a measure does no good against antiradar missiles.

ANTIWEATHER FEATURES

Radars at sea are exposed to weathering – salt spray, even solid water for equipment mounted directly on superstructures. Moreover, many parts are by nature delicate: servos which move antennas, joints permitting radio-frequency energy to pass, even antennas requiring very precise dimensional relations among their parts. Exposed antennas are also subject to distortion by wind, and to ice loading. Hence the wide use of radomes, particularly in those navies (British, French, Dutch) subjected most often to bad-weather conditions.

Yet another feature is the balancing vane, which

Radome enclosure insulates a complex antenna system from wind and weather, obviating, for example, balancing wind vanes, as shown in this photograph of the French missile frigate Suffren*; her single massive DRBI-23 three-dimensional radar also illustrates one solution to the problem of superstructure crowding. The two automatic 3.9in (100mm) guns forward are controlled by a director with a DRBC-32A radar, forward, and the Masurca launcher aft by a pair of DRBR-51 directors, Terrier-style. The radome surmounting the short mast aft is presumably a French equivalent of TACAN, and she has a navigational radar on her bridge (DRBN-32) and a surface search set aloft (DRBV-50).*

J A JEDRLINIC

helps to balance asymmetrical air forces on large antennas. Vanes were introduced into US service with SPS-6 and SPS-8; small square vanes also appear on the SPS-28/29/37 series, although they are not always visible. Abroad, vanes are most prominent in Soviet radars, where they may balance relatively weak motors attempting to train heavy antennas. It seems unlikely, on the other hand, that the end plates of SPS-48/52 have any aerodynamic function.

The problems of the shipboard environment extend well beyond weathering and corrosion from stack gases. Even when she is not working in a seaway, a ship is a mass of vibrations, many of which subtly influence radar performance. Such influences may include combustion fans, even crewmen using their tools. Some fraction of radar signals bounce off the ship structure back into the antenna. Masts and antennas lie in the direct path of the signal, and their vibrations feed back into the antenna. The effect of such vibration is to smear out radar signals, and it is felt more and more as radars rely increasingly on coherence, *ie* on the detailed relationship between subsequent pulses, to overcome clutter and noise. That is, as radar signal processing comes closer to making full use of the information contained in each train of pulses, spurious information injected into those pulses by the ship herself becomes more and more significant.

The effect of a vibrating whip, for example, in the path of a radar beam is additional to the interference of the ship structure as a whole, which reduces radar performance from its theoretical level, particularly in all-around coverage. Moreover, by reflecting radar signals into side lobes, the ship structure can materially reduce the effective accuracy of radar signals. The flexing of the ship structure can have a similar effect, so that the radar receiver signal does not quite correspond to the geometry of incoming signals. In recent years attempts have been made to calculate the effect of ship structure on radar efficiency in terms of the time delay imposed on a combat system by radar inaccuracies, including the smearing-out of radar beams.

5. Radar Functions

The earliest radar function was to search,* *ie* to sweep out some volume of space to discover, in a crude way at least, its contents. Search requires a relatively broad beam, so that large volumes can be swept rapidly. An air search radar like SPS-6 has a beam 10° high and 3.5° wide, which means that at a range of 50nm the beam covers all altitiudes up to about 53,000ft over an area about 3nm wide. In this way most of the space around a ship, out to the (S/N-limited) range of the radar, can be covered on every sweep of the antenna, at several rotations per minute.

AIR SEARCH AND SURFACE SEARCH

A problem in wartime air search radar development was that antennas like SK were large in both vertical and horizontal dimensions, and hence did not produce very high search beams. Lack of high-altitude coverage became a serious problem late in the war, and some observers actually preferred SC-series radars to the big SKs because of the less well-defined (vertically) beam of the former. To some considerable extent the SR set developed late in the war was an attempt to match the more diffuse beam of the SC to the range of SK by a sharp increase in power. A 1945 Pacific Fleet Board on Radar Lessons suggested that the SR beam, 20° wide and 55° high, ultimately give way to one only 4° wide (for very accurate bearings) and 60–70° in height. However, so diffuse a beam implied very limited range: hence the performance of SPS-6 previously noted. The latter type of beam shape assumes that effective radar range will be so great that even a narrow beam will cover all practical altitudes.

It was British experience that the beamwidths required for effective air search were so extensive that accuracy was far too poor for gunlaying. The Royal Navy therefore developed a class of intermediate Target Indication radars, such as Type 293, which had fan beams narrow in bearing but broad in elevation (typically 1–2° in bearing, 30° or more in elevation). They searched in bearing and passed on their data to fire control tracking radars which could search in elevation and then lock onto the target. The elevation search was generally so quick that it was not worthwhile designing the Target Indicator to carry it out. In US practice, a pencil-beam three-dimensional radar supplements the long-range air search set in most missile installations; the original motive appears to have been some equivalent of the British Target Indication role.

It is important to distinguish *air search* from *surface search*. In principle the same radar might do both; but in fact, in view of the fading problem, surface (or even low-level flyer) search requires a different range of wavelengths. Thus US World War II surface search sets, such as SG, operated in the S-band, whereas the air search sets (SA, SC and SK) operated in the P-band. S-band pencil-beam radars originally designed for other functions were also used against low flyers (SM and SP). Another type of surface search radar is *navigational radar*: a high-resolution, short-range type. There is some ambiguity between surface search and navigational radars, but the latter will generally employ lower power and a shorter wavelength (X- and even K-band).

Surface search antenna design presents special problems: the US parabolic reflectors (such as SG) and the British 'cheese' are alternative solutions. The 'cheese' differs from the more conventional radar antennas in that its beam is extremely wide *in elevation*, although it is narrowly defined in azimuth. Hence a 'cheese' pointing horizontally pours much of its radiation into the sea, and the beam splits up due to reflection. However, a small ship rolling deeply benefits from the great vertical width of the beam, which continues to sweep the sea surface throughout the roll. In such circumstances the roughness of the waves will reduce reflection and beam-splitting. On the other hand, a narrow-beam radar such as the SG may lose its effectiveness as its beam is alternately swept up and directed into the sea during each role. The later American (narrow-beam) surface search radars, such as SU, were therefore mounted on special stabilized platforms which could compensate for ship motion and so keep the beam parallel to the surface of the sea; the Royal Navy needed no such complications, but paid by using a more diffuse beam more of the energy of which was wasted.

It might be imagined that the 'cheese' would, by virtue of its wide beam, sweep out a large volume of sky, but in fact reflection from the sea surface would so break up its wide beam that there would be many gaps at high angles of elevation. However, the Royal Navy did use the wide-angle beam of the 'cheese' for air search, mounting Type 293 at an angle to bring its beam clear of the sea. In this mode it operated as a short-range precision air search (target-indication) set.

It should be noted that the earliest British sea search radars had to perform aboard much smaller vessels (deeply rolling corvettes) than did such US products as

*British writers often refer to search radars as air or surface *warning* radars; this designation may reflect the low resolution of early British search sets.

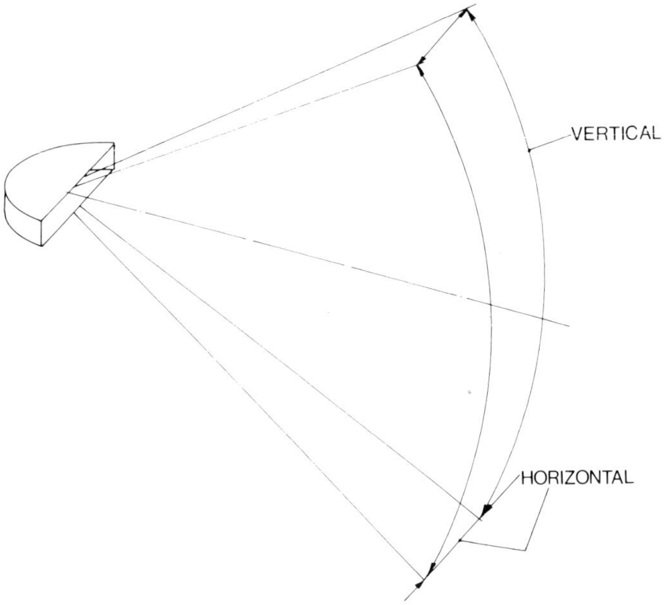

◀ *'Cheese' beam form.*

▼ *At the end of World War II, major efforts were underway to improve radar coverage immediately* over *a task force: 'zenith search'. The battleship* North Carolina, *photographed in June 1945, shows two such attempts: an SCR-720 night-fighter radar, in an upward-pointing radome, on a platform forward of her forward funnel; and and SC series radar on her mainmast, arranged so that its antenna could tilt upwards. Also evident here is the result of the other major late-war problem: low-flyers. A Mk 22 'orange peel' fire control radar, designed to improve performance against low-flying aircraft, is visible just forward of the first funnel; it is mounted on the Mk 37 director on the other side of the funnel. This particular problem was only solved postwar, with the advent of the pencil-beam Mk 25 (dish) fire control radar.*

SG; the British, moreover, far more concerned with Arctic (bad weather) operations. In contrast to the sea search radars, the wartime US S-band jammer, TDY-la, benefited from ship motion, which, in effect, wiped out the nulls (fades) in its antenna pattern.

Long-wave air search radars do not require stabilization. Their antenna patterns show large nulls near the sea surfaces, and fat lobes above it; it takes a very large roll indeed to bring sea clutter into the display of a long-range air search radar. However, as we shall see, stabilization of some type is essential for height-finders.

Both conventional air and surface search radars provide a *two-dimensional* picture; altitude enters only in the sense that fades will occur at different ranges for aircraft at different altitudes. Moreover, both types provide very poor coverage in the cone overhead. Towards the end of World War II this latter problem began to assue great importance, because it was sometimes possible for Japanese aircraft to approach undetected at high altitudes, over land, and then dive on an unsuspecting ship out of the 'dead' cone directly over her. Radars were therefore designed specifically for zenith (*ie* overhead) search (*eg* SPS-4, SG-6).

Zenith plus two-dimensional search was at best an interim solution; in 1945 a requirement for a 'hemispheric' search radar, effective up to 80° elevation, was formulated. A variety of ingenious solutions was proposed, but the first operational US hemispheric search radar was SPS-39, which scanned electronically in elevation but mechanically in bearing, so that all altitudes at each bearing could be scanned rapidly. This set and its successors were generally described as three-dimensional (altitude, range and bearing).

SM was the first US pencil-beam height-finder, developed specifically because fighter control was difficult when the altitude of incoming raids was not known. There was a tactical means of overcoming poor altitude data, 'stacking' Combat Air Patrol (CAP) aircraft, but it was inefficient at best. Even the pencil beam generated by a radar such as SM was not sufficient, but it helped, particularly as raids became much more intense late in World War II.

HEIGHT-FINDERS AND FIGHTER CONTROL

Search radars provide relatively imprecise location data – hence the existence of two other classes of sets, height-finders and fire control sets. The former were developed for fighter control: friendly fighters needed an enemy's altitude as well as his bearing and range. A variety of precision altitude-finding radars were developed during World War II; the SM and SP of the US Navy used very narrow ('pencil') S-band beams to define accurately the altitude *angle* to a target. This angle could be meaningful only in relation to the surface of the sea rather than the deck of a rolling ship. SM and SP had to be gyro-stabilized to compensate.

One difficulty encountered by pencil-beam radars was that at low angles they tended to fix on an image reflected by the sea rather than on a low-flying target. The dual-antenna SX was an attempt to solve this problem by displaying all echoes on a given bearing on a range–height indicator; false echoes would then show clearly. More recent height-finders (SPS-8 and SPS-30) have dispensed with the associated search antenna of SX by means of remote displays connected to other air search radars aboard ship, as with the SPA-8 series of

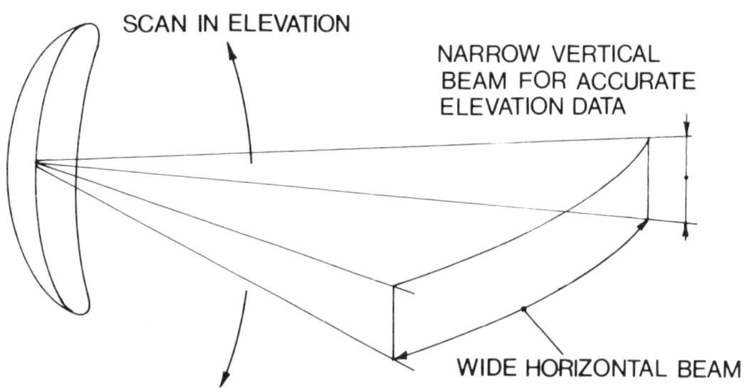

A typical height-finder ('orange peel' antenna). The considerable horizontal beamwidth is limited somewhat by the requirement that nearby targets at different altitudes be distinguishable. Note that in some radars the entire antenna nods; in others (eg SPS-8), only the feed moves.

SM was too heavy for many applications, and it was ultimately replaced by SP, which with a smaller antenna could be placed aboard ships as small as radar picket pestroyers and destroyer escorts. This full-size SP is shown aboard the carrier Yorktown *(April 1945). It required a special precision IFF (BO) to allow fighter controllers to distinguish friendly from enemy aircraft. This particular set shows the dipoles of both wartime IFF systems, Mk III and Mk IV. The former are the long dipoles mounted on frames protruding from the dish itself. The mattress above carries the shorter dipoles of Mk IV, a standby IFF system provision for which was built into many US wartime naval radars. The fat horizontal dipole below the SP pedestal was an ECM radar receiver, AS-56/SPR-1; the narrow dipole is probably a test antenna.*

PPIs with correlated range–height displays.

SM and SP generally did not scan in height, at least not continuously. The natural development was automatic scanning in elevation (SP-2), but the great weight of the antenna precluded very rapid motion. In 1945 SX employed instead a moving (Robinson) feed, the lesser weight of which permitted very rapid vertical scanning while the heavy antenna remained fixed. Unfortunately the Robinson feed had a complex shape, so much so as to make manufacture extremely difficult, SPS-8, the postwar US height-finder, continued the Robinson feed, but added an antenna elevation feature which permitted variation in the 11° fast-scanning sector. The dish-shaped SPS-8B introduced a new organ-pipe feed, retaining the fixed antenna feature of earlier versions, and the present SPS-30, which is probably the last US height-finder without electronic scanning, resembles SPS-8B in operation.

Other countries differed in their choice. Only three manufactured conventional height-finders: Britain (Types 277/8 and 983), France (DRBV-10) and the Netherlands. Both the British and the Dutch preferred to 'nod' the entire antenna – in fact Type 277/8 did not scan at all in the vertical, it merely pointed on command – but the French combined Robinson feed with a wide antenna to achieve a crude 3-D pattern as the antenna rotated and scanned vertically in comparable periods.

An alternative means of height-finding was to 'stack' several narrow beams angled at various elevations. In effect such a 'stacked-beam' radar consists of several narrow-beam sets operating simultaneously, and rotating together. Such a system provides a form of hemispheric scan – a mechanical alternative to the electronics of SPS-39. The main example in US naval service was the enormous, and enormously complex, SPS-2.

Another system developed during World War II was the 'vee-beam'. Two air search radars were operated with their beams at an angle. Coincidences between the two sets of returns gave a measure of target height, although not so precise a measure as might be provided by a pencil-beam height-finder. Systems of this type were widely deployed on land during World War II, but their inherent weight made the simpler pencil beams preferable at sea. However, well after the war the

Stacked-beam height-finding, as in SPS-2 (schematic).

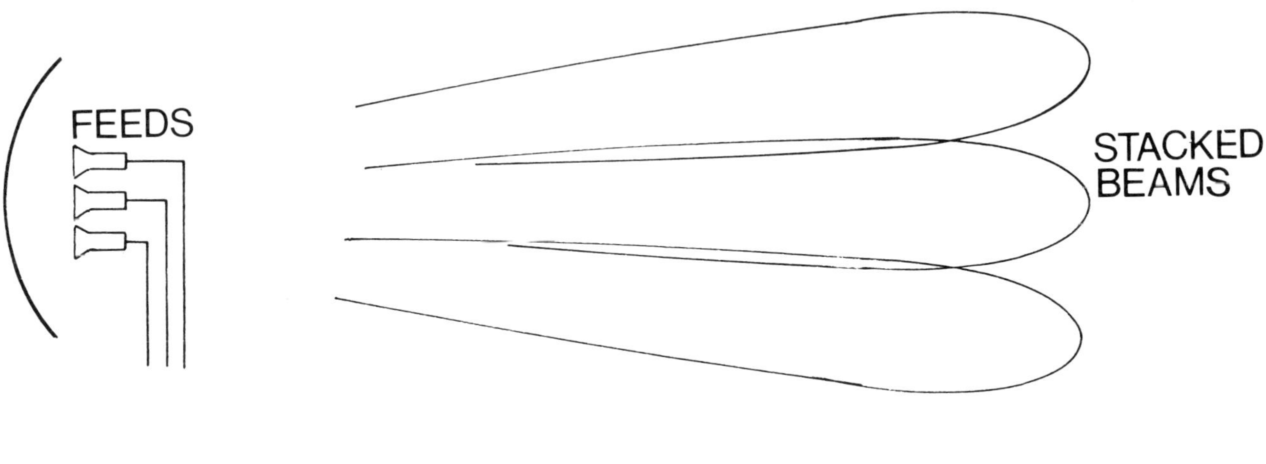

Soviets deployed a 'vee-beam' search/height-finder system, 'Headnet-C', in which two antennas were mounted back to back. Operation is somewhat cumbersome: the operator finds his target with the horizontally-arranged search set, then fixes a range and bearing and turns on the second, angled, set.

The 'hemispheric' radars are also height-finders, the FRESCANS in particular (SPS-39, -48, etc) performing this function in many cases. Their great advantage over pencil-beam radars is that, because their beams are electronically formed and steered, they can be electronically stabilzed merely by altering pulse shape, *ie* they can respond far more rapidly to ship motion.

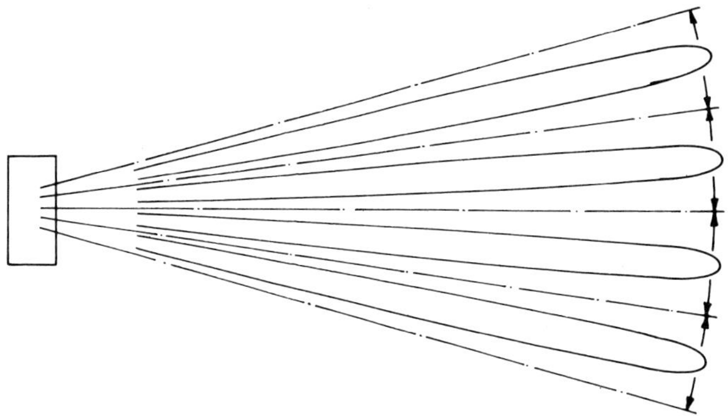

Stacked-beam height-finding, as in SPS-48, where the beams scan and do not overlap (schematic).

THE SATURATION PROBLEM

The success of the fighter or weapon control to which a height-finder contributes depends upon the efficiency of the system which binds the radars to the weapons which are to be assigned to prospective targets. Such a system begins with the extraction of plot data from the radars, so that the officer(s) or computers of the system can perceive the battle as a whole; the same data must also be stored accurately so that it can be used as the basis for fire control or for fighter interception assignment. In World War II plot extraction was a manual function; men in a CIC (or, in the Royal Navy, an AIC) would decide which images on radar scopes were

Unlike the US SM and SP, the British pencil-beam radar, Type 277, was relatively light in weight and was suitable for alternative use as a small-ship surface search set. It could be scanned in elevation from its control panel, but did not normally scan automatically. Here it is installed as the sole radar of a frigate, HMS Loch Alvie *(April 1954).*

CONWAY PICTURE LIBRARY

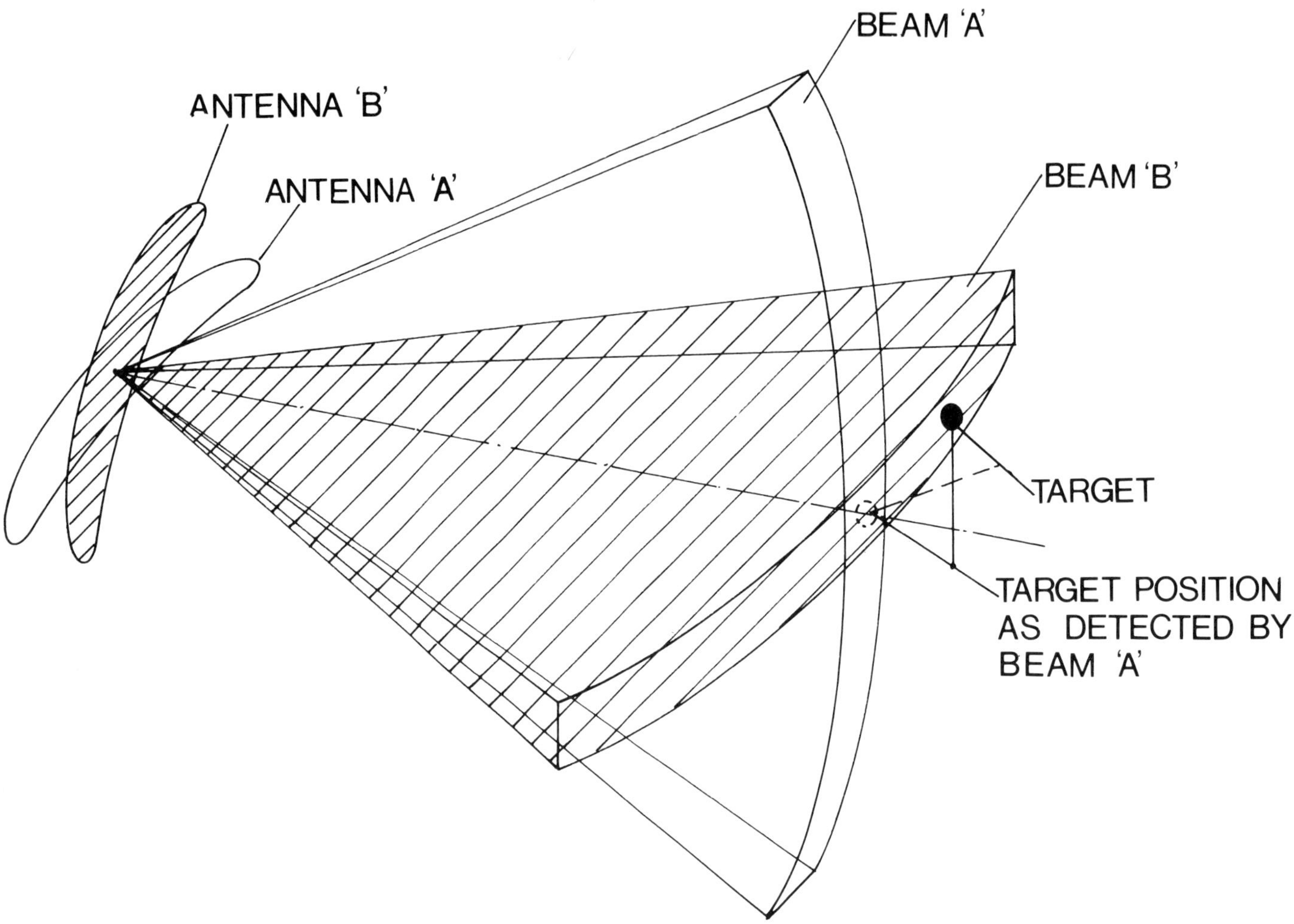

Vee-beam, as in Soviet 'Head Net-C'. Slanted beam is shaded. Difference between bearing as measured by the two antennas gives altitude of target.

targets, and would call out their data so that others could plot them on displays, from which evaluators could make their weapon or aircraft assignment decisions. Specialized fighter control officers, for example, would vector fighters or groups of fighters onto enemy contacts on the basis of these plots. The only important airborne equipment was VHF voice radio and a good compass.

This system was subject to saturation in mass raids. After World War II higher aircraft speeds began to increase the potential rate at which enemy attackers might approach a target, and by the late 1940s saturation had become a serious concern to air defense experts both at sea and ashore. The attack generally visualized was a 'stream' of enemy aircraft similar to, but much faster than, the Allied bomber raids of World War II.

Postwar experiments were discouraging at best. For example, it appeared that a wartime-type vertical plot could barely accommodate twelve tracks, and that with twenty it was hopeless. Even with twelve, errors of four or five miles were common, and some tracks were omitted entirely. Low-level saturation would not even be apparent to those running the plot. There were other problems as well. Inaccuracy bred an inability to distinguish tracks. Then there was control saturation, an inability to decide the particular attackers at which to direct fighters or gunfire. It was estimated, for example, that one intercept officer could not control more than four targets. Some of these problems could be traced to operating practice, but others were necessary consequences of the manual nature of the system. For example, it was estimated that the time lag between the reporter at a radar display beginning his report and the filtered plot appearing on the front of the plot written in by the air plot officer might be twenty seconds. At 600mph that is an error of about three miles.

About 1949 the Royal Air Force has a potential solution to the control saturation problem: 'broadcast' control. In this system the controller broadcasts only enemy course, speed and position. Individual aircraft are provided with computers allowing them to calculate vectors for themselves. This scheme required a precise means of position determination as well, often by way of a second crewman to navigate, *ie* to work out an attack vector. Figures prepared in 1950 for the US

Weapon System Evaluation Group give some indication of the potential of broadcast control. Operational Development Force experiments suggested that conventional control techniques could deal with six or eight separate (simultaneous) raids approaching from a narrow sector. This was considered equivalent to 25 fast aircraft per hour. Broadcast control was expected to be able to handle up to 60 raids per hour, but an automatic system – which became NTDS – could do even better.

Broadcast control was a major factor in the US adoption of TACAN (Tactical Air Navigation), an aid permitting just such precise position-finding. Broadcast control did not in itself solve the problem of saturation presented by attackers arriving from many different directions at once. It did, however, permit a very limited number of controllers to handle many more attacks. Moreover, since the mid-1950s much of the saturation problem has been alleviated by the rise of computers capable of generating intercept vectors automatically and then of passing course instructions to aircraft without taxing scarce fighter-director personel. The goal of the US Air Force SAGE system was a fully automatic intercept from take-off onwards, and missile systems perform a somewhat similar function, albeit generally at short range.

The Royal Navy was also very much concerned with the plot saturation issue, and sought some means of storing radar plots in a computer. This it called a Comprehensive Display System (CDS). Individual radar operators entered plots into the system by pointing instruments at blips on the faces of their displays, and these were collected and displayed together on a central PPI. The US Navy took over this program, and in 1951–52 NRL demonstrated an air defense system utilizing electronic memory and switching to track targets and display them selectively. This shore system employed three radars, and stored a track identifying number, range and bearing, identity, height (high, medium or low), and a raid size (single, few or many). Up to 96 track numbers were available. CDS proved effective, but it was far too bulky for shipboard installation. In 1953, therefore, NRL developed a more compact Electronic Data System (EDS). This remained in limited service as late as 1968, the first of twenty being installed aboard the frigate (DL) *Willis A Lee* in 1956. Later installations included the four radar-picket destroyers of Destroyer Division 262, and missile cruisers.

EDS features included a simplified means of extracting plots from the radar display. Special circuitry permitted the comparison of new plots with earlier ones, so that target speeds could be estimated and tracks more easily established; NRL estimated that a single operator could track up to eight targets, compared to two in the non-electric CDS. The central data display was an automatic plotting board, 5ft square, on which markers were positioned by electrical data in the system memory. EDS also was the first data system to provide for automatic tactical data exchange among ships of a force, tests being conducted by Destroyer Division 262 in 1959. The consequences of this development are discussed in the next chapter, but what is important here is the evolution of the individual ship's CIC towards greater automation and in particular towards greater automation of the transfer of data from radar to central plot. The step beyond EDS was the Naval Tactical Data System, NTDS, which remains in widespread use in improved form as this is written. EDS was primarily an analog system, but with NTDS all data was handled in digital form. Among other things, this made the combination of data from different shipboard sensors particularly efficient, and permitted the use of central computers which could process the sensor data and even make decisions on its basis. For example, the combat system of a ship like the frigate *Perry* maintains a file of target tracks, decides which tracks represent urgent threats, and aims and fires weapons on that basis. The digital character of NTDS also facilitates data exchange within a fleet.

There was also a severe data saturation problem given the limited data rate available from the radars. Conventional air search radars rotate at up to 15rpm, and therefore 'look' in the same direction up to 15 times each minute. It takes several such 'looks' to provide any clear picture of a target's course and speed. For example, NTDS enters a target track when an operator enters three succesive target plots. Vectoring, or broadcast control, requires height-finding as well, which means a dedicated antenna pointing at the search radar's target to obtain course and speed data in the vertical. All of this takes time. Systems such as ADT save time by deciding more quickly that a target exists, and by using the character of the returning echo to begin to estimate target course and speed. They are essential as warning time shrinks, for example as the Soviets begin to deploy 'pop-up' weapons such as the SSN-7 missile, and sea-skimmers which appear very suddenly on the horizon. Moreover, the integration of all of a ship's radars through a central computer permits both the long-range search set and the height-finder to scan simultaneously, with data from both stored and associated continuously for use when needed.

SHORT-RANGE CONTROL

EDS and its progeny were essentially long-range systems concerned with fighter control. At the same time the US Navy sought to develop a better means for controlling short-range antiaircraft fire, a problem studied by Bell Laboratories under the designation Gun Fire Control System Mk 65. Bell concluded that the problem was saturation, and that what was needed was a better means of evaluating incoming threats and assigning available weapons to them; Bell called a system for this a Threat Evaluator and Weapon Assigner (TEWA) or, in automated form, an ATEWA.

Requirements included very rapid target acquisition by a search radar with a beam sufficiently narrow to permit quick designation to fire control radars. Bell identified three phases of operation: I, detection and tracking; II, attack analysis, with the assignment of specific threats to specific directors; and III, assignment of weapons to these directors to counter the most urgent threats. This was a revolutionary approach: previously, directors had generally been permanently linked to specific weapons. Detectors and tracking, moreover, required an effective three-dimensional radar with a high data rate whose output would be of sufficient quality to permit Track-While-Scan operation.

A simulated ATEWA was built at the Bell Laboratory in Whippany, New Jersey, in 1949–51. It included a novel display device, a vertical screen on which lighted characters representing up to ten raids, and up to four directors, could be projected. Tests showed that automatic director assignment was essential against heavy raids, but that manual assignment was almost as good against lighter ones. The next step was an operational prototype, Weapon Designation System Mk 3, which was installed aboard the command cruiser *Northampton*. It employed the prototype 'hemispheric scan' radar, SPS-3. In fact Bell tests with SPS-3 prior to its installation aboard ship showed some of its defects, leading to a Bell proposal for a new radar, the CXRX of the second and third ATEWAs, aboard the missile cruisers *Boston* and *Canberra*. Much of the ATEWA effort went into the development of effective displays. For example, the scan of each fire control radar was graphically displayed, so that when it was assigned the operator could see its numeral moving towards the letter indicating the target.

The *Northampton* installation proved both effective and reliable; in its offical history, Bell claimed that during the entire period of the test, the equipment was never out of service during any scheduled operation period. Subsequent Designation Systems controlled short- and medium-range missile fire: Mk 7 in the *Bostons*, Mk 9 in many early Terrier ships. In each case Phase II and III were manual for the missiles, and in Mk 9 they were manual for guns as well. Later systems were designated Weapon Designation Systems (WDS): Mk 1 in Tartar ships, Mk 3 in three carriers with Terrier missiles; Mk 4 in one carrier and ten Terrier frigates (*Leahy* class); and Mk 8 in the first two *Belknap* class missile frigates (now officially referred to as cruisers). In most of these ships the three-dimensional radar was an SPS-39 electronic-scanning type.

The final stage was to merge WDS and NTDS to form a single combat system capable of accepting data both from onboard and from external sources, and capable of moving automatically from target detection through to engagement. Within such a modern combat system, the Weapon Designation System is the element which actually control missile fire. In effect, the merger

The Mk 4 (or FD) dual-purpose fire control radar epitomized the lobing technique; here one is seen aboard the destroyer Nicholas, *January 1944. Upper and lower parts of the antenna produced two separate beams, slightly different in angle; the strength of the return from each could be compared, and the radar pointed at its target when they were equal. The related Mk 3 surface gunnery set employed a similar technique, in which the two halves of the antenna, split either vertically or horizontally (depending upon the version) produced alternative beams. In many later systems, the entire antenna switched rapidly between alternative beams so as to generate an error signal when it was off-target.*

▶ *Lobe-switching. Radar is switched among alternate beams pointed in slightly different directions. With target antenna axis, returning signals differ in strength; on axis, returns are all the same strength, giving high precision in target-tracking. Returning signals are thus equal in intensity only when target lies on line bisecting angle of intersection of the two beams.*

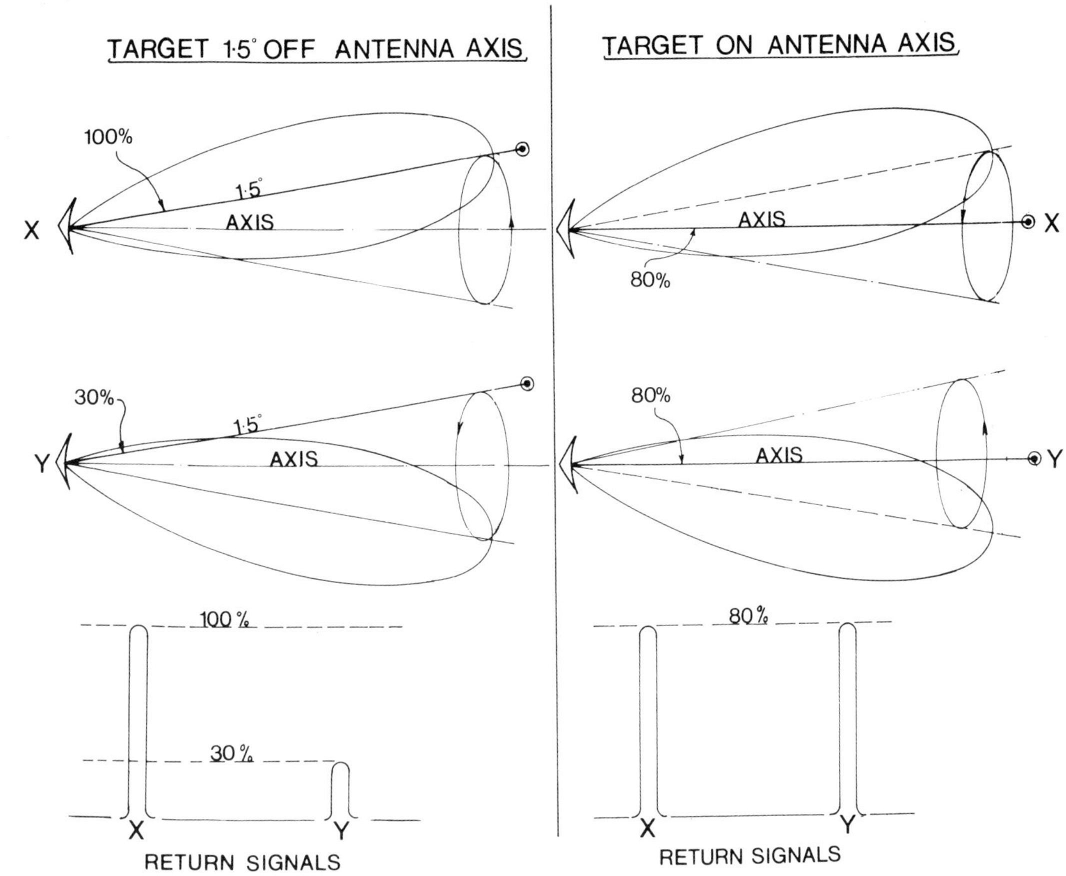

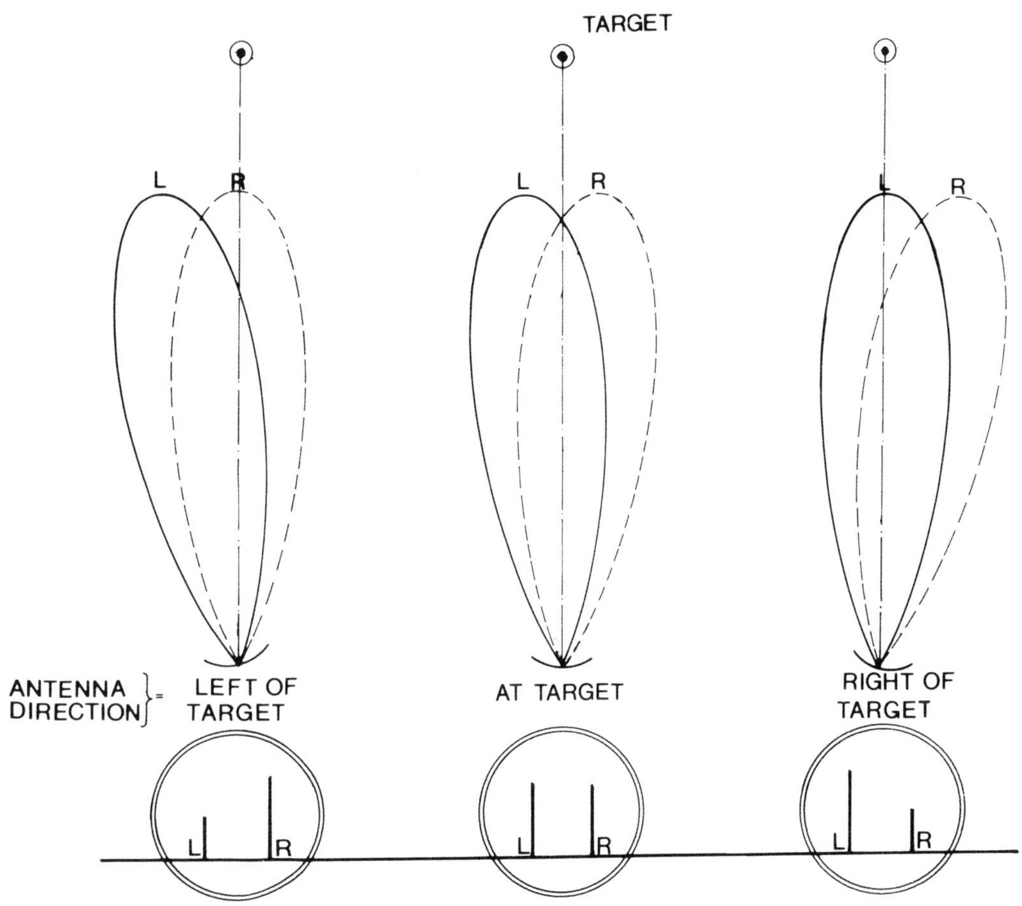

RETURN SIGNALS FROM TWO BEAMS

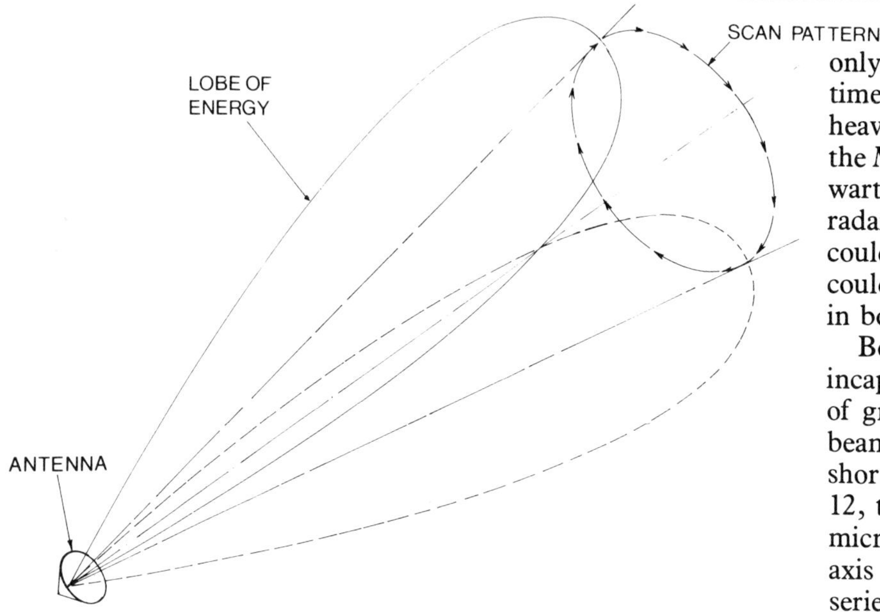

Conical scanning. Beam is nutated about axis, and variations in return measure extent to which axis is off target.

of NTDS and WDS represents a meeting of the two ends of the saturation problem first perceived at the end of World War II. At the same time, the single, high data rate radar is no longer the primary sensor of the WDS, as that system now accepts data from all shipboard search radars, evaluating that data automatically through the resources of NTDS.

FIRE CONTROL AND LOBING

The usual search radar also provided bearings insufficiently precise for gunnery use. Long before height-finders were developed, there were specialized fire control sets such as Mk 3, to provide accurate range and bearing *given search radar or other data as a basis*. Because the need for such radars pre-dated the development of effective S- and X-band transmitters, their designers had to show very considerable ingenuity. Even with the advent of very well-defined radar beams, fire control demanded the ability to search for a target in the relatively small volume of space defined by a search radar, as well as to track its motion.

The solution adopted in several US wartime radars was very ingenious: *lobing*. In effect one antenna system was used to project two radar signals in alternation, the signals directed at a small angle to each other. Generally the radar would not be pointed quite directly at the target, so that one sub-radar or the other would receive a stronger signal. Only when the entire system was pointed *directly* at the target would both signals match; as a dividend, the relative strengths of the two signals would indicate the direction required for corrections to point the system more effectively.

This technique was first used in the Mk 3 main-battery radar, which consisted (in effect) of two L-band radars side by side. This main-battery system required only accurate bearing and range data, but at the same time accurate elevation angle data was required for heavy AA directors. The same solution was adopted in the Mk 4 radar, which surmounted the great bulk of US wartime 5in DP gun directors. Mk 4 consisted of two radar antennas one above the other; and each of the two could be operated in halves, so that the Mk 4 *system* could produce, in effect, *four* beams, *ie* attain accuracy in both bearing *and* elevation.

Both Mk 3 and Mk 4 were L-band radars, clearly incapable of any very great accuracy except by the use of great ingenuity. However, the idea of alternating beams in various directions suvived the transition to shorter waves and hence narrower beams. In 1948 Mk 12, the successor to Mk 4, began to be replaced by a microwave dish called Mk 25; the latter spun about an axis just off-center, so that it produced a continuous series of beams angled slightly with respect to each other ('conical scanning'). Unless the radar were pointed accurately at its target, the echoes generated by these slightly different beams would differ, and the returning stream of signals would vary in a way indicating how far *off*-target the radar was pointed.

Lobing was used to improve the accuracy of several wartime search radars, including the pencil-beam height-finders SM and SP. It could also compensate partially for broad beams powered by small antennas (*eg* SA). However, in such cases precision could be attained only when the antenna ceased to scan.

A rotating-beam radar is somewhat complex mechanically. An alternative developed since World War II is *monopulse* operation, in which a single antenna serves several radar feeds and specialized circuits compare the sets of returns. A major consequence of monopulse operation is that rapid fluctuations in target strength, such as these due to rapid changes in presented target area, do not confuse the radar. From an ECCM viewpoint, monopulse does not present to a would-be deceiver a regular sequence of signals which can be systematically distorted to present a false picture of target motion. It would also appear far easier to achieve frequency agility in a monopulse than in a lobing or conical-scanning radar.

Monopulse radars can often be identified by their multiple feed horns, which in parabolic dishes meet at the focal point. Each element of the feed adds in its own off-center signal, and in many cases the radar carries out its comparison literally by mixing those signals by means of waveguide 'plumbing'.

Both lobing and monopulse present a wider possibility, automatic target tracking, by following small changes in target position. In effect the target is always making the radar position incorrect. Lobing, of whatever type, automatically generates these corrections which make tracking a practical proposition and such corrections, translated into rates of target motion, can be fed into fire control computers to estimate future

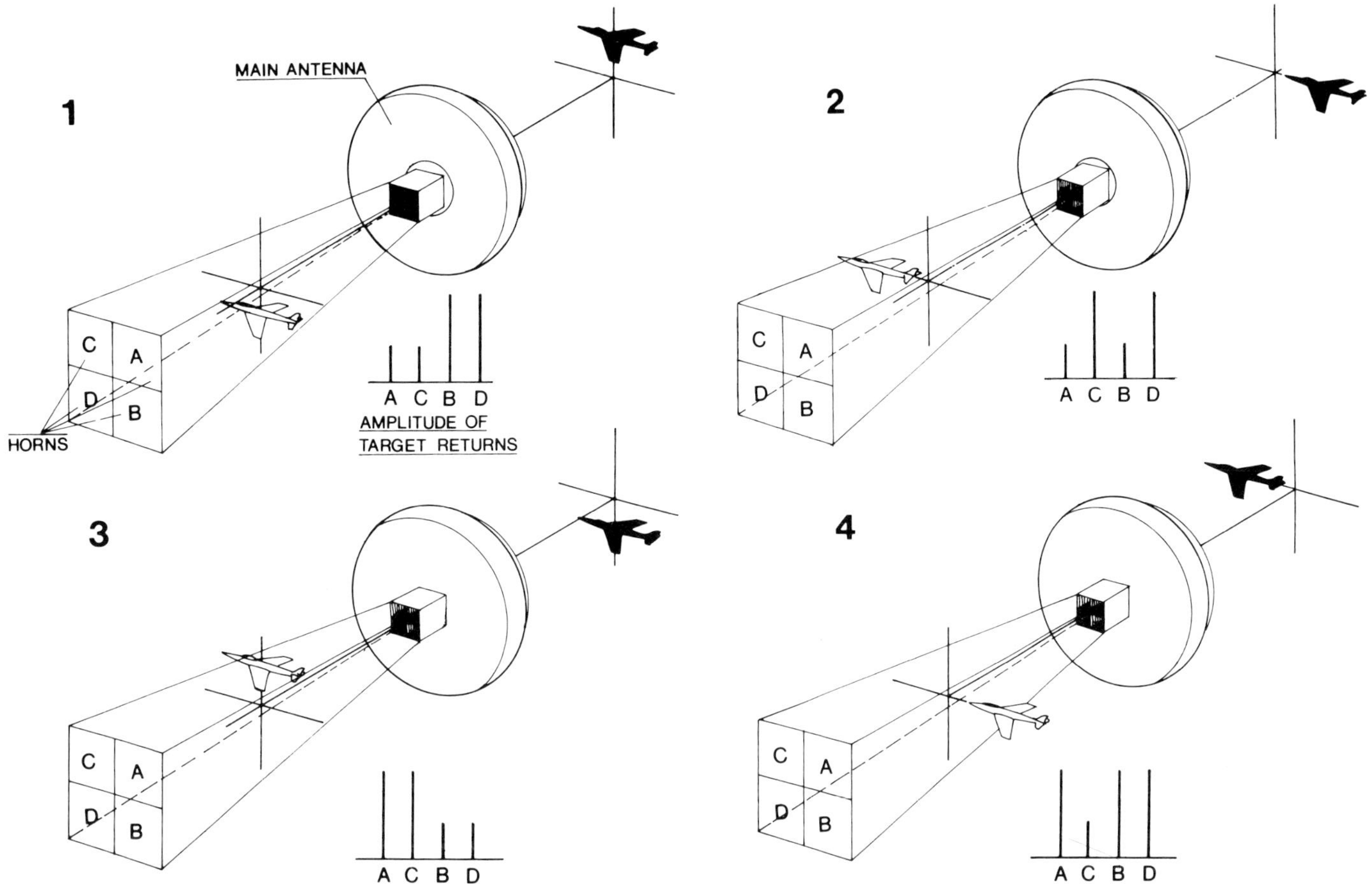

The principle of monopulse (simultaneous lobing). Returning signals are broken down into alternate (usually four) channels (corresponding to alternative antenna positions) and compared.

target position. Serious radar-aided AA predictors first came into existence late in World War II.

MISSILE CONTROL

The introduction of guided missiles for air defense enlarged and merged height-finding and fire control functions. In effect, a missile is a pilotless fighter requiring particularly good target data, since it is not nearly as resourceful as a human pilot once it reaches the vicinity of its target. Moreover, its own radar-homer – if it even has one – has not nearly the power of an airplane's. In addition, missiles were originally devised in large part because of the speed with which air engagements might develop.

A typical control sequence might be: (a) target detection at long range by an air search radar; (b) more precise target location by a height-finder; and (c) as the target comes into range, tracking by a fire control radar. What follows depends on the method of guidance of the missiles. All missile control systems must continuously track both target and missile, and somehow command the missile into its target. Radar offers several possible shortcuts. In one of them, beam-riding, a radar tracks the target and the missile is directed to fly *along* the beam tracking the target, a maneuver probably not particularly efficient, especially at long range. Moreover, at very long ranges, a target, however high, will be at a low angle of elevation – at 50nm, 50,000ft is only 10° – which means that a direct beam-rider of very long range may run into problems of sea clutter.

At low angles, the wide initial capture beam could reflect from the sea surface to cause problems; in this respect the X-band of the first Terrier installation (*Boston*) was superior to the C-band of subsequent ones (SPQ-5) because it reflected less readily from the sea surface (minimum elevation angles were, respectively, 2.5° and 6°). C-band had offsetting advantages, however.

This was the thinking behind the development of the first generation of US antiaircraft missiles, consisting of a medium-range (Terrier) and a long-range (Talos) type. Terrier employed simple beam-riding, in which the guidance radars produced a wide capture beam to gather the missile into the guidance beam. The width of the capture beam and the time required for it to take hold defined a minimum range for the system. Once the missile had been 'captured' it rode the tracking beam. Talos represented a somewhat different problem. It had so long a range (ultimately over 70 miles) that direct beam-riding was not practical. Instead, the Talos system comprised paired SPW-2 missile control radars, computer-driven on the basis of data produced by the SPG-49 target-tracking sets. In effect, then, Talos would ride a secondary beam.

BY COURTESY OF NORMAN POLMAR

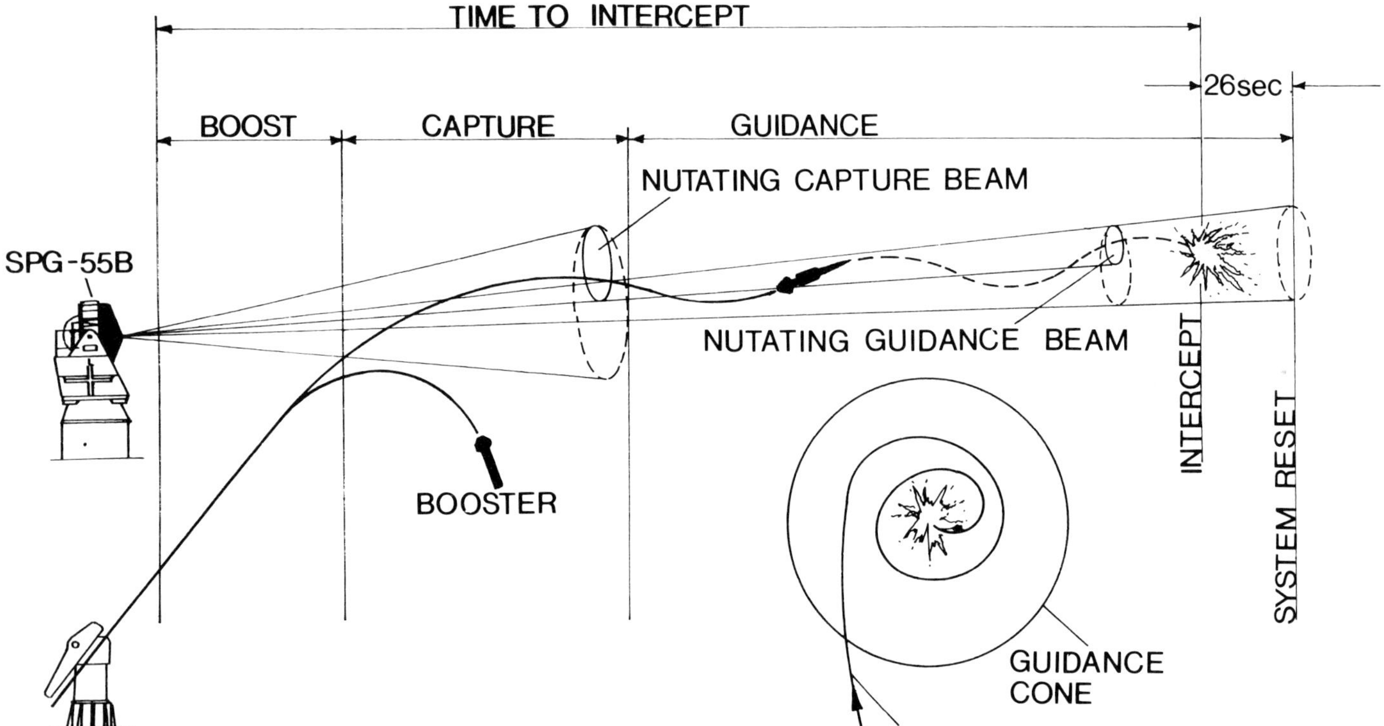

◀ *Terrier was the first US radar-guided shipboard missile; here it is shown aboard the test ship* Mississippi. *Guidance was by a much-modified Mk 25 fire control radar, a type later used aboard the missile cruiser* Boston. *In this photograph the radar atop the mainmast is SX, the predecessor to SPS-8.* Mississippi *served as test ship for many of the US naval radars of the first postwar decade, down through SPS-8B.*

▼ *Talos was actually the progenitor of Terrier, although it entered service somewhat later. It employed a combination homing system, a guidance beam being generated by the massive SPG-49 radar. When Talos appraoched its distant target, it switched over to semiactive homing. The small dishes of SPW-2, seen above and below the SPG-49s, tracked a beacon on the missile, and the SPG-49s followed a programmed course. In later versions of the missile, they also generated a CW signal for terminal semiactive homing.*

GIORGIO ARRA

The earliest US beam-riders used conical-scan radars; the nutating (slightly off-axis spinning) motion of the beam both kept the radar on target and provided deviation information to the missile. In one system, the pulse repetition rate varied as the beam nutated, so that a PRF counter aboard the missile could sense the quadrant of the beam in which the missile lay. The systematic variation in beam strength, as the beam spun, provided an indication of how far from beam center the missile was. Only one operational conical scanning tracking radar appeared, the Mk 25 Mod 7 of USS *Boston* and of the experimental missile ships. *Canberra* and other early Terrier ships used instead the monopulse SPQ-5, and subsequent missile control radars (SPG-49, -51, -55 and -56) have been of this type. The multiple monopulse feed is evident in the feed horn of SPG-51.

◀ *Flight sequence of Terrier beamriding missile (antiaircraft mode). System reset is where computer automatically adjusts system if manual reset has not been performed.*

Semiactive homing was a simpler alternative. Once more, a radar tracks the target. The missile is fired towards the target but ultimately steers itself toward the radar energy reflected by the target. Hence the radar beam need not be nearly so well defined as for beam-riding. Generally the missile illuminator is a specialized radar distinct from the target tracker which controls it. Semiactive homing might, for example, extend the effective range of systems originally designed for beam-riding. Moreover, since the missile steers itself toward the target rather than follow a path dependent on the motion of the launching ship, its path can be more efficient. This system was adopted for Tartar, the

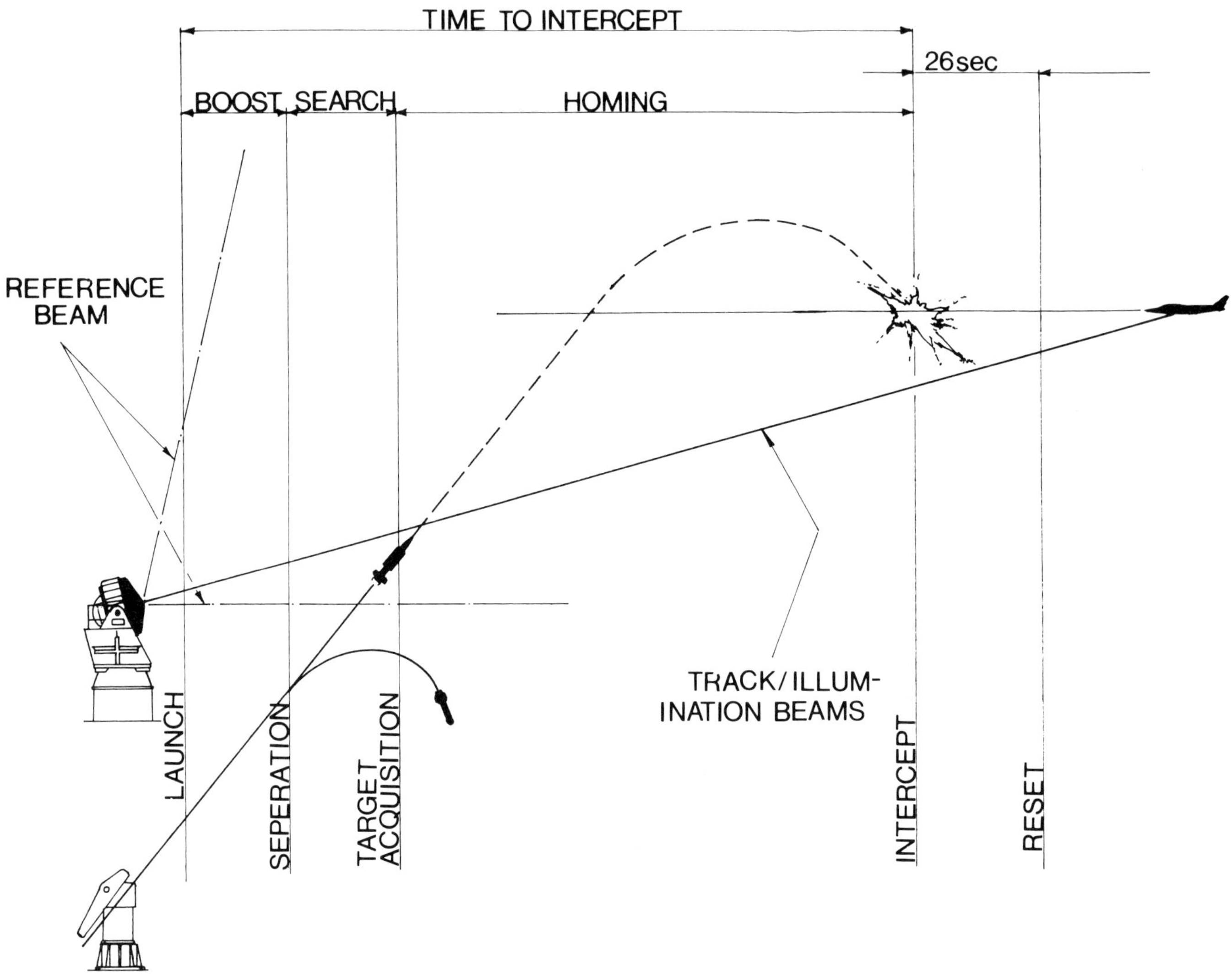

Flight sequence of Terrier/Standard semiactive homing missile (antiaircraft mode). After target acquisition missile follows 'up-and-over' trajectory while maintaining constant collision angle. Computer automatically resets system if manual reset has not been performed.

short-range second-generation derivative of Terrier, and ultimately for Terrier as well.

US semiactive missile guidance systems require target velocity (*ie* doppler) data as well as direction. Generally the missile obtains its doppler information by comparing the reflected beam with a beam transmitted directly by the tracking radar. The latter can be either a pure CW (continuous-wave, *ie* non-pulsed) type as in the land-based Hawk system, or else it may be a more conventional pulse radar with CW illumination injected through the same antenna, as in the Tartar and the Sparrow system. In the early 1960s systems based on pulse-doppler radars without separate CW illumination were introduced. Pulse-doppler provides accurate speed data, with data-processing to filter out the frequency spread introduced by the pulsing. The Advanced Sparrow seeker, for example, was designed to operate with either CW or pulse-doppler illumination.

Several developments of these ideas are possible. One is ICW, Interrupted CW illumination. If a missile can be designed to coast between bursts of illumination, then a single tracker-illuminator can share its time among several targets and thus among several missiles. In addition, recent developments in K-band active radars suggest that it will soon be possible to provide useful active seekers: the Phoenix air-to-air missile already incorporates an active terminal seeker. Such a device could not guide the missile from ship to target, but it could take over the illuminator function during the final portion of missile flight, reducing the fraction of time each illuminator would have to spend on each engagement, even in the absence of ICW.

AUTOPILOTS AND TRACK-VIA-MISSILE

Recent development has followed two paths. One is an attempt to counter the vulnerability to target saturation

inherent in any semiactive system: a radar illuminator must be held on target through much of the missile's flight. The advent of phased array radars such as SPY-1 made it possible to track many targets (and several missiles) simultaneously and with great accuracy. Thus the system can detect a target and direct a missile towards it by setting an autopilot. Both missile and target are tracked, and the missile can be given mid-

▲ *Later versions of Terrier were guided by a new radar, SPG-55B, which could operate in alternative beam-riding and pure semiactive modes. Beam-riding was retained because it permitted positive control of nuclear-armed Terriers, in accordance with US national policy on the control of nuclear weapons. Here two SPG-55s and a fully-loaded Terrier share the after end of the missile frigate* Dahlgren.

▼ *Continuous-Wave (CW) Illumination (Talos).*

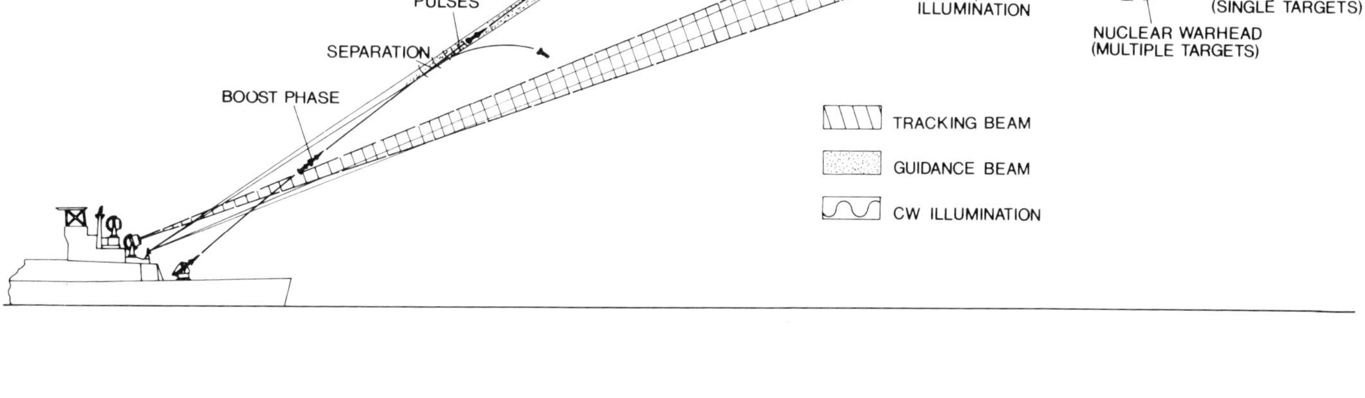

course correction commands to bring it into position. In the last moments of flight, it can be directed into its target by an illuminator controlled by the central system computer receiving search radar data. Meanwhile other missiles can be launched, and other targets tracked and engaged. This system and the SM-2 missile incorporating an autopilot are the basis of Aegis, which incorporates the SPY-1 phased array radar.

The extremely precise measurements possible with radars like SPY-1 suggest that a future system can be designed which does away with the illuminators. That is, if the missile itself is provided with a terminal seeker, then it can be command-guided into seeker range of the target, and left to home independently. Such a system should exploit to the fullest the rapid target-handling capacity inherent in a phased array radar controlled by a fast computer. Although no such system now exists, its components should be in existence in the near future, particularly given increased interest in self-homing ('fire-and-forget') air-to-air missiles.

The other major development is TVM, or Track-Via-Missile. Pulse-compression allows a search radar to achieve good range resolution at long range without excessive transmitter peak power. However, it is impractical to operate a semiactive homing missile on the basis of a pulse-compression radar: the level of signal processing required is far too high, if the size of the missile is to be kept down. In a TVM system, the missile receives the reflected radar signal from the target, but retransmits it down to a computer which processes it and then transmits commands back up to the missile. As in Aegis, a TVM system in principle allows simultaneous control of several weapons, since missile guidance is not directly linked to continuous target illumination. The Typhon system of the late 1950s was to have employed TVM, although its successor, Aegis, does not. At present the only US TVM system is the Army Patriot tactical air defense missile. A phased array multifunction radar tracks and illuminates several targets in rapid sequence. To some considerable extent, too, the use of command guidance permits the system to fly its missiles along the most efficient paths. Finally, TVM (or, for that matter, Aegis) achieves better accuracy than a pure command system because terminal homing is used; the accuracy of the system inproves as the missile approaches its target. It has been reported that the Soviet SA-10 land-based air defense system employs TVM; a naval version, provisionally designated SAN-6, is mounted aboard the new Soviet nuclear cruiser *Kirov*

At the other end of the spectrum, the Basic and Improved Point Defense Missile Systems employ a ship-launched version of the Sparrow semiactive homing air-to-air missile. The original BPDMS used a simple radar illuminator which was pointed at the target like a searchlight, but the Improved version (Nato Sea Sparrow) employs a powered tracking/illuminating radar, often linked to the dual-mode TAS (Target Acquisition System).

Long-range/fleet air-to-air defense missiles underwent a similar evolution. Thus Sparrow I was a beam-rider, but the current Sparrow III is a semiactive homer. The heavier Phoenix carried by the F-14 is an air-launched equivalent of Aegis, which is why an F-14 equipped with it is said to be capable of destroying six widely separated airborne targets simultaneously.

For a time the US Navy used radars to guide its surface-to-surface missiles, Loon and then Regulus. In the latter case a transponder beacon on the missile was used, and the radar tracking beam was coded to give guidance commands – hence the multipurpose ('Q') designator of Regulus radars (SPQ-2, BPQ-1 and -2).

CARRIER AIR CONTROL

We have already made reference to specialized fighter-control radars (height-finders). In fact carrier warfare requires considerably more: a system of homing beacons, and, after World War II, a system to make landing possible even in bad weather, much as instruments work at major airports.

Early carrier homing arrangements consisted of beacons plus airborne direction-finders – an unsatisfactory arrangement. The next stage (1937) was a homing beacon broadcasting directional information: YE (airborne equipment ZB).* YE was first tested aboard the carrier *Saratoga* in 1938, and during World War II its radar-like mesh antenna surmounted the highest mast of every US carrier. YE remained in use as late as 1960, and was adopted by the Royal Navy. YG was a simpler, standby system.

As YE rotated through each of twelve 30° sectors, its transmitter was automatically keyed, an assigned morse-code letter being broadcast twice. The beacon was keyed to the ship's compass, so that its headings

*The Y series began with fixed equipment. YA was a 300–500kc/s beacon incorporating a radio-telephone. YB was the ground station of an instrument-landing system, comprising YC (localizer and beacon in truck trailer) and YD (marker beacon in a motorcycle sidecar). YF was an abortive semiportable marker beacon. YK was a radar beacon (AN/CPN-3). YL/YN was an unsuccessful landing craft homing system, of which YL was the transmitter. YL was later tried in CCA experiments, about 1950. YO was an abortive instrument low approach system; YP was a ceilometer (reassigned as AN/GMQ-3); and YR was a nondirectional radio-telephone beacon, a combination control tower radio and radio compass locator station. The missing designations were all ship-related: YE and YG, the radio beacons; the racons (radar transponder beacons) YH, YJ and YM (AN/CPN-6); and YQ, mounted on the AEW ship so that the airborne radar could transmit a picture centered on that ship. It operated at S-band, using three horizontal dipoles 120° apart. All of these wartime types were superseded postwar by AN series devices, but they are listed here to provide some indication of the direction and breadth of wartime development.

were accurate, and it could be detected at about 90nm by an airplane at 5000ft. The less sophisticated YG required course corrections cranked by hand. Both systems operated at 246mc/s.

Such a system was unsatisfactory in that it provided, at best, only directional data. In particular, homing beacons were sufficient until it became necessary to vector fighters at considerable distances, where reference range and directional data would be essential. The alternative was a transponder which could react to radar signals – YJ. This system was broad-tuned to respond to a variety of P- and L-band radars, but its response would be at a frequency slightly off the radar frequency, so that the receiver of a radar set communicating with the beacon would have to be de-tuned. In effect an airborne radar could seek beacons or targets but not both. Special switches were provided for airborne radars, the idea being to avoid confusion and interference – and, for that matter, clutter from sea returns. YJ operated at sub-bands A (IFF Mk III) and B (515mc/s). It succeeded an A-band land or ship transponder, YH, and was itself originally intended for land operation. YJ-2 was the carrier version.

Most airborne radars fell into the S- and X-bands. A new transponder, YM (later AN/CPN-6), was developed for X-band. Beginning with installation aboard USS *Yorktown* in December 1944, it replaced YJ. As compared to the latter, it responded instantly instead of 30 seconds later, and had a far more sophisticated coding. Both YJ and YM were considered effective at 70–150nm with an accuracy of a quarter of a mile or 5°

AN/CPN-17 used CPN-6 components for S-band. None of the radar beacons proved popular during the war; according to the 1945 Radar Board, 'pilots are often reluctant to trust someone else [*ie* a radar operator] to find their home ship and prefer to listen to the YE code with their own ears'. There were also serious maintenance problems. In 1945 L- and P-band radars were no longer in wide use aloft, and carriers mounted YE and CPN-6 postwar.

The greatest advantage which could be claimed for a transponder was that it automatically gave range as well as bearing. In some sense the transponder would 'echo' back the radar signal, albeit with some slight delay. Work therefore proceeded on a better beacon system after World War II. The result achieved was TACAN (Tactical Air Navigation), which provides both range and relative bearing. Its heart is a big transponder operating in the UHF band about which nine parasitic aerials, which provide directional information, rotate. So much information requires a certain ingenuity, as the aircraft using TACAN must distinguish distance data from directional signals. Moreover, a means of identifying the individual TACAN beacon is necessary.

This means that the antenna must put out a special signal marking a reference direction, plus auxiliary signals marking intervals around the circle in which it rotates, plus distance interrogation replies, *plus* an identifying signal. The array of parasitic antennas rotates at 15 revolutions per second, and TACAN is claimed to be accurate to 1000yds at a slant range of 195nm. The system is thus a cross between YE and YM in that bearing data are transmitted continually, whereas range data are by transponder. The designation of early TACAN, AN/URN-3, suggests its wide use outside the fleet, and TACAN is of course a feature of all ships working closely with aircraft.

CARRIER-CONTROLLED APPROACH

Beacons were one essential element of the development of all-weather carrier operations, including very high speed air defense. Another essential was blind landing. This began as Carrier-Controlled Approach (CCA), *ie* a blind system to bring an approaching airplane to within 100yds or so of the carrier, where the Landing Signal Officer (LSO) could take over using his eyes. The system was pushed forward with some urgency, since at the end of World War II many senior officers considered that the greatest threat to the Fleet was radar-controlled bombers launching guided missiles in weather too poor for flying. Both the postwar missiles and CCA were attempts to solve this problem.

CCA was inspired by the wartime success of GCA, or Ground-Controlled Approach. Work began in 1944, and the initial problem was to find a radar precise enough to provide a good picture for an LSO. AN/APQ-7, a precision bombing radar in a wing-like radome, was selected. NRL provided a new B-scope display showing the sector from about 170° to 230° relative to the carrier, *ie* her port quarter. The X-band APQ-7 (0.4 × 30° beam) scanned a 30° sector every 1–2 seconds by an electrical technique similar in theory to FRESCAN. It was to be located under the after end of a carrier's flight deck, supplemented by a pair of APG-17 dishes, which provided air speed indication.

The entire system, controlled from a space beneath the flight deck near the LSO station, was installed aboard the escort carrier *Solomons* (CVE-67) between mid-June and 10 July 1945. Initial daytime trials proved quite successful, and night landings were made in August. At this time the system was considered effective with visibility as low as 500ft, in rough seas with winds of up to 43kts and moderate rain. Final approach was controlled by an LSO using illuminated wands – hence the need for 500ft visibility.

The procedure at this time was for CIC to guide the plane around the landing circuit, using SG. As the pilot passed between ½ and 1 mile on the port beam going down-wind (*ie* in the direction opposite to that of the ship), CIC ordered a 95° turn to port and handed over control to the CCA crew, who controlled his heading so that he could see the LSO at 100–300yds. Note how closely CIC, CCA and the LSO had to cooperate.

Moreover, the *Solomons*' experiment did not take into account a need to control large returning air groups.

However they were so successful that installations on *Essex* class carriers were ordered, the first two to be USS *Franklin* (CV-13) and *Antietam* (CV-36). It appears that *Franklin* had already been modified as of early 1946, although she may not have operated in this configuration. As of 1946 the CCA system was to have used an auxiliary CIC under the flight deck aft near the LSO's platform; the main innovations were remote repeaters for SX and SG, so that CCA could control aircraft from the first.

An incoming strike would be picked up by CIC and then passed to a 'pick-up controller', who would break it into groups which he could station 15 miles from the carrier. As planes landed on, he would move these groups into a Traffic Area 5–10 miles astern, from which the Traffic Controllers could pick up planes one by one, using SG, and vector them to lower wheels and flaps and reduce speed and altitude to 100kts and 300ft. The pilot would then be vectored into the 'groove', a path about 10° on the carrier's port quarter. At 2000yds he would reduce speed to landing speed and altitude to 90ft – and await the appearance of the LSO.

This extremely ticklish system was greeted enthusiastically as a means, for example, of avoiding the non-combat losses of a latter-day Marianas 'Turkey Shoot'. Of course it also presented a serious potential for bad-weather task force defense, and as such it was pressed forward.

For some reason, however, neither *Franklin* nor *Antietam* carried out tests. By March 1947, the Material Improvement Program called for CCA aboard all active carriers: the three *Midway*s, nine *Essex*s, two *Saipan* class CVCs and nine escort carriers. *Valley Forge* (CV-45) and *Philippine Sea* (CV-47) were to have prototype installations. They were evaluated by the Operational Development Force that fall. Tests aboard *Philippine Sea* continued through the fall of 1948. By now APQ-7 (modified) had become SPN-3, and APG-17 redesignated SPN-2. The system generally followed the 1946 proposals, except that the CCA station was now adjacent to the CIC, and a new precision radar, SG-7, was now used to vector aircraft into the field of the approach radar (now SPN-3).

MARSHAL CONTROL

The 1947–48 tests included attempts to use escorting DDRs to define the 'holding' patterns. On problem was arranging the random incoming aircraft in a smooth pattern in the 'marshal area', *ie* the holding area. In 1948 this was the only holding area; aircraft were picked off, handed over to traffic control and landed, without milling about astern. A maximum rate of about one landing per minute was achieved, with a 200ft ceiling and a 500yd visibility. In this system SX provided marshal control, handing over to SG-7 for traffic

The carrier Philippine Sea *tested Carrier-Controlled Approach (CCA) techniques in 1947, using a modified SG radar, visible here sponsored out over her flight deck from her funnel. This SG-7 formed the basis for SPN-6, the standard US Navy 'marshal radar'.*

control, and then to SPN-3 for final approach. This terminology carries over to more recent systems. What is worth noting is that only SPN-2 and -3 were essential to CCA; the rest of the system might as well have consisted of standard carrier radars.

One serious shortcoming of the system was the requirement that weather be good enough for formation flying, *ie* for the formation and break-up of the initial holding pattern. Capacity during one operation was said to be eight aircraft or eight sections of two aircraft each; and 25 per cent of all landings would be wave-offs. Nonetheless, the system was considered worthwhile.

The precision demanded of control radars implies X- or S- band and very short range. In 1948–49 the success of the prototype system led to proposals for a more advanced version aboard the rebuilt carrier *Oriskany* (CV-34) and the large carrier *Midway* (CVB-41), to use

The carrier Hornet, *refitted at San Francisco in January 1965, shows her Carrier Controlled Approach system. The SPN-6 marshalling radar (***87***), high up her mast, is just visible against the dark paint or the mast proper. However, the SPN-35 Aircraft Approach Control Central (***67***), usually radome-enclosed, shows both of its antennas; it occupies the same position as the much smaller thimble-shaped radome of SPN-8, in earlier configurations. Above it the SPN-12 air speed indicator (***68***) is visible on its platform. Other unusual antennas visible are the whip for* Hornet*'s ASW Command and Control (ASCAC) link (***88***; Type 66046 for radio type TCS); a radio-sonde (weather information) antenna (***75***; AT-775/SMQ-IA), the long dipole of an ESM antenna (***63***; port and starboard: AT-1025/SLR), and ECM whip (***65***; AT-924/SR; both* **63** *and* **65** *are duplicated to starboard), an old AS-45A/APR-6 'Y' shaped ESM antenna (***77***), and the slave IFF antenna (***83***; AS-1065/UPX) for her SPS-30 height-finder (***80***);* **85** *is an IFF test antenna (AS-1065/UPX). The small radome just visible below* **90** *is an AS-1154/SLR-10 ESM antenna. TACAN is* **99** *(SRN-6). Not visible here are the deceptive ECM antennas of ULQ-6, and the slotted waveguide of a Raytheon Pathfinder navigational radar on the forebridge. The large air search antenna (***81***) is SPS-43A, and the usual radio antennas, including 'Phasor-90s' (such as* **69** *and* **70***) are in evidence.*

a pair of new radars, the S-band SPN-6 for 'marshal control' and traffic control, and SPN-8 for approach control, including air speed indication. This system could land one airplane every three minutes. By the mid-1950s a third radar, SPN-12, had been added as a specialized air speed indicator: a 2ft parabolic dish emitting a continuous signal, the doppler shift in whose echo would accurately measure the speed of an approaching airplane.

Late in the 1950s a new SPN-10 replaced SPN-8. Systems incorporating it were now referred to as automatic CCA, SPN-10 being the *landing control central*. It consists of two radars capable of tracking two aircraft at the same time, so as to achieve a landing rate of two aircraft per minute. In order to keep the size down, the designers went to the Ka-band, about 1cm in wavelength, and a 4ft parabolic dish with a spinning beam then sufficed.

SPN-10 made possible completely automatic landing in any weather, and it was first tested aboard USS *Antietam* on August 1957. The automatic lander used a data link connected to the airplane's autopilot to transmit up the precision tracking data generated by a pair of SPN-10s.

During the 1960s a new generation of CCA radars appeared: (a) SPN-35, for precision approach, displaying azimuth and elevation to a radar operator who guides the pilot to a transition point a mile from his ship (this X-band radar generally occupies a radome); (b) SPN-41, to provide approach guidance to an approaching pilot, in effect giving him an electronically marked landing path; (c) SPN-42, a replacement for SPN-10 using more advanced technology; (d) SPN-43, a replacement for SPN-6, described officially as an air traffic control suveillance system with an effective range of between 250yds and 40nm (30,000ft) and, like its predecessor, an S-band parabola; and (e) SPN-44, a solid-state replacement for SPN-12. The complete modern system would, therefore, consist of SPN-35/41/42/43/44, but in a few cases no SPN-35 appears to have been fitted.

NAVIGATION AND LORAN

Radar has contributed to navigation in two important ways. In the small, it has been used to provide very detailed pictures of the area immediately surrounding a ship, so that approach to a harbor in poor visibility is practical. Several wartime S- and X-band sets were considered very suitable for such tasks as buoy detection. In a slightly broader sense, surface search sets were valued for the navigational data they could supply. Modern navigational radars, as opposed to surface search radars, are small sets optimized for very great precision *at short range*. The Raytheon Pathfinder 1500 is typical of modern commercial navigational radar. A comparison with a standard surface search radar, SPS-10, is shown in the accompanying table.

Comparison between Pathfinder 1500B and SPS-10

	Pathfinder 1500B	**SPS-10**
Band	X	C
Peak power (kW)	7	285
Pulse-length (microseconds)	0.14	0.25 and 1.4
PRF	750 and 1500	625–650
Antenna width (in)	48	120
Beam size (degrees)	2 × 20	1.5 × 16

One of the few conspicuous gaps in US naval radar development was navigational radar, both for large and for small ships. In the early 1960s many standard commercial sets (*eg* the Pathfinder 1500 and Marconi LN-66) were bought and deployed, but Vietnam experience suggested that their reliability was inadequate. However, no new navigational (AN/SPN) radar has appeared.

The other face of navigation is position-finding in the open sea, perhaps under conditions of such low visibility that even stars cannot be seen. A variation of radar, LORAN (Long Range Navigation), was devised during World War II for this purpose; it and such variations as Omega continue in wide use today. The basic LORAN unit consists of a pulse emitter ('master') and a transponder ('slave') some considerable distance (usually about 300 miles) away. Both pulses are omnidirectional. A ship using LORAN listens for both the initial pulse and the slave pulse, the latter beginning later than the initial pulse owing to the distance of 300 miles. The first pulse must cover to reach the slave in the first place, and the remaining time difference measures the difference in paths from master and slave stations to the ship. What the ship actually hears is a delay, a net time difference – which is equivalent to a *distance* difference. The equivalence of distance and time used in LORAN is what makes LORAN a kind of radar.

In any case, all that the ship knows is that it is at some point the distances to which from two known points are *some fixed distance apart*. That difference specifies a hyperbola on which the ship must be, a second fix, perhaps using a second slave station, gives a second hyperbola, and the ship is at the intersection of the two. This 'hyperbolic navigation' has the great virtues of simplicity and radio silence on the part of the navigator.

A great deal depends on how far the pulses of master and slave propogate, and on the 'base line' between the two. In effect, LORAN is measuring the angle at the ship corresponding to the base line, so that the longer the base the better the accuracy of LORAN at very great ranges. Clearly base line, too, is a matter of propagation. For these reasons the tendency in LORAN has been toward *lower* rather that higher frequencies. In current parlance the LORAN of World War II is LORAN-A, the more recent type LORAN-C, and a VLF type (which can penetrate water and is thus useful to submarines) is Omega. Some typical characteristics are shown in the table.

LORAN Characteristics

	LORAN-A	LORAN-C	Omega
Frequency (kc/s)	1700–2000	100	10
PRF (cps)	20–35		
Base line (nm)	300	500	500
Range day (nm)	700	1000	5000
night (nm)	1400		
Error line (nm)	0.2–5		
fix (nm)	0.08–10		

IFF

IFF (Identification Friend or Foe) is not, of course, a radar, but it is so universal that it deserves mention in any account of radar system. Probably the most important point about IFF is that it is part of an integrated system necessarily embracing many kinds of radars – and radar targets. For example, an airplane would have to identify itself to radars of widely different frequency, such as SK and SP, not to mention the set of Allied navies, especially the Royal Navy. Somehow the very wide variety of Allied radars would have to share a common IFF *system*. All would have to understand its reply – which would have to be coded to prevent the enemy use of the system.

The need for some means of positive identification predated radar. Once radar had been devised, US and British workers independently produced very similar systems, which were brought into correspondence when both nations fought as allies. The earliest US system was the NRL XAE of 1937, a shipboard yagi antenna mounted on a rifle stock which could be pointed at an unknown airplane. The pilot would turn on his omnidirectional identification beacon, and the ship would transmit back an acknowledgement, which flashed a light on the airplane, visible from the ship. The system worked on a frequency of 500mc/s, or 60cm. The air-to-ship system was tested aboard USS *Ranger* in 1938, and production for operational use began in January 1939.

Radar required automatic identification. For its first radar (XAF) NRL devised a rotating reflecting antenna, producing a coded echo when it was illuminated. During the winter exercises of 1938–39 which tested XAF, this device was mounted in a destroyer, and it proved useful. Quite independently, Watson-Watt conceived an analogous device for aircraft identification. Unfortunately this interaction between echoes from two or three nearby aircraft could have a similar effect. What was needed was a more positive means of identification – a *transponder* which would reply to each radar pulse it sensed.

This Mk I system was introduced into British service in 1940, and its details were disclosed to the United States that fall. However, even then it was obsolescent. By late 1940 Britain had a variety of air defense radars in service on quite different frequencies. It seemed unlikely that any practical airborne system could search through such a range quickly enough to react in time to incoming radar signals. An attempt to do so, for the 10–15m of CH, the 7m of a system called MB and the 3–5m of early AA fire control radars, was designated Mk II; it was not very successful.

Once again, quite independently, NRL had hit upon a very similar system. NRL had a pulse transponder in 1939; to challenge, the radar was switched to a special PRF which triggered responses. In US service, the CXAM IFF system (BD and BE transponders) was considered equivalent to the British Mk II. As yet, the radar *itself* was the interrogator. Interrogator and transponder operated on the radar frequency, and such operation could not be satisfactory when many different radars were used.

Mk III was Watson-Watt's solution, the precursor of modern IFFs. IFF challenge and response were to occupy a separate, specialized band (A-band). It was adopted as the standard Allied IFF of World War II (1942) and remained in US service for some time after the war. An important question was how to ensure that returning signals gave an accurate indication of target *bearing* and not merely of target range. The solution adopted was to locate the IFF interrogator *on* the radar antenna rotating with it to give directional indications. Generally the same device would receive the response; it was then an 'interrogator-responsor'. BL interrogator-responsors were built into SC-2 and SK search radars, and a specialized interrogator-responsor was built into the pencil-beam fire control sets. On the other hand the SG X-band surface search set did not incorporate an integral IFF antenna. It benefited from the operation of omnidirectional interrogators, BM and BN. Ships were also equipped with responders (BK) to answer the challenges of other ships and of aircraft.

The band selected formed a 30mc/s slice of the P-band, the use of so wide a band (157–187mc/s) being dictated by problems of maintaining stable frequency. In addition, a higher frequency G-band was adopted specifically for fighter control, in which the greater precision of shorter wave signals would be important. In practice a Mk III IFF transponder would tune its receiver over this band every 2.5 seconds, responding almost instantly to any signal received; then there would be a 'dead time' corresponding to resetting the sweep, and the cycle would begin again. The slight delay in response could contribute to position ambiguity, and it was sometimes argued that the need to sweep through many frequencies would cause the transponder to miss a large percentage of interrogating signals. The means of avoiding position ambiguity was stopping the radar with its interrogator while both pointed in the appropriate direction. This of course cut

Existing IFF systems require a dedicated interrogator for each search antenna, since the radar pictures from the different sets are not automatically integrated. The missile cruiser Little Rock, *shown here as built, was typical, with the usual Mk X array antennas visible atop the SPS-17 mattress on the foremast and at the base of the massive SPS-2 on the short lattice aft. SPG-49 and SPW-2 Talos guidance radars are visible right aft. Note, too, the waveguide running up the lattice mast amidships, not yet topped by an SPS-39 FRESCAN radar. TACAN has already been installed, as well as a set of short dipoles for ship-to-air communication. Also visible amidships is a crane for Talos missile strike-down during underway replenishment.*

the data rate very badly.

Meanwhile NRL has designed an alternative system, designated Mk IV after US adoption of Mk III. It differed from Mk III in employing separate frequencies, 470 and 493.5mc/s, for challenge and reply. Mk IV was held in reserve against possible compromise of Mk III, although a few were used in the Pacific at the end of the war. A major argument against Mk IV in Europe has been its closeness to the frequency of the German *Wurzburg* radar, 550mc/s. The Mk IV system on board ship consisted of BA transponders and BG, BH and BI interrogator-responsors; BP was a kit to convert BN to Mk IV operation; and BE was a recognition transmitter and BF a corresponding receiver. Mk IV had greater directivity than Mk III, due to its higher (G-band) frequencies; the usual array on an SA, SC or SK had a beamwidth of 7–10°

Mk III was clearly an interim measure. In September 1942, the Radar Committee of the combined Communications Board, combined (British and American) Chiefs of Staff, directed NRL to produce a new system, which became Mk V/UNB (United Nations Beaconry). It followed Mk IV in using separate (high frequency) interrogation and reply signals with replies associated with radar echoes. Transmitter and receiver frequency separation permitted the use of higher gain antennas, and higher frequencies (950–1150mc/s) made for better directivity. Twelve channels were made available within this range as an antijamming measure. Moreover, signals were coded to permit, for example, identification of one among several friendlies. On a PPI, transponder coding would be displayed as a dot-and-dash elongation (radially) of the target pip. Although the first Mk V systems, AN/CPX-2 (shipboard) and AN/APX-2 (airborne), appeared in August 1944, Mk V did not complete service evaluation until 1947–48.

Mk III was distributed to all Allied forces, including those of the Soviet Union, so that from a postwar point of view it could be considered thoroughly compromised. Even so, postwar financial stringency required that it remain in service, a 1950 report indicating that stocks would probably remain until 1951.

Mk V was considered successful, but few were produced. Installations were confined to the CVBs and to several fleet carriers, in which it was important to be able to track fast targets (jet fighters). According to the 1950 report, 'These installations will work with transponders of the new Mk X and will permit limited IFF and excellent tracking and control features. Jet types have been tracked in excess of 175 miles at above 20,000 feet . . . Stocks of IFF Mk V are exhausted. There are no spares and little test equipment to maintain IFF Mk V/UNB installations'. That left the new IFF system, Mk X.

This was *not* the tenth Allied IFF system. Mk VI had been an abortive wartime simplified Mk V; but at the end of World War II NRL had begun to investigate 'X', the IFF of the future – which naturally became Mk X. Free of the press of wartime, the NRL group was able to consider more deeply the character of an ideal IFF. Mk V had used a universal 'code of the day' to distinguish friend from foe, but NRL considered this feature a serious potential security risk. Another danger inherent in Mk V was *enemy* use of our IFF to identify our own craft. Ideally some cryptographic system could have answered these problems, but by 1948 the onset of the Cold War (and of fast aircraft) had made the new IFF system more urgent. Mk V was to be adapted, but pressure on frequency allocation forced NRL back to two-channel (interrogation and response, 1030 and 1090mc/s) operation. Production began in 1949.

SECURITY AND RANGING PROBLEMS

The Navy planned extremely wide use. As of August 1950, partial operation aboard Mk V/UNB carriers was to begin within a month, with 50 per cent of active ships and naval aircraft using Mk X by 1 July 1952; the balance of the fleet would be operational by 1 January 1954. Mk X was to be operational by fire control IFF, surface-to-surface IFF and air-to-air IFF by 1 January 1956. Advantages were held to include improved directionality (3°–7° beam, depending on the antenna), correlated PPI presentation, greater reliability, and an uncompromised system. However, air-to-air IFF (except in AEW aircraft) would still be nondirectional. Moreover, Mk X had no inherent security features, just as Mk V had not.

NRL was well aware of the latter problem. It collaborated with the Federal Aviation Administration to produce universal identification systems for military and commercial aircraft, the system in use at major US airports today as ATCRBS (Air Traffic Control Radar Beacon System). The method used was coding, a total of four modes each allowing for 8192 codes. This system compared with the three modes of Mk V (general IFF, one friendly from another, and flight leader identification); the fourth mode being reserved for altitude information. In this Selective Identification Feature (SIF) form Mk X became available in 1959. An alternative designation for SIF, Mk XI, was not used.

Mk X (SIF) did not provide real security; its interrogation pulse was not coded. Thus there was always a real possibility that an enemy might use Mk X interrogation pulses to induce US aircraft to identify themselves, then use their IFF systems as beacons. As early as 1951, NRL had a vacuum-tube binary coder, effective but far too massive for IFF use. Transistors improved matters, and by 1956 a tri-service group had been formed to implement what has become Mk XII, the current US system. Mk XII is also sometimes referred to as AIMS, the Air Traffic Control Beacon System for IFF Mk XII System. The great advantage of a fully cyptographic system is that even an enemy with full

As the radars have been integrated, the need for independent IFF interrogators has decreased: after all, each radar feeds a single combat system. Thus the new US Aegis ships will have a single IFF antenna, a fixed array UPX-29. This is the RCA mock-up of the superstructure (and, more importantly, combat system) of an Aegis striker cruiser, located at Moorestown, New Jersey, with the IFF system surmounting its lattice mast. One face of its phased array is visible below the bridge windows, with two slaved illuminators above: the SPY-1 radar is so precise that they do not have to track their targets at all, but need only obey its tracking commands. An SPS-49 long-range two-dimensional radar is located abaft the lattice mast, and an SPS-55 surface search set is mounted on a platform forward of it. The land test site itself is a consequence of the growing importance of integration among radars: it is a test of the ability of the Aegis computers to operate properly with the Aegis hardware, particularly the search and tracking radars.

knowledge of the design of the system cannot use it, because he cannot then guess correct interrogation or response codes.

IFF operation still presents considerable problems. For example, for many years the US Navy developed longer range air-to-air missiles for Fleet Air Defense. In Vietnam, however, it proved almost impossible to exploit the long range of such weapons as Sparrow, because IFF was never sufficiently trusted. Instead, rules of engagement called for visual identification before engagement, and sometimes Phantoms had virtually to pull alongside targets before they could fall back and fire their missiles. There were always, for example, fears that target aircraft might be civilian (hence without IFF transponders) or might be friendlies damaged in battle (and hence with inoperative transponders).

One solution is to use inherent aircraft characteristics for identification, in effect imposing the entire burden on the observing radar. Such a development is practicable in view of advances in computing power per unit weight and volume, so that complex radar data processing systems can now fit aboard aircraft. There are several possibilities, the most mature development begin direct observation of the jet engines of an approaching aircraft. Its compressor blades impose modulation on the returning echo, and this modulation cam be used to identify the particular engine type. This is JEM, Jet Engine Modulation. Other techniques seek something closer to an image of the target: high range resolution (HRS), using very short pulses to identify particular parts of the target; multifrequency (MFS), a sequence of short frequency steps, making use of the variation of a radar cross-section with radar frequency; target motion resolution (details of target motion are characteristic of its configuration); and inverse synthetic aperture radar (ISAR) techniques, from which an image of the target can be generated directly.

At another remove, target track itself can be used as a form of target identification. One of the great advantages of NTDS-equipped warships off Vietnam was their ability to keep track of the very large number of friendly, neutral and enemy aircraft in their surveillance areas, although this was a task greatly complicated by the small number of enemies and also by the extraordinarily high cost of erroneously identifying a neutral (such as an airliner) as an enemy. Track IFF data is also used by modern naval combat systems to rank incoming threats, so as to engage the most urgent ones. However, it is not sufficiently reliable to displace all existing systems, and development of new IFF systems continues.

The current effort is proceeding along two complementary lines. As of February 1980 it was named USIS, the US Identification System; it had formerly been named the NATO Identification System, or NIS, and the Air Force is the lead agency. One line is direct: development of the Mk XII system as well as Non-Cooperative Target Radar (NCTR) using mechanisms such as JEM. The other is indirect, a new concept. As combat systems, both shipborne and airborne, become more highly integrated, information in those systems should become increasingly available to all users on a more and more timely basis. A good deal of that information should permit target identification in real time, and it seems unwise to duplicate it unnecessarily. In particular, in a task force at sea there will be many interrogators operating almost continuously, and all will be at very similar frequencies in L-band. In effect the task force becomes an L-band beacon, even though much of the information received in return will be redundant. This will be the case even within one ship equipped with several radars operating simultaneously.

The new Aegis cruiser goes one step in this direction by substituting UPX-29, a single, electronically scanned IFF antenna for the IFF antennas more conventionally provided for each radar. However, IFF integration within a battle group is still a future hope.

IFF installations have varied considerably in size and shape, as one might imagine from their very different frequencies. For example, BL (Mk III), affixed to many mattress antennas, was itself a small mattress with four vertical dipoles projecting from it. BL also used a series of omnidirectional antennas: quarter-wave dipole with horizontal 'group plane' reflector ('steering wheel') and half-wave dipole ('stovepipe'). The latter was preferable for larger ships in view of its better performance. BK, BM, and BN were similar dipoles, generally mounted in pairs on masts, BK on one side and BM on the other. The IFF installation in SK-2 was, unusually, a second dipole mounted in front of the feed dipole. Mk III and Mk IV versions could be distinguished by the size of this dipole. Mk III being longer than the SK feed element, Mk IV considerably shorter. Generally mattresses (SC and SK) were built with dipoles for both Mk III and Mk IV installed, but only one set wired. SRs delivered at the end of World War II had only the former, and no auxiliary mattress was visible, the Mk III dipoles being integral with the mattress proper.

The antennas of Mk IV were generally similar. The BA transponder was an omnidirectional dipole, shorter than those of Mk III because of the higher frequency employed. The original interrogator-responsor, BI, used a small (48in × 36in) mattress, as did BH and BE-1 (20in × 28in). BE used a mattress plus a yagi, to achieve better directionality. In all 34 BEs, 400 BE-1s, 10 BHs, 3 BLs and 20 BI-1s were built. A total of 400 BGs were incorporated directly into the frames of their antennas. Of these equipments, BI (presumably 'I' for 'Interrogator') was the first – a rare example of the use of that letter in a designation.

Mk V continued the trend down in wavelengths. Its main visible manifestation was as a replacement for BL mattresses mounted on air search radars like SR-6. In a typical mount on the slotted-waveguide version of SR-6, the mattress of BL was actually larger than the main antenna. Mk V showed instead eight much shorter dipoles, and its sub-reflector was of much finer mesh.

As in the case of SK-2, postwar air search radars such as SPS-6 were not really suitable for mattress IFF arrays. At the same time the character of Mk X suggested that all, or nearly all, radars, even the short-wave surface search types, should have integral IFF. The solution adopted was ingenious: an IFF dipole inside the horn feeding the antenna itself fed in IFF pulses. In this way relatively sharp IFF patterns could be achieved. Of course, when the Navy returned to P-band mattresses such as SPS-28, a more conventional IFF antenna was once more practical – indeed, essential, given the wide mesh of the main antenna and the short IFF wavelenghts. Recent IFF antennas are enclosed broadside arrays, probably printed dipoles. They appear at the top or bottom of many recent radars, including the FRESCANs.

In some cases, such as the SPS-30, the IFF broadside antenna is mounted separately, near the radar, and rotates in coordination with it. Such remote directional IFF antennas are often mistaken for suface-search radars, which they resemble.

6. Integrated Fleet Systems

A Task Force is more than a random arrangement of warships. The developments of the past two decades have converted it, via the Naval Tactical Data System, or NTDS, into the equivalent of a *single multihulled ship*, *ie* an integrated system. Here the main fruit of integration is that all the hulls of the Task Force partake of all (actually only most) of the sensor information *each* ship obtains. This integration is an essential response to the high speed and great stand-off range of modern threats. It is the culmination of a long history of shipboard and fleet system integration, beginning with the integration of facilities on board each ship.

The radars aboard a ship are elements of a more or less well integrated system that provides information to the ship commander and his subordinates. This apparently trivial point took some time to be well appreciated. Its great emblem was the creation of Combat Information Centers, or CICs, aboard warships: spaces in which information from warships, and even pilots aboard aircraft, could be combined to provide a plot of an overall situation. Indeed, the classic photograph of a CIC shows plotters at work correlating this information on a large vertical chart. CIC was an essential element of any useful fighter control system; the idea originated with the British air defense system evolved just prior to World War II. There it was called plotting of enemy aircraft in a 'filter center' into which radar and visual reports were fed.

RADAR-PICKETS AND AIRBORNE SYSTEMS

Pacific warfare emphasized carrier operations, in which it was essential to be able to control fighters defending a Task Force. Such control required more precise information, including accurate altitude data – hence the emergence of height-finders such as SM and SP. But fighter control also required the CIC, since only in the CIC could the character of an ongoing air battle be appreciated. Even so, the necessarily crude communication between ships meant that a carrier would have to depend largely on her own radars for detailed CIC inputs. It must have been clear as early as 1945 that such a system would find it difficult to deal with high-speed aircraft. For example, a radar like SK-2 could be relied upon to detect medium- and high-altitude aircraft out to 80 or 100nm, which meant as much as 20 minutes from the carrier. SM and SP were effective out to a lesser distance, so that fighter control could really be exercised over no more that 50 or 60nm. Matters could become even less satisfactory near land, where the big air search radars would be distracted by returns from the ground.

Two solutions were attempted. The first, which actually went into practice before the end of the war, was the use of lesser units as radar-pickets and, more importantly, as fighter directors. Twenty-four destroyers were equipped with SP radars, aircraft homing beacons and enlarged CICs; groups of fighters could be assigned to their control. These ships could also perform an early warning function, and they could interrogate homebound streams of aircraft to separate off enemy intruders attempting to slip by defensive patrols. This latter function has survived as the present PIRAZ (Positive Identification of Returning Aircraft Zone). At the very end of the war seven destroyer escorts were similarly converted, and attempts were made to use submarines as pickets, on the theory that they alone could submerge when attacked. Many (unconverted) destroyers on early warning picket duty off Okinawa were sunk or very severely damaged.

However, radars on the surface were useless against low-flying aircraft at any great range, and the Japanese were great exponents of torpedo attack from about 500ft. The solution chosen was to move a big search radar aloft, using a carrier plane. This was Project 'Cadillac', in which TBM torpedo-bombers were converted to carry a big X-band search radar (APS-20) with an 8ft antenna. At 20,000ft the APS-20 could detect a battleship at 200nm and a low-level (500ft) bomber at 75; one 1946 document referred to Airborne Early Warning (AEW) as 'presently recognized as the only satisfactory long range search radar'. The TBM-3W could do little beyond carrying the big radar and a VHF data link back to a 'terminal ship', at first a carrier although later on radar-picket destroyers were adapted for such duty. It could, however, act as a radio relay link to carrier fighters beyond the carrier's radio horizon. Typical shipboard installations (1946) included a

The radar-picket symbolized the divorce between individual ship radars and shipboard weapons. This is the destroyer escort Lansing, *in November 1963. She was converted to support Continental Air Defense, but that requirement led to much of the work which evolved into NTDS. Her air control role is reflected in her large TACAN aft; an SPS-8 height-finder (fighter control) radar is just visible abaft that tripod. Her foremast carries the usual destroyer comnbination of an SPS-28 or -29 mattress with SPS-10 above it, and the mast is crowned by a URD-4 radio direction-finder. Her 3in/50 gun forward carries the antenna of an SPG-34 fire control radar. More subtly, the closed-in area amidships, just visible here, indicates the increased requirement for internal space, for an enlarged CIC and its operators.*

388

During World War II, twenty-four Gearing *class destroyers were converted to radar-pickets, tripod masts replacing their amidships banks of torpedo tubes. This is USS* Duncan, *in June 1950, showing her small-antenna SP, with its characteristic IFF dipoles (BO), with a YE aircraft homing beacon above and ECM search antennas on the yardarm. A TDY jammer is visible on the platform below. The foremast carries SU-2 and SPS-6B, and is of the tripod type which had only recently been introduced. Twelve more destroyers were converted under the FY52 program.*

BY COURTESY OF A D BAKER III

The submarine radar-picket was devised to reduce the rate at which conventional surface ship pickets were being lost to Japanese air attack off Okinawa; two were converted as part of the crash anti-kamikaze program of 1945, Threadfin *and* Remora, *and four more were under conversion as of VJ-Day. None of these conversions was very complex, and these submarines were not redesignated. However, the idea was considered valuable enough for two more submarines,* Spinax *(shown here, 28 July 1954) and* Requin, *to be converted at Portsmouth Navy Yard under Project 'Migraine'. Their predecessors had no specialized radars, but these two units had SR-2 and an SV-2 height-finder on the main deck over the after torpedo room, which served as a CIC. These experiments encouraged further conversions suitable for operational use ('Migraine II'); in 1948 the conversion of six, at the rate of one per year, was envisaged. The next pair,* Tigrone *and* Burrfish *had their radars on masts, and half their batteries were removed to provide further space aft.*

BY COURTESY OF A D BAKER III

▲ *After a total of ten wartime submarines had been converted to radar-pickets, two new ones were built. This is* Sailfish, *newly completed, in 1956, showing her three air search radars: the massive (and collapsible) BPS-3 height-finder, the large BPS-2 medium-range air search radar, and the small BPS-4 for short-range air search. The third specially built picket was the huge nuclear submarine* Triton, *large enough to be considered for a time as an alternate National Emergency Command Post Afloat. Interest in radar-picket submarines declined in the late 1950s, partly because of the general decline in the value of pickets which also killed off the destroyer radar-pickets. Both of the* Sailfish *types reverted to attack submarines in 1961, their height-finders being removed. They underwent FRAM modernization as attack submarines in 1964.*

▶ *'Cadillac' was an attempt to provide the fleet with a high-altitude radar capable of detecting low-flying aircraft at a great distance; the TBM Avenger shown carried an APS-20 radar in its belly radome, with a peak power of 750kW at S-band. Its 8ft antenna scanned at 6rpm, generating a 3.5° beam (2-microsecond pulses, PRF about 300). Design goals were detection of aircraft at 50–60nm, and of ships at 200; in an early test, a TBM at 5000ft detected two bombers (JMs, Martin Marauders) at 500ft at up to 70nm. The APS-20 radar continued in service, in much-modified form, through the 1950s and early 1960s aboard modified Skyraiders. Power was up to 2MW in the late model APS-20F.*

GRUMMAN AEROSPACE CORPORATION

LOCKHEED CALIFORNIA CO

'Cadillac II' was a larger land-based airplane with an enlarged APS-20 aboard, as well as a CIC organization of its own. The first were converted Flying Fortresses (PB-1Ws) which carried at least two fighter-direction officers, a CIC watch officer, and a radar control officer: they could operate over a fleet, controlling their own fighters, and were the distant ancestors of the current AWACS. During the Korean War several were fitted with primitive height-finding radars, which proved quite successful. The converted Lockheed Super Constellation (WV-2 or EC-121) shown was the result, carrying a two-dimensional search radar (APS-20, generally with a larger antenna giving a 1.6° beam and with up to 2MW power) under its belly, and an APS-45 height-finder in the fin-shaped radome above (X-band, with a maximum intrumented range of 120nm).

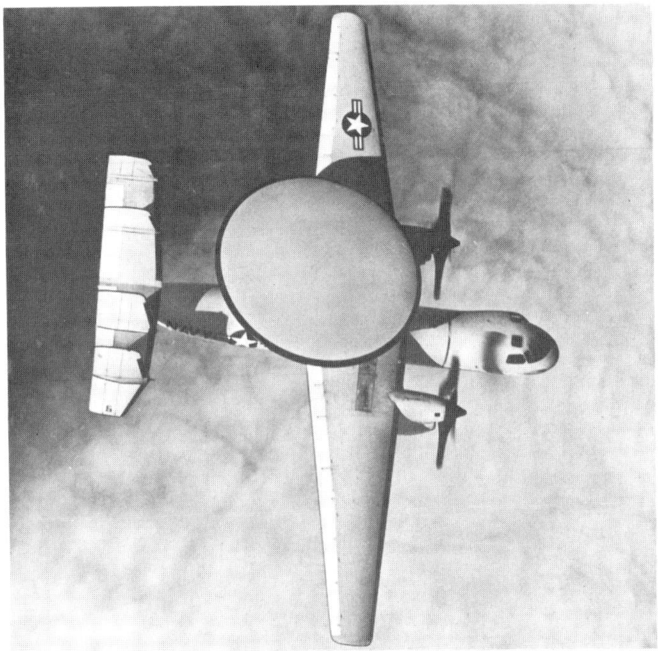

Large carrier aircraft could take over the airborne CIC function of the land-based Super Constellations; this Grumman E-2C carries the rough equivalent of a destroyer air search radar in its UHF-band APS-120. Indeed, a shipborne version of the experimental APS-70 'Cadillac II' radar was nearly chosen for production in place of SPS-40, and was tested aboard a destroyer. Like SPS-40, APS-120 (and its close relatives, APS-96 and -125) employs pulse-compression. The E-2 was the first US airborne early warning (AEW) airplane to be fully effective at high altitude, above 30,000ft, and it introduced a unique height-finding system. A radar signal could travel to an airborne target either directly or it might reflect off the sea. There were, then, three alternative paths: direct-direct, direct-reflected, and reflected-reflected, the time difference between them giving the altitude of the target airplane. Pulse-compression permitted the radar to use so short an effective *pulse that these differences could be measured.*

special YQ radio beacon and specialized IFF. In effect, the antenna on the TBM-3W was connected to a shipboard PPI by the data link.

'Cadillac' was first proposed early in 1944, its goal a combination of long-range surface search and low-flyer warning. The first Navy demonstration followed at AA Field, Bedford, Massachusetts, on 20 October 1944. Tests at Brigantine Naval Air Station early in 1945 showed detection of two JM medium bombers flying at 500ft by a TBM at 2000–5000ft at ranges of between 45 and 70nm. The system now (March) went to sea aboard the carrier *Ranger* (CV-4); that fall AEW was tested aboard fleet carriers (*Bunker Hill* had the first complete installation; *Hornet* was ready only at the end of 1945; it appears that *Enterprise* was also fitted for AEW at this time); and in November the CNO approved its installation aboard the AGC *Adirondack*. In March 1946, the net cost of installation was given as 5.5 tons. Two 5in guns were to be removed as compensation.

The great significance of 'Cadillac' was that for the first time one ship was able to use directly the data collected by another platform, in this case the TBM-3W. The AEW data link presaged the data links of NTDS, which convert a fleet into a multihulled yet integrated unit.

It proved difficult for fighter-director teams to use the shifting radar picture presented by the airborne set; one solution was to move the fighter-directors into the air: 'Cadillac II', the airborne CIC. Initially it was a PB-1W, a converted B-17; later, converted Constellation airliners were used. By 1960 it had become possible to incorporate much of the airborne CIC installation in a large carrier airplane, the Grumman E-1 (originally WF). More recent and far more elaborate airborne CIC equipment fills the new E-2 Hawkeye.

THE COMMUNICATIONS REQUIREMENT

Another factor leading to fleet integration was the grow-

Different naval strategies demand alternative uses for airborne surface search radars. In the case of the Soviet Navy, what is important is information about enemy naval formations provided from beyond the horizon of surface or submarine strike platforms. For about fifteen years that has been the function of the 'Bear-D' maritime reconnaissance bomber, which transmits back to a ship or a surfaced submarine its radar picture of a potential target area. A related airplane, the 'Moss' (based on the transport version of the 'Bear'), was the first Soviet AEW platform, successfully guiding Indian aircraft during the 1971 Indo-Pakistani War.

BY COURTESY OF A D BAKER III

Specialised fleet flagships emphasize the growing role of sensors and command and control, as opposed to shipboard weapons, in modern naval warfare. This is the command/missile cruiser Oklahoma City; *she has surrendered her superfiring triple 6in mount forward in favor of additional command spaces. Amidships she carries an SPS-39 three-dimensional radar, part of her Weapon Designation System; presumably the SPS-8B aft was needed for aircraft control, as that (as well as missile operation) was an important role. The large discone/cage radio antenna forward emphasizes her need for good communications. This photograph is dated December 1960; the ship had not yet been fitted with the large (SPS-37A) air search radar forward, nor with the flat antenna for her SPS-39 or the improved intercept radar (SPS-30), nor did she have active electronic countermeasures gear installed.*

ing need for communications, which by 1945 meant that the best flagships were special units, *not* capital ships. The earliest units of this type were the amphibious command ships., AGCs, which required carrier-level radar suits in order to control air cover over landings.

However, even the Task Forces were not immune to this trend. Advances in communications and the great range at which all battles were fought made it less and less obvious that a Fleet Commander had to ride a capital ship. In 1945 the incomplete heavy cruiser *Northampton* was selected for flagship conversion. She exchanged the great bulk of her armament for extensive radio systems – and the most powerful air search radar in the world, SPS-2. In effect a carrier operating with *Northampton* would operate on the basis of the latter's radars; in some sense *Northampton* and the carrier would operate as two physically separate parts of the same integrated unit. Although a similar conversion of the incomplete large cruiser *Hawaii* was canceled in 1954, the concept survived in a series of specialized missile cruiser conversions in which gun armament was sacrificed for staff spaces.

The means of integration, NTDS, flowed directly from the postwar problem of carrier group air defense, the same problem which had produced SPS-2. The basic point, which can already be seen in AEW, was that as aircraft speed (and lethality) increased, inefficiencies in Task Group operation due to nonintegration of the units of the Group became more and more serious. In other terms, it became essential to smooth data flow through the Task Group.

BY COURTESY OF A D BAKER III

The command cruiser Northampton *was the epitome of the specialized flagship, almost entirely unarmed, but fitted with both a massive air search radar (SPS-2, on her tower foremast) and the first of the Bell Laboratories Threat Evaluator and Weapon Assigner (TEWA) systems, served by the SPS-3 radar on her after radar tower. Her hull was raised a full deck level to accommodate massive control facilities, including a two-level CIC. To some extent it appears that the rationale for her completion in this form was that the new flush-decked carriers planned in the later 1940s would be unable to carry effective long-range radars; many task force air defense studies of the early 1950s envisaged* Northampton *or a similar CIC as the primary task force radar platform, located in the center of the formation. Here she is shown soon after completion in 1953, with all of her new radars aboard. They included two zenith/surface search sets, one on the port forward side of the forward tower. An SPS-6B was later mounted on the starboard side of the forward tower, and SPS-8 is visible aft. Also visible are the new sleeve broadband antennas, as well as a very tall forward radio mast supporting a 'conical-monocone' broad-band antenna and one end of a flat-top.* Northampton *served as a Fleet Flagship in 1954–55 (Mediterranean) and 1955–61 (Atlantic), and then as a National Emergency Command Post Afloat (NECPA) 1961–70.*

The first of the specialized fleet flagships were the amphibious flagships, or AGCs: Eldorado *is shown, refitting at Mare Island in November 1944. The dark spots near her rigging are antenna connectors, and their number suggests the degree to which she functioned as a floating antenna farm. Operationally, these ships functioned more as coordinators of activity afloat and ashore than as air control centers; usually the air control elements of their CICs coordinated Combat Air Patrols under the immediate control of destroyers anchored offshore, enjoying better radar coverage. The radars visible here are an SK-2 'dish' and, above it, the usual SG for surface search. The after kingpost carries, to port, a YG aircraft beacon, used on carriers as a back-up for the much larger YE. On the forebridge, a new Mk 52 fire control system with its Mk 26 range-only radar has been circled. An additional SG was generally mounted forward, and by the end of the war most AGCs had SP aft, as well as an ECM system; the circled item on the after kingpost appears to include an ECM dipole antenna. Radio suits were extremely complex, including long-range ship-to-shore links, tactical links (the antenna of a TBS ship-to-ship radio is visible at the extreme port end of the after kingpost), and Army radios for ship-to-shore (short-range) communication: in 1946 the standard suit included 112 receivers and 34 transmitters, plus portable units to be carried ashore.* Eldorado *was the last AGC to remain active, ending her career only in 1973. By that time she had SPS-30 (amidships) and SPS-37A. Her enlarged deckhouse enclosed, among other spaces, a CIC and a Joint Operations Room.*

DATA-LINKING AND AUTOMATED CICs

In 1945, the Task Force depended largely upon fighters for its air defense, and fighters were not effective in very bad weather. German and Japanese successes with, respectively, bomber-launched missiles and bomber-launched rocket suicide aircraft suggested that the main future threat to the Task Force would consist of radar-guided bombers launching guided missiles in weather too foul for carrier aircraft. CCA was one approach to this problem; the other, pushed very vigorously, was long-range ship-launched missiles: Terrier, Talos, later Tartar, and now Aegis. Such weapons required very considerable integration *within* the CIC, leading ultimately to such electronic systems as NTDS. In addition, a missile ship might well find radar data collected by another ship or by an AEW airplane essential. The carrier would *not* be the missile ship, so that the fighter control ship might not be the ship firing missiles, and there would thus be a coordination problem. Ideally, indeed, the entire Task Force would have to be able to feed into a centralized CIC from which commands might originate.

Automated data exchange was therefore an important element of EDS and NTDS development from the mid-1950s onward. The EDS system included a crude data link, and in a 1959 test two of the picket destroyers of Division 262 exchanged data over a distance of over 400nm, using a high frequency (HF) radio link. By that time NTDS was well along in develpoment, partly as a result of Air Force efforts towards automated air defense systems for North American Continental Defense, which ultimately became the SAGE (Semi Automatic Ground Environment) System.

Project 'Lamplight' was part of the air defense effort, an attempt to extend the ground SAGE network to seaward aboard specialized picket ships. The pickets had to be able to transmit their radar pictures ashore in a form suitable for SAGE computers, the essence of SAGE being the computer-processing of radar information in order to present to a battle commander the shape of an air battle developing over a very wide area, so that he could properly marshal his forces. As envisaged in 1954–56, 'Lamplight' would use EDS in Phase I and some new digital-orientated system in Phase II. In fact the Phase II system, NTDS, would prove too massive to fit aboard radar-picket DERs, but nonetheless it was successful in other naval applications.

A mid-1955 Office of Naval Research study on Tactical Data Processing led to a Bureau of Ships Technical Requirement late that year; Characteristics and an Operational Requirement were approved by the CNO the next spring. Very fortunately, the Navy could benefit from early solid-state computers which made shipboard installation at least conceivable. NTDS would function both as a single-ship, automated CIC and as a means of data exchange (through an HF radio link) between ships, or between ships and aircraft. An

48 OP CONSOLE

BY COURTESY OF NORMAN POLMAR

◀ *The Combat Information Center (CIC) was the key to effective utilization of radar information during World War II: this one was installed aboard the carrier* Yorktown. *The three blackboards were status boards, to summarize the condition of combat air patrols and offensive strike aircraft (including how much fuel they had left); the large vertical screen was used to plot a summary of the air and surface situation, the plotters standing behind it. The two tables were used for tracking and for the arrangement of interceptions. Radar monitors, such as the one in the foreground, fed data into the CIC, which could communicate with ship lookouts and with the CICs of other ships in the Task Force. Such organizations functioned magnificently against large raids (which could be tracked as a unit). However, in 1944–45 the Kamikazes generally attacked singly, along different bearings, and at widely different altitudes, and CICs such as this one were saturated. The Naval Tactical Data System, NTDS, was an attempted solution.*

BY COURTESY OF NORMAN POLMAR

◀ *A modern CIC is highly automated. This one was aboard one of the first NTDS-equipped missile frigates, USS* Wainwright *(shown in September 1968). The consoles here include radar operator positions and summary plots at which tactical decisions can be taken. In effect, the operator at a radar console decides when his radar has detected a valid target and enters that target into the computer; to a considerable extent he has been automated away by modern ADT (Automatic Detection and Tracking) systems. The functions of this particular CIC included aircraft control and weapon designation for the ship's Terrier missile battery.*

▲ *The development of Weapon Designation Systems was a parallel effort to overcome raid saturation, this time against the shipboard weapons of a single ship. This is the assignment console of such a device, descended from the original Bell Laboratories ATEWA/TEWA. The crosses and circles are a schematic representation of potential targets around a ship, and of the capabilities of shipboard weapons and directors. NTDS consoles, by contrast, show a modified radar map of the area around the ship.*

NAVAL RESEARCH LABORATORY

As conceived by Bell Laboratories, the Threat Evaluator and Weapons Assigner (TEWA) or Weapons Designation System required the services of a high-precision search radar rapidly scanning the hemisphere around and over the ship. The first such system was SPS-3, shown here in its initial test form at the Chesapeake Bay Annex of the Naval Research Laboratory in the early 1950s. It was not successful, although it did go to sea for a short time aboard the command cruiser Northampton. *Ultimately the FRESCAN radars (such as SPS-39) replaced it.*

experimental system was tested at the Naval Electronics Laboratory between April 1959 and November 1961, and service tests aboard USS *Oriskany* and two missile frigates (*King* and *Mahan*) began in September 1961.

Both the data link and the automated CIC concepts have proven useful. Major US combatants have both, but even the austere *Perry* class frigate has what amounts to an automated CIC (combat system), necessitated by the character of the threat it faces, the 'pop-up' submarine-launched missile. In its case, economies have been realized by the deletion of the NTDS data link, with provision for its later installation. Data-linking in general has become far more important within the Fleet with the increasing emphasis on passive detection systems. Two or more ships which can exchange data freely can rapidly triangulate to locate an emitter.

Effective data-linking demands very precise navigation. Any one ship in a data-linked group must be able to determine its position relative to the others accurately and continuously: it must be properly 'grid-locked'. Errors in grid-lock can, among other things, generate false target tracks as two or more ships detecting the same target fail to correlate their detections. In addition, real targets may be missed. The great advantage of data-linking is that it vastly improves on the detection capability of any one sensor by adding up the detections and many equivalent sensors. Poor grid-locking loses single detections, and breaks up apparent tracks consisting of series of detections by different radars. Thus navigation, perhaps satellite-assisted, becomes an essential element of successful integrated Task Group operation.

Externally, NTDS shows in the provision of particular HF antennas, but they are not very distinguishable against the forest presented by all modern major combatants. Internally, it consumes weight and volume and electric power. For example, missile frigates (now missile destroyers and cruisers) fitted with it as part of their AAW modernisations in the 1970s had deckhouses added to house NTDS-related equipment. Much more importantly, the success of NTDS and of similar but less powerful systems means that the radar suit of a group of ships operating together consists of all the radars of that group feeding into what amounts to a single CIC. The same holds for sonars, and for passive detection devices.

Moreover, the US development of NTDS is typical of Western navies in general, most of which operate with a combination of some type of computer-driven combat system and some type of digital data link. Examples include the Royal Navy's Action Data Automation Weapon System (ADAWS) and Computer-Assisted Action Information System (CAAIS, in *Amazon* class frigates), the French Navy's SENIT, the Dutch SEWACO (Sensoren Wapens Commando), and the Italian SADOC.

7. Radar Development

The idea of radar occurred independently to scientists in several countries almost simultaneously as a natural consequence of advances in radio technology. What differed was the speed of development, once the idea was in hand. For example, it appears that American scientists conceived a radar defense system substantially before the British, yet the latter had by far the more urgent need, so that by 1940 Britain had an enormous lead. For that matter, it appears that at first Germany led both of the Allies in producing naval range-finding radars, although by 1941 she had fallen far behind. From the account of radar operation already presented, the major developments have been (a) the idea of radio echo-ranging; (b) the 'TR box' (duplexer); (c) the PPI display; (d) microwave radar; (e) electronic scanning; (f) pulse-compression; and (g) Automatic Detection and Tracking (ADT).

Radar echo-ranging was proposed as a navigational aid in patents issued by several countries as early as 1904; indeed, Heinrich Hertz, the discoverer of radio waves, had observed their reflection from metal as early as 1886. Even so, military and naval authorities do not appear to have tried radar-like systems during World War I. In June 1922 the idea was revived by Marconi, who suggested the use of high frequencies, but nothing appears to have come of his comments.

Subsequently a number of radio researchers detected echoes (which they thought of as interference) from ships and aircraft. For example, in the fall of 1922 Dr A Hoyt Taylor and Mr Leo Young, both of the then (US) Naval Aircraft Radio Laboratory, suggested that a scouting line of destroyers could detect ships passing through by observing distortions in signals from ship to ship. Although both were later prominent in US radar development, their proposal at the time was deemed impractical.

BEGINNINGS

In 1930 Dr Taylor was superintendent of the Radio Division of the Naval Research Laboratory. One of his subordinates detected aircraft (accidentally) using a long-base radio direction-finder, while listening to 32.8mc (9.1m) radio signals. Taylor recognized this phenomenon as a potential means of aircraft warning; echo experiments were pushed to higher frequencies, up to 100mc (3m). By 1932 NRL had proposed a system to provide warning of aircraft entering an area 30 miles in diameter, to consist of a circle of 100mc transmitters, with receivers in another circle 15 miles further out, and experiments showed that aircraft could be detected and located at a range of 50 miles.

Before World War II many of the world's navies employed wire cage antennas slung between high topmasts, as in this view of the heavy cruiser HMS Berwick. *In such installations the height of the topmasts dictated effective radar range.*

The first US experimental shipboard radar was a 200mc set installed at the Washington Navy Yard aboard the destroyer Leary *in the spring of 1937. Its antenna was a yagi mounted on the starboard 5in gun to permit training on objects, and it incorporated a duplexer (TR box) which had been introduced in July 1936. Two 1-week trips were made, down the Potomac and into Chesapeake Bay and briefly out into the Atlantic; aircraft were detected at up to 20nm in April 1937. Sea clutter due to ocean waves was also observed for the first time in these tests.*

The system was not yet quite a radar, since it operated by comparing at the receivers radio transmissions via the air with those transmitted along the ground.

True radar required pulse-transmission, which had been tried as early as 1925 in attempts to gauge the height of the ionosphere. At NRL pulsing was seen, in March 1934, as the solution to the problem of co-locating transmitter and receiver. A special Research Section was formed to push forward radar developments, and during 1934 a 60mc/s (5m) pulsed radar was built. Later the 28.6mc/s radio antenna used for the 1930 experiments was converted to a radar which in the spring of 1936 detected an airplane at 25 miles. The Bureau of Engineering, responsible for NRL, decided in June to classify this project 'Secret' and to press it forward with the highest priority.

'Highest priority' was still not very great; indeed, both the US and the British radar efforts of this period were carried out on very much a shoestring. The latter had begun in earnest when a British radio physicist, Dr Robert Watson-Watt, had been asked about the practicality of a radio death-ray for AA purposes; his reply, that radio might better be used for detection and location, effectively began British radar work. At this time (1935) Watson-Watt was himself astounded that, of all the possible 'signatures' an airplane might make or might be made to make, radio had been entirely neglected in favor of sound and even infrared signals.

The first US air search radar was CXAM, based on the XAF prototype shown aboard the battleship New York. *It was installed in December 1938 and tested at sea through March 1939, when it was removed. After the* Leary *tests the Commander-in-Chief US Fleet asked that 'radio detection and ranging' equipment be provided for the Fleet, and a February 1938 conference decided to install a set aboard a major ship as soon as possible. Work on XAF began in March; after tests in the Atlantic Squadron, its commander, Admiral A W Johnson, declared it 'one of the most important military developments since the advent of radio itself'. In Fleet exercises in Caribbean waters, XAF detected aircraft at 100nm, surface ships at 15, and 14in shells in flight at 7; navigational buoys were picked up at 4nm, and birds in flight at 5.5. XAF operated at 15kW (5-microsecond pulses) at 200mc.*

With the support of BuEng, the NRL team began work on a shipboard system. They felt that the keys to success would be a shorter wavelength (1.5m, or 200mc, was chosen) and a means of switching receiving and transmitting functions so that a single antenna could be used. Young and Dr R M Page of NRL developed a duplexer suitable for the 1.5m signals, and in August 1936 NRL had a laboratory model in operation. An April 1937 demonstration before the CNP, Admiral William D Leahy, and the Secretary of the Navy, gained essential support: a primitive antenna was affixed to the starboard 5in gun of the old 'flush-decker' *Leary* for tests.

Ranges were limited by the power output of available 200mc tubes. NRL considered longer wavelengths preferable where space was not a problem, and when radar (80mc/s) was demonstrated to the Army Signal Corps in November 1936, that service was urged to use 100mc/s. The first Army long-range search set, SCR-270, used this (3m) frequency, for which tubes of higher power were already available.

The British also operated at lower frequencies. From the first, Watson-Watt had espoused the 'cult of the third-best', *ie* the choice not of the ideal (far future), nor of the next best, but of what could be built – and used – at once, in other words what could fight in a war he feared might well come by 1938. Watson-Watt therefore began with the lowest frequencies suitable for aircraft detection, *ie* the highest available power and the largest antennas, his problem being aircraft detection from land sites. Moreover, he did not need a duplexer. The Royal Navy adopted this land-based technology, and it began the war with 7.5m (40mc/s) air search radar, each set requiring separate transmission and reception antennas.

By way of contrast, in the United States it was a *naval* agency, NRL, which originated radar systems. Moreover, without the sense of urgency felt in Britain, NRL saw less point in compromise – hence the immediate adoption of 200mc/s. A characteristic feature of the US effort was the enlistment of the major radio laboratories: Bell Laboratories began work on a 700mc (40cm) set. NRL persevered at 200mc and had a system, XAF, ready for ship installation late in 1938. An alternative type, CXZ (400mc, 75cm), was developed by RCA. Both were tested in Fleet maneuvers that winter, XAF aboard USS *New York* and CXZ aboard her sister *Texas*. CXZ had been developed in some haste (RCA had been brought into the program only in March) and proved unsuccessful, but XAF did extremely well. The captain of the *New York*, for example, recommended immediate installations on all carriers.

INTO PRODUCTION

The first six production models, CXAM, were delivered in May 1940. CXAM and its offshoot, CXAM-1, were too large for installation in ships smaller than heavy cruisers, so NRL designed a smaller antenna set with higher power (XAR, 330kW) which went to sea aboard the test destroyer *Semmes* in July 1941. Production derivatives were built by both RCA (SA) and General Electric (SC series). This began the Navy practice of having NRL develop radars for commercial redesign and production, NRL being the primary US Navy radar developer.

XAR was the Naval Research Laboratory prototype for the major World War II long-range air search sets: SA, SC and SK. It provided a twenty-fold increase in peak power and an 11db gain in receiver sensitivity, and its receiver had the first antijam features in any Navy radar. The transmitter power of 330kW was achieved with an oscillator tube specially designed for pulse operation at NRL request, and receiver sensitivity was attained with a new type of tube called Lighthouse, specially developed by General Electric. It was tested aboard the experimental destroyer Semmes *in July 1941.*

From 1942 on, SC became the standard air search radar of US destroyers, cruisers and some capital ships. It was also employed as a standby set for large carriers. SA, originally fitted to destroyers, was soon relegated for the most part to lesser craft. SK, which became the standard wartime large-ship radar, was essentially SC with a much larger antenna: in effect the major wartime air search sets can all be traced back to the NRL experiments of 1936. There was also, at a metric wave-length, SD, an omnidirectional air warning set for submarines.

Meanwhile a parallel series of L-band fire control radars was developed in conjunction with Bell Laboratories. Western Electric, the production arm of the Bell System, began work on a prototype, CXAS, in April 1940. It became FA, but it did not see much service. However, by the spring of 1941 Bell had built a prototype lobing radar, FC (Mk 3), and the FD, or Mk 4, dual-purpose set soon followed. FD proved enormously successful, and 667 were built. Its main deficiency was that at low angles its operator tended to track the image of a target reflected from the sea surface – a problem also encountered with the height-

BY COURTESY OF NORMAN POLMAR

◄ *The Mk 3 surface gunnery radar was one of the earliest in US service; this Mod 1 version was mounted aboard the battleship* Washington. *The mattress in the background is CXAM-1, in effect the predecessor to SK. Note, too, the antenna for a TBS ship-to-ship tactical radio, visible just to the left of the fire control radar antenna.*

finders. A nodding 'orange peel' X-band height-finder was developed to solve this problem. FD was succeeded by the very similar Mk 12, and by 1945 nearly all Mk 4/12 radars operated in conjunction with Mk 22. In main batteries, NRL and Bell Laboratories developed the phase-scanning Mk 8 'polyrod' radar, and later the more conventional S-band Mk 13.

The radars would have been of limited value had it not been for extensive work on radar displays, and on means of transmitting radar data to remote display units. The earliest sets used A-scopes, but both British and US scientists, independently, soon hit upon the

▼ *Submarine radars were among the first developed by the Navy during World War II, both for surface search (SJ, a microwave system) and for air warning, which could permit a submarine to crash-dive before it could be attacked. The earliest model of the air-warning radar, SD, operated at P-band and was omnidirectional. Unfortunately, The Japanese soon produced a search receiver by means of which their aircraft could home on SD transmissions. Efforts were made to improve the directionality of the SD radar, but in any case it was ineffective against the primary threat, the low flyer. This is the submarine* Permit *(January 1943), and SD is the antenna on the tall mast to the right, with SJ at left.*

idea of the PPI. A recent NRL history notes that both the PPI and the idea of remote displays came out of experience with the XAF radar in the 1939 Fleet Exercise. The US PPI was devised in 1939–40, the first example being demonstrated aboard *Semmes* in April 1941 in connection with the SG prototype. By that winter there was a PPI for the P-band air search radar as well, although many of the latter entered service without it. One problem of early installations was that indicators such as PPIs were incorporated directly into the radar transceivers, which made for crowded CICs (or precluded them entirely). The solution was remote repeaters, which in wartime were prefixed 'V': VC, VD, VE, and so on, followed by an AN/SPA series. Wartime display technology included a great variety of special types, from the B-scopes of fire control systems to the AEW PPIs.

Even before the war, it was appreciated that much more efficient radars could be built using microwaves; but NRL and its industrial colleagues were unable to produce sufficient output at such high frequencies. The British were far more successful. In the summer of 1940 they decided to disclose to the United States their recent scientific advances, on the theory that sooner or later the Americans must enter the war. These included the magnetron, the key to microwave radar. Between disclosure in September 1940 and the late spring of the following year, a prototype microwave surface search set, the forerunner of SG, was conceived and built.

The initial generation of S-band surface search sets comprised SE, for auxiliaries and small combatants such as the old destroyers; SF, for MTBs; SG, for cruisers and capital ships; SH, for large modern destroyers; and SJ, for submarine surface search. In fact both SE and SH proved unsuccessful, and SG was light enough to fit in destroyers. However, SF proved overweight, and so it was used for such small combatants as patrol frigates. A new set, SL, was designed specifically for such applications (*eg* for destroyer escorts), and a new SO series was evolved for MTBs. In a few cases SO appeared aboard larger ships, however. The final wartime S-band series was SV, an air search radar intended to replace SD. The two pencil-beam height-finders, SM and SP, also operated at S-band, but they have little in common with the surface search sets.

The next step was X-band,* which produced, in search radars, SS and ST for submarines and SU (plus some late SO-series) for surface ships, specifically destroyer escorts. However, SU proved good enough for larger units, where its installation could save considerable topweight. Thus we see SU in postwar carriers, balancing off the weight increase associated with SX.

*The progression from P to S to X is a progression in relative difficulty of power tube manufacture, so that it is characteristic of radar development in other countries.

By late in World War II US submarines were equipped entirely with microwave radars. The most prominent antenna shown here is SV (air search) with a late model of SJ (surface search) beside it. The loop antenna proved capable of receiving radio signals at periscope depth. Above it is the short whip of an SPR-1 countermeasures receiver. The submarine is USS Grouper, *July 1945.*

Microwaves were also essential to lightweight radar for heavy machine-gun fire control, although the ultimate goal, radar-direction integral with the gun mount (GUNAR) was not reached until postwar. Wartime progress was limited to GFCS Mk 63, whose radar dish was mounted with the guns. The level of wartime effort is attested to by the variety of systems produced, though not necessarily deployed. Only GFCS Mk 57, a group concentration director, entered service (October 1944), being employed for the first time at Okinawa. Postwar a more sophisticated system, Mk 56, was widely deployed. Both used a variety of X-band dish radars.

EARLY BRITISH RADAR DEVELOPMENT

Britain was the first nation to use radar operationally, and the first to equip much of her fleet with it. Although the British and US governments exchanged radar data very freely from 1940 on, Royal Navy systems had and have a very distinctive character due in large part to the fact that the Royal Navy pressed into service in 1939 equipment one technical generation more primitive than that fielded by the Americans one to two years later. The British systems, then, tend to illuminate the consequences of advances in radar technology.

The earliest US policy on radar for carriers was to provide a single high-powered air search set, originally CXAM and then SK. In the Yorktowns *for example, it was mounted atop their tripod foremasts, which were heavy enough to support it adequately. However, sometimes the single radar failed in battle with unfortunate results, and by the fall of 1942 CinCPac had asked that all carriers receive an auxiliary air search set. The extemporized nature of this addition shows in a photograph of the new carrier* Essex *at Pearl Harbor (August 1943), with SC-2 on a new lattice mast built out from her funnel and exposed to funnel gases. By way of contrast, SK is clearly in the ideal position, except that in some arcs SC-2 blocks it. A short topmast carries the vital YE aircraft beacon and, below it, an SG for surface search; a second SG was mounted in the shielded enclosure just abaft the SC-2 tower. At this point, it could be argued, all available real estate for radar installation had been taken. However, within a year most* Essexs *would have a third major radar, for fighter control (SM and then SP), and later many would receive a jammer (TDY) and even a CCA marshaling radar (SPN-6).*

One reason wartime British ideas have persisted so far beyond 1945 has been the low state of postwar British finances. As far as possible, the Royal Navy used its 1945 technology, replacing sets on a one-for-one basis; the great exceptions were a series of large radars evolved for aircraft carriers, and probably begun before the end of the war: Types 982 and 983 and their successor, Type 984, and the Type 965 long-range air search radar, which probably was a genuinely postwar response to new conditions of AA warfare. There were also new missile-guidance systems. One problem in evaluating British radar since 1945 is that problems in the British economy have seriously delayed developments: a radar like Type 984, which appeared in service in 1957, was the design contemporary of the US SPS-2, which entered service three years earlier. Throughout the war British radar design was considered well ahead of US standards, at least in many areas; but the gap between prototype and production was generally far greater. That situation persisted, for a time at least, into the postwar period.

The main effect of inefficiencies in early British air search ('air warning') sets was to encourage the development of *integrated* systems by means of which the crude indications of the warning sets were refined into data suitable for gunnery sets. Patterns set as early as 1943 have persisted down to the present in the

categories of British naval sets, which do not quite correspond to the classes assigned by the US Navy.

The British started from a very different set of conditions. They felt compelled to use current ('off-the-shelf') technology in view of the imminence of war, and often they could not place advanced devices in production simply because the consequences of stopping production were far worse than were those of continued use of less-than-optimum equipment. In fact one of the great surprises of the war was that systems regarded as belonging to the distant future, centimetric radar, proved quite practical during the war.

Unlike their US counterparts, the British radar scientists were concerned first with land-based air defense

Type 286, shown here aboard HMS Whitshed *(1941), was the first British destroyer radar, adapted from an airborne surface search (ASV) type. It was fixed, with three sets of yagi antennas: one set pointing dead ahead for transmission, two angled to the sides for reception and for crude direction-sensing. It was replaced by the far more directional Type 291.*

systems, which were largely unconstrained by size. Moreover, they did not have a vast electronics industry upon which to draw – hence Watson-Watt's early decision to try for the lowest practical frequency, which would be easiest from a power-generation point of view. 'Lowest' was a matter of *target* cross-section, not, as in the US, of permissible radar *antenna* dimensions. Con-

The British escort carrier Attacker, *completing at Mare Island, shows standard British air search radars: the four dipoles of Type 279B on her mast, with the wires between clearly visible through their spreaders, and Type 272, the centimetric surface search radar, in its 'lantern.' The mast also carries the lightweight US aircraft beacon, YG.*

sequently the first of the CH (Chain Home) early-warning sets operated at 50m as completed, although much shorter wavelengths were soon achieved to improve accuracy. Moreover, there was little point in working towards a 'TR box', and as late as 1943 many British warships used separate transmitting and receiving antennas.

The pressure under which Watson-Watt and his colleagues worked must have been enormous. He first conceived of radar air defense at the beginning of 1935, and on 26 February had already demonstrated radar of a primitive sort. By June aircraft were being tracked, and by July consistent tracking to about 40 miles was reported. By the end of 1935 Watson-Watt was reporting accuracies of a quarter-mile in range, 2° in bearing, and 15 per cent in altitude. By the next March aircraft were being observed regularly at 80 miles. Radars, which then had the cover name RDF (for 'Radio Direction-Finding') participated actively in an April 1937 British air defense exercise.*

Meanwhile, frequencies were forced up. Experiments at 4m in February 1936 were successful, but instead 13m was chosen that June for an Army mobile set. However, in April 1938 7.5m was selected. At the same time a 1–1.5m ASV (Air to Surface Vessel) radar was devised, but at this time too little power could be achieved at such a high frequency to make it useful for naval work, a problem not encountered in the US Navy at this time.

SYSTEMS FOR THE ROYAL NAVY

Hence the first naval sets were 7m types, and sea trials with them were held in March 1937. At such wavelengths no reasonable antenna could be very directive. The British chose to use dipoles with dipole reflectors – in this very early antenna, two such sets, one above the other, turning with the mast passing vertically through both. This arrangement was similar in principle to such contemporary US sets as CXAM, but the much smaller number of radiators decreased directivity very sharply. In free space each dipole would produce a very fat beam pointing roughly ahead, as much as 90° wide. Effective vertical coverage was set by sea reflection and consequent cancellation, but there would be gaps through which low- and even medium-altitude aircraft might slip. In addition, the lack of a TR box meant that two antennas were necessary, one on the foremast to transmit and one on the mainmast to receive. Even so, the 7m radar provided potentially useful warning of air attack, and it was adopted as Type 79, in the numerical sequence the Royal Navy used for radio sets (the aircraft carrier beacon was Type 72). About 1940 a separate radar series 200, was begun, the first being Type 279, which was essentially 79 plus a barrage predictor for long-range AA fire. The 100 series had already been taken up by asdics (sonars). This example suggests, correctly, that British practice in designations is far from systematic: the first radar was

*The US Navy was responsible for the word 'radar'. The two officers responsible for radar procurement, Lt-Commander (later Rear-Admiral) F R Furth and Lt-Commander (later Admiral) S M Tucker, were the first to use the acronym RADAR, for 'radio detection and ranging', and by order of the CNO, Admiral Stark (18 November 1940) this term superseded such rivals as 'Pulse Radio Equipment' (whence the 'P' in radar designations) and 'Radio Echo Equipment'. 'Radar' became a navy codeword, to be used 'in nonclassified dispatches or correspondence . . . and also in conversation'. Meanwhile the Royal Navy preferred 'radio-location' or RDF ('radio direction-finding'), adopting the US term only in 1943.

HMS Ramsey, *a former US destroyer transferred to the Royal Navy, carried the early Type 286 fixed air search radar at her foretop; this view shows clearly the three sets of dipoles, one pointed dead ahead, the other two angled to either side for reception. Atop the bridge is the radar office and weatherproof 'lantern' of the Type 271 surface search radar, essential for convoy operations, if only for station-keeping at night. In practice it appears that HF/DF was a more effective submarine detector. Note, aft, the substitution of a single centerline set of triple torpedo tubes for the usual beam tubes of US Navy ships of this class which the Royal Navy considered excessive topweight.* Ramsey *was photographed at the Charleston Navy Yard, June 1942.*

279, not 200 or 201, and many sets with lower numbers than 279 were designed much later.

The early Type 79 was soon succeeded by Type 279, which incorporated a ranging unit, and then by an interim Type 280, an Army gunlaying radar. The latter introduced the shorter wavelength (half the 79/279 figure) which characterized the major British wartime air search radar, Type 281. Early developments also included a series of fire control radars with a common transmitter, Types 282 through 285, some of which remained in service postwar. They operated at a much shorter wavelength, 50cm, having been developed directly by the Navy for gunnery; they were not direct descendants of Watson-Watt's air defense work. Given their relatively simple yagi antennas, most had very broad antenna patterns, and were therefore effectively

▼ *Centimetric radars were in great demand when they were first introduced, and many ships could not receive them; this is the 'Hunt' class destroyer* Melbreak, *October 1942, outfitted entirely with longer-wave sets: Type 291 for air search and Type 285 (facing aft) for dual-purpose fire control. The crosses on her yardarm are for Type 85 tactical radio, soon to be replaced by the simple dipoles of Type 86.*

▲ *The British cruiser* Penelope *demonstrates the consequences of dual-antenna operation in this September 1942 photograph: one antenna on the foremast to transmit, one on the mainmast to receive. This particular view is unusual because the two antennas of her Type 281 air search (warning) radar are not turned in the same direction. Each consisted of an insulating frame carrying four pair of dipoles at its ends; the forward antenna also carries an IFF interrogator (Type 241) associated with the radar, since the interrogator had to transmit. It was, in effect, a smaller version of the same antenna. The directors fore and aft carried Type 285, with a main-battery director, presumably Type 284, just visible forward.*

▼ *The heavy cruiser* Norfolk *shows a standard British cruiser radar suit of December 1942: dual-antenna Type 281, Type 273 aft for surface search, and Type 284 on the main director; her secondary directors carry Type 285. Atop her transmitting antenna forward is the aerial, barely visible, of a Type 243 IFF set with vertical dipoles (rather than the small horizontal one of Type 241).*

The carrier Indomitable *(1943) shows several British radars characterisitic of the early war period, particularly, on the bridge front, the 'lantern' of Type 271; above it is a medium-frequency radio direction-finder. The tripod mast is surmounted by the 'dustbin' of the Type 72 aircraft homing beacon, the British equivalent of the US YE. Above the bridgework are the aerials for three Type 282s, atop pompom directors, and the main battery director carries Type 285. A mast atop the homing beacon carried aerials for Type 281B, with Type 79B (barely visible here) on the mainmast.*

range-only radars requiring some auxiliary means of tracking. This appeared only with the advent of the Type 293 Target Indicator late in World War II.

The principal British air search radars operated at such long wavelengths that their beams were very imprecise; Type 293 effectively interpolated between them and the fire control sets. A similar philosophy, applied to carrier interception, led to the development of Types 980 and 982. Long wavelengths did make for fat lobes which added to the range of such radars as Types 279 and 281. Once the Royal Navy had obtained TR boxes for single-mast operation of these sets, it often combined the two in its carriers, as their different lobe patterns (due to their very different wavelengths) combined to give solid coverage. In capital ships, on the other hand, the elimination of one long-range air search set permitted the fitting of Type 293 on the mast thus freed. Finally, long wavelengths made for poor low-level coverage. During World War II the Royal Navy adapted its new Type 277 surface search radar for limited height-finding, and also obtained SM height-finders from the United States, whilst postwar it developed a nodding-beam height-finder, Type 983.

THE MAGNETRON

Both 293 and 277, which used the same transmitter, were practicable only because of the development of the magnetron, the high-powered S-band transmitter. In fact, as soon as the latter appeared, it was pressed into the most urgent service, surface search against submarines. Thus radars such as Type 293, which had been designed relatively early in the war, could not be produced until the needs of the ASW force had been met. Until the advent of Type 271, the S-band surface search set, these ships had used Type 286, a small-ship set developed from the airborne ASV radar and relatively unsatisfactory, given its low power and long wavelength. Thus it might properly be said that Type 271 was the most important British naval development of the Battle of the Atlantic.

CONWAY PICTURE LIBRARY

The British light carrier Theseus, *shown in May 1951, retains her wartime radars, directly descended form the earliest British types. Her mainmast carries Type 279B, her tripod foremast Type 281B, and her bridge carries Type 277, a pencil beam set serving alternatively for surface/low-flyer search and for height-finding.*

▼ *The battleship* Vanguard *incorporated most of the operational British naval radars of her time; she is shown, newly completed, in December 1946. She was, for example, the first ship to have Type 960 air search radar (on her mainmast); her foremast carried the Type 293 Target Indication (short-range air search) set, with Type 277 below on the fore starfish, and the double aerial of Type 275 (secondary fire control) below that. The main battery director carried the Type 274 microwave set (a 'double-cheese') and the small vertical 'cheese' of the Type 930 splash-spotter.*

CONWAY PICTURE LIBRARY

It still did not provide small ships such as destroyers with effective air cover. That fell to Type 291, a P-band air search radar, in appearance a miniature Type 281 with all the latter's faults of broad beam. In addition, it had a relatively short effective range. By the end of World War II it was typically combined with a Type 293 S-band target indicator; the latter functioned in practice as a short-range air search radar. Early in the 1950s Type 291 began to vanish from the destroyer force, as Type 293 functioned well enough as a combined air and surface search radar. This practice survives in modern British frigates, in which the Type 992Q target indicator, a lineal descendant of Type 293, performs the same role.

The Royal Navy early appreciated the value of the TR box, but the need to produce new sets in quantity limited its application. For example, by 1941 it was possible to reduce both Type 79 and 279 to single-antenna (single-mast) operation, and sets so modified were given a 'B' suffix. Meanwhile the US Navy supplied its 200mc TR box, and by 1942 a Type 281B existed. However, the war effort precluded production of many B series sets before 1943–44, and even after the end of the war there were cruisers operational with separate transmitting and receiving antennas.

In 1945 the Royal Navy was beginning to enjoy the fruits of the centimetric radar revolution in all types of radar: in surface and short-range air search (293), in fire control, even in shellsplash-spotting. To a far greater degree than the United States, it emphasized the S-band in its postwar program, avoiding the L-band entirely. For example, the three principal new carrier air radars, Types 982, 983 and 984, all operated in the S-band, precision apparently taking precedence over the greater range generally associated with longer wavelengths. Type 984 in particular was a great achievement, combining all carrier radar functions in a single massive instrument. In the carrier *Hermes*, for example, it was combined only with a navigational radar and with a 'quarter-cheese' Type 993 for gunfire control, a successor to Type 293. It was, however, extremely expensive and must have been difficult to maintain, and it was outlived by simpler systems such as Type 982.

Postwar, too, the Royal Navy began to investigate the large-scale use of radar-picket ships. For a time the design of a Fleet Air Defence Escort (FADE) with massive radar equipment was pursued, the conversion of a fast minelayer being seriously considered. Then attention turned to the conversion of destroyers; part of the fleet modernization program which produced the many destroyer ASW conversions was a large-scale picket program. Although the latter took some years to accomplish (it began about 1952 but destroyers were not converted until 1958–59 and 1961–62), it produced a major new radar, Type 965, which was related to the US SR/SPS-17 series.

CONWAY PICTURE LIBRARY

The cruiser Mauritius, *shown in December 1951, still retained a mixture of mid- and late-war British naval radars. Her main battery director carried the late-war Type 274, with Type 277 and Type 293 above it, but her secondary-battery directors carried Type 285 radars with multiple-yagi aerials, and on the face of her bridge she carried a pair of Type 283s for blind-fire control of her main battery against air targets. Her light directors are not visible.*

GERMAN WARTIME DEVELOPMENTS

German interest in naval radar actually predated that of the Royal Navy, but German naval radars were not highly developed, owing to German operational ideas and national priorities – like the British, the Germans considered land-based air defense the primary task. Moreover, it appears that at first they thought of radar in terms of gunlaying and accurate target-tracking rather than of warning or search. In the German Navy radar was considered to be more effective as a rangefinder than as a means of evading air attack, but it must be kept in mind that the Germans built their capital ship fleet with little expectation (at least at first) that it would be subject to serious air attack.

The German radar story begins with a Dr Rudolf Kuenhold of the German Navy Signals Experimental Establishment (*Nachrichtenmittel-Versuchs-Anstalt*, or NVA), a sonar expert who in the summer of 1933 experimented with 13.5cm signals using a magnetron of 100 milliwatts power, *ie* a tenth of a watt. His experiments were sufficiently successful for the German Navy to establish a special company (GEMA, *Gesellschaft für*

CONWAY PICTURE LIBRARY

The cruiser Tiger *(December 1966) shows standard British postwar surface-ship radars: the 'cheese' of Type 992 forward, with the Type 960 air search radar aft. Below it is the typical array antenna of Mk X IFF systems, rotating with the light frame of the 960. The two MRS 3 directors forward each carry the dish of Type 903, the postwar standard gunnery radar. Note, too, the surviving cage IFF (which appears to be Type 253) on the lattice foremast.*

CONWAY PICTURE LIBRARY

HMS Kent *(October 1963) shows typical British postwar missile-ship radars: the 'cheese' of Type 992 forward for short-range precision search, Type 965 aft for long-range search, and Type 277Q or 278 operating as a height-finder in support of the Seaslug long-range missile system aft, the latter beam-riding using a Type 901 barely visible aft. The 'candlesticks' at her yardarms are characteristic of British naval design practice, and are the antennas of short-wave radios. ECM equipment surmounts the foremast, and forward she carries the usual Type 903 fire control radar. The Royal Navy never developed a high-data-rate three-dimensional radar system comparable to the US SPS-39/52 or -48, apparently finding existing equipment, combined with an automatic data system, sufficient.*

MoD(N)

HMS Birmingham *(September 1976) epitomizes recent British practice, which involves far fewer radars – or radios – than does US practice. She is, in theory, equivalent to a US missile cruiser or destroyer, carrying a weapon (Sea Dart) probably the rough equivalent of the US Standard Missile of even of the Standard Missile (Extended Range), given its ramjet powerplant. There is, however, no integral three-dimensional radar;* Birmingham *has only a double 965 forward and the slotted-waveguide 992Q aft for target acquisition. The small directors, later covered by radomes, carry Type 909 radar tracker-illuminators, and there is also a small navigational radar (Type 1006) at the foretop, with HF/DF at the foretop and, probably, ECM radomes below that. The two cones pointing out at angles from the side of the forward superstructure are probably wire-wound conical antennas, achieving wide bandwidth and covering all polarizations; similar antennas are commonly used for ESM, particularly pointing upwards at mastheads.*

C & S TAYLOR

HMS Phoebe *(June 1977) was converted to carry four Exocets in place of her previous twin 4.5in guns; however, she retained her former gun fire control radar and director, forward. Presumably it controls the single Seacat point defense missile system forward of the Exocets.* Phoebe *shows a transition from active to passive radar operation in missile ships: her most prominent array is the ESM system at her foretop, consisting of the omnidirectional cone and four log-periodic 'arrows' which presumably function together as a direction-finder. In theory, such a system can detect the radar emissions of an enemy ship, which 'leak' over the horizon. To some extent, too,* Phoebe's *helicopter should assist her in engaging potential targets beyond her horizon.*

The German destroyer Z39, *photographed in August 1945, shows the standard German* Seetakt-*type mattress antenna, in this case almost certainly FuMO 24; the small object at the masthead appears to be* Berlin-S *(FuMO 81), and the smaller aerials are ESM devices.*

Elektro-Akustiche und Mechanische Apparate, January 1934). Presumably the sonar-like name was intended to cover Dr Kuenhold's new idea. Like the British, the Germans chose a 'cover' name: decimeter-technology, or D/T. GEMA found that practical systems could be built at decimeter wavelengths, and German naval fire control radars operated at 80cm (*ie* 8dm, or 368mc/s). An operational rangefinder, *Seetakt*, went to sea aboard the 'pocket battleship' *Graf Spee* in 1936 – the first naval radar in the world.

Seetakt (7kW, PRF 1000 or 2000, ultimately 500) was capable of detecting large ships at 10nm, and cruisers and 6, and it could give bearings accurate to 0.2°. The antenna was a mattress fixed to the rangefinder, its upper part used for reception and its lower for transmission. The upper half was divided into two receivers, to improve bearing accuracy. Clearly *Seetakt* could be used for surface (later even for air) search, but that involved turning the rangefinder with its operators inside. In fact German capital ship operations (*eg* the *Bismarck* sortie) suggest that they did not think of sets like *Seetakt* as search devices. One reason was their short range: *Bismarck*'s improved set was credited with 75 per cent of optical ranges.

Seetakt continued to be produced in gradually improved versions through the war, virtually all German surface units from destroyers up carrying one of their gunfire directors, fore and aft aboard major units. In addition, in 1943 German capital ships began to mount a third 80cm mattress on a mainmast bracket, where it was free to rotate. A special radar room appeared at the foot of the mainmast, but there was no PPI display.

What is striking is that the Germans concentrated through the war on radar for fire control, and did not develop the type of air search sets which were common in Allied navies. Where they had to adopt radar for antiaircraft fire control, as in flak ships, they generally used land types too massive for combat at sea.

However, much of the electronics effort expended by the German Navy went into radar warning devices, as a natural counter to the massive use of radar-equipped aircraft against the U-boats which constituted the bulk of the German Navy. The reasoning for such measures was exactly paralleled by US developers waging a pro-submarine battle in the Pacific. However, the Germans paid heavily for their disbelief in the S-band, as for some time Allied bombers equipped with microwave search radars faced German submarines with P-band search receivers. Even after the recovery of H2S from a crashed Lancaster bomber, the Germans were unprepared for the advent of X-band airborne radars. As in other sections of German industry, scientific ingenuity rather outdistanced productive capacity. Thus in 1945 Germany had prototypes of C-band radars, and even an L-band fire control system mounted on its own director, but her heavy ships were still equipped much as they had been four years earlier. Similarly, the impressive array of U-boat systems is somewhat deceptive, as in many cases only prototypes ever appeared. What productive capacity there was went primarily to the *Luftwaffe*, for air- and ground-based air defense.

JAPANESE AND ITALIAN NAVAL RADAR

Japan differed from Germany in three key ways: she emphasized naval systems far more, at least for much of World War II, than land-based air defenses; she was not blessed with a scientific corps of microwave doubters; and she found the mass-production of precision equipment far more difficult. Thus Japanese naval radar appeared in great profusion but showed relatively little sophistication, as well as low power.

Japanese naval officials interrogated postwar indicated that the first Japanese information on electronic detection was derived from an ultra-short-wave iceberg detector aboard the prewar French ocean liner *Normandie*, inspected in New York by a Japanese engineer. In fact the first real stimulus to action was the receipt of German information on radar principles early in 1941. Presumably the material provided included the principles of the German TR box; many Japanese sets had duplexers. As in the case of much Axis military inform-

BY COURTESY OF A D BAKER III

The Japanese escort destroyer Maki, *shown here in Manila Bay awaiting former Japanese prisoners of war (October 1945), displays two of the principle wartime radars: the twin horns of the S-band Mk 2 Mod 2 Kai-4 on a stub foremast, and the vertical 'ladder' of Type 3 Mk 1 Mod 3 aft, carrying four pairs of horizontal dipoles and operating at 2m. The two horns of the S-band set were separate receiver and transmitter.*

ation exchange, only outline ideas were provided, the Japanese doing their own engineering.

It should be emphasized that by no means did the Japanese merely copy the Germans. Their own electronics experts were quite competent, and they included the inventor of the yagi antenna. In addition, the Japanese benefited from the capture, in 1942, of early Allied types, a British searchlight control radar and a US AA set. On the other hand, they appear to have failed to recover the radars sunk with HMS *Repulse* and *Prince of Wales*, although attempts were certainly made. Postwar records fail to indicate the extent to which the Japanese Army and Navy were able to cooperate on radar development.

Throughout the war, Japanese naval sets were plagued with very low power: 30kW in their last P-band air/surface search sets (Type 2 Mk 2 Mod 1-Kai-3), and up to 2kW in 10cm equipment. Land-based radars were somewhat more powerful, a Mk 1 Mod 4 (6m) producing 100kW, which was the best Japanese performance.

Design was limited by production facilities. The postwar Naval Technical Mission found that 'antennas, except microwave, were extremely simple prefabricated systems . . . elements were copper tubing cut to length with the pinch left at both ends; masts were of wood-lattice construction; feeders, simple open wire lines; and disconnect-connectors were simple clips. These components were made by school children and had to be simple . . .' The kinds of systems produced reveal naval priorities not very different from those of the US and British fleets. Like those fleets, Japan began with air search sets, only progressing to surface fire control late in the war, although several surface search sets were in use.

By the end of the war, battleships generally had three air search (*eg* one Type 2 Mk 2 Mod 1, 1.5m type and two Type 3 Mk 1 Mod 3, 2m radars) and two surface search (Mk 2 Mod 2, 10cm) sets, as well as intercept devices for P- and S-bands. Carriers and cruisers were similarly equipped, but lesser craft such as destroyers had a single 2m air search set and a surface search set. The latter had originally been designed for submarine use, and employed a pair of horn antennas.

Japanese technology did not permit the development of height-finders nor indeed of operational surface fire control sets; and there were not even prototypes of AA control sets. However, there was a wide variety of shore-based AA and searchlight control radar, operating in bands from 60cm up.

The Italian Navy, too, developed its own radar devices in the period immediately before World War II; during the war it benefited to some extent from German assistance. Experiments in the 1920s showed that distance could be measured with radio waves, and in 1936 a special section of the Navy's Electronic and Communications Institute (RIEC) was formed, led by Prof Ugo Tiberio, who headed it through World War II. Progress was limited by the state of Italian industry, and (as in Germany) it appears that radar was seen at first primarily as an improved rangefinder, most commonly mounted on the director of a battleship or cruiser. Indeed, the performance of British naval radar at Matapan (28 March 1941) proved an unpleasant surprise, as intercepted messages indicated detection of the immobilized cruiser *Pola*, in the dark, at a range of 6nm. One result was increased official interest in radar.

DUTCH AND FRENCH SYSTEMS

Dutch naval interest in radar dates from before World War II. HSA (Signaal) was known as the Hazemeyer company prewar, and was deeply engaged in the development of conventional fire control systems. In 1934 a Dutch scientist, Jhr Ing J L W C von Weiler, was

The cruiser De Zeven Provincien *illustrates most of the initial Dutch postwar naval radars. Fore and aft, she has an M45 director for her 57mm battery; a ZW-01/SGR-103 surface search radar; an M25 director for her 15cm main battery; a VI-01/SGR-104 height-finder; and a DA-02/SGR-105 target indication radar. The large tripod mainmast carries at its head an LW-01/SGR-114/06 S-band air search set.*

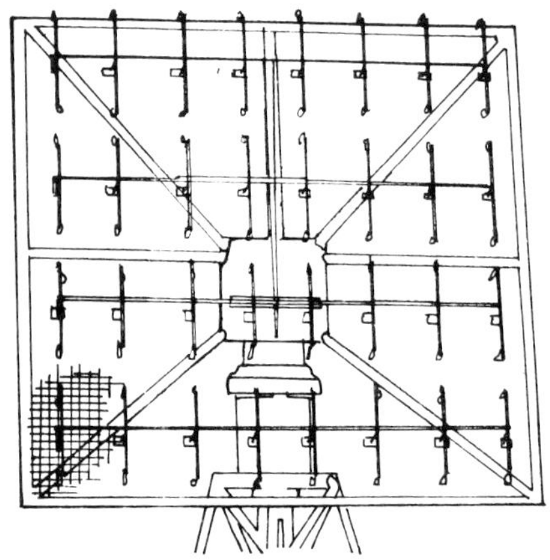

Dutch radar antenna fitted to Issac Sweers, 1941

asked to study the application of ultra-short radio waves to telecommunications. He developed a workable device, and almost immediately found that passing sea-gulls reflected his waves. Echoes from buildings were so clear that the distances between them could be measured. Partly because the output of the device was through an earphone, the Dutch called their system an 'electrical listening apparatus'. Power was low, and the range on an airplane was about 15km (about 8nm). Even so, it was planned to adapt it to Army artillery fire control, coupled with a searchlight. One of the four produced saw service during the German invasion in May 1940. It appears in retrospect to have been a rangefinder rather than a search set, which would have been natural given Signaal's fire control orientation. As Holland fell, von Weiler and his assistants were asked by the Dutch Navy to go to England; they arrived at Portsmouth aboard a destroyer with two of their sets.

The Signaal history describes the Admiralty as interested but cool, which seems natural in view of the low power of the Dutch 70cm system. Ultimately it was decided that both sets would be married to the Hazemeyer twin Bofors mountings of the incomplete Dutch destroyer *Isaac Sweers,* which had been towed to England and which was completed in July 1941. The Royal Navy designated the system 'Type 289 of Dutch origin', and it proved effective in navigation as well as in fire control. According to the Signaal history, von Weiler was active in developing fire control systems for a British radar establishment which had initially concentrated on long-range air warning sets; the Dutch group was responsible for some of the British 50cm artillery fire control antennas as well as ECCM devices.

France, like the Netherlands, had begun to develop some naval radar systems even before World War II; indeed, as we have seen, the Japanese were most impressed by the CW iceberg-detection system installed aboard the liner *Normandie,* which they examined in New York Harbor. The first French military efforts were intended for the protection of land areas against air attack, and consisted of metric-wave CW *barrages électromagnétiques* similar to those which had been proposed in the United States about 1930. The iceberg detector, which operated at 16cm, was first tested aboard the liner *Oregon* in 1935 and installed in the *Normandie* at the end of that year. Power, as elsewhere, was very low: for example, the original 16cm pulse transmitter produced a peak power of 10W, and in 1939 the best that could be done was a 300W magnetron.

The French destroyer Cassard *displays most of the early generation of postwar French naval radars in this July 1960 photograph. She carries DRBV-20 on her foremast, and the two-beam target indication radar DRBV-11 on her lattice mainmast. The antenna just below the big mattress of DRBV-20 is the associated IFF, presumably for a Mk X system, and below that (and forward of it) is the DRBV-30 surface search radar.*

Finally half a kilowatt was achieved, but that was just before the fall of France, and all equipment was destroyed to keep it out of German hands.

The principal French effort was in metric wavelengths, and it continued even after the Armistice. Thus in February 1941 a metric set was installed aboard *Richelieu* at Dakar, capable of detecting aircraft at 50nm and surface ships at 6–12nm. Similar sets were fitted to the battlecruiser *Strasbourg* and to the *Jean Bart*. There was no TR box and no attempt at a rotating directional aerial. In *Strasbourg*, for example, the aerials were fitted to the four bridge tower yardarms, each angled 45° to the centerline. The starboard forward and port aft antennas transmitted, and the other two received. This equipment was destroyed when the Germans occupied southern France in November 1942. In any case, it was quite primitive compared to the British and US technology which France could freely acquire at the end of World War II and which formed the basis for the subsequent growth of the French radar industry. From a naval point of view, that industry is synonymous with the Thomson-CSF company, responsible for all French naval radar systems since the war.

US DEVELOPMENTS SINCE 1945

In 1945 the US Navy operated a vast, even bewildering, variety of radars, a variety due more to the large number of different manufacturers than to the spectrum of requirements. This problem shows particularly in the relative characteristics of surface search sets. As early as 1943 work had begun on SR, a new series of air search radars which would use standardized components and displays. The SRs were also said to be the first properly shock-mounted US radars. SR and SRa were rough equivalents of the late-model SC with a similar wavelength, but in 1945 a complete series, SR-2 through SR-7, was contemplated.

Most of these sets were intended to fill requirements perceived before the war. For example, the prototypes of the major US search radars (SC/SK and SG, were both under test in mid-1941, as were the Mk 3 and Mk 4 fire control sets. During the war two entirely new categories appeared: emergency (portable) search sets, SN and SQ, and fighter control (height-finding) sets, SM, SP and SX.

SN and SQ proved very useful aboard several aircraft carriers in 1944–45, but the use of emergency sets was discontinued after World War II, on the theory that major combatants carried so many radars (including gunnery sets) that the disablement of all was very unlikely. The height-finders, and indeed the entire fighter-direction organization, proved more useful and have continued down to the present.

War experience revealed two great gaps in air coverage: low flyers, typified by Japanese torpedo planes, and very high flyers, which might appear overhead quite suddenly since they were outside the cover of SK at long range. Both problems came into prominence as Fast Carrier Task Force operations began in earnest in 1943. By 1945 solutions had appeared: Airborne Early

By the end of World War II, the standard US cruiser or battleship radar suit included both air and surface search radars and a height-finding or fighter control set. In some cases the latter was used to track air targets before handing them over to fire control radars (as in the British target indication system); in others, its presence made the ship more effective in directing a combat air patrol. There were also massive ECM suits. This postwar photograph of the light cruiser Juneau *shows both, with SPS-6B on her foremast, and the SP fighter control set aft. The topmast above the SP carries the usual pair of DBM radar direction finders, with the short dipoles of ship-to-air radios below it. ECM search antennas ('swords') and Mk III IFF transponders ('ski-poles') share the yardarm below the SP. The foremast carries the heavy antenna of a TDY jammer below the air search radar, and the wartime Mk 12/22 fire control radar has been replaced by the dish of a Mk 25 on the forward director. The 'sword' before the bridge is intended to 'look through' during jamming, to measure its effect. Shielded from the TDY jammer, it could sample enemy radar counter-countermeasures. Large cruisers generally received the SPS-8 height-finder on their mainmasts during the 1950s, as did the surviving battleships.*

Warning (AEW) for long-range detection of low flyers, and zenith search against very high flyers, the latter ultimately to be replaced by a more satisfactory 'hemispheric radar'. Meanwhile the new generation of long-range air search sets was designed for greater high-altitude coverage, a trend begun with SR as early as 1943.

It is possible to discern four major stages in wartime radar technology. The first was the exploitation of P-band, which had been very high frequency indeed when NRL began work in 1936. Coincident with the 1.5m air search sets was a series of what would later be called L-band fire control radars, using relatively long wave magnetrons.

British disclosure of the 10cm magnetron opened up S-band radar, most of it used for surface search or for airborne systems. The next phase was X-band (3cm), which was still relatively new at the end of World War II. As shorter wavelengths were pursued, system components had to shrink further and further in size, and it became harder and harder to achieve the requisite powers.

The greater efficacy of shorter wavelengths suggested yet a further step: K-band (1.25cm). Unfortunately this was close to the resonant wavelength of water, and ranges were very short. Ultimately useful K-band radars were realized in two sub-bands outside the water-absorption peak: Ku (2.5–1.75cm) and Ka

(1.25–0.75cm). An early Ku-band radar, 'Cindy', was delivered at the end of the war.

As in the case of radio, there were pressures on the Navy to choose radar bands to accommodate civilian users, in particular television stations (P-band). At the end of World War II the Navy was also beginning to use P-band (VHF) radios, and there were complaints of interference. Hence the shift to L-band in the postwar SR-2, -3 and -6 search radars – in fact in a general distribution of frequencies in 1945 the Navy agreed as an ultimate goal to vacate P-band. L-band was considered preferable in any case because of the greater gain, greater frequency diversity, and better directivity possible.

However, tests in 1949 and 1950 showed considerable failings. For example, streamlined jet aircraft showed very small cross-sections at L- as compared to P-band. At the latter wavelength the radar signals reflected off the whole aircraft, whereas at the former the returning signal was made up of reflections from individual aircraft surfaces – and there were far fewer corners to reflect strongly in a jet than in the older propeller aircraft. It also appeared that higher frequencies were more subject to atmospheric effects, or 'trapping'. NRL had fought for the retention of P-band systems; now it developed prototypes for a new generation of 1.5m and 75cm air search radars, the current members of which include SPS-37, SPS-40, and SPS-43.

The carrier Yorktown, *shown in April 1945, was typical of US late-war radio and radar practice, and well illustrates the crowding which had to be accepted. Unlike* Hornet, *she carried her main search radar (SK) on a heavy mast sponsoned out from her funnel, with a secondary air search set (SC-2) on the other side. The foremast carries an SP fighter-control radar, with its BO IFF array above it, and an AS-56/SPR-1 ESM dipole forward of it. The platform above carries the SG surface search set and three distinct carrier beacons: YE above; and two back-ups, CPN-6 forward, and YJ aft. In the photograph, YJ is superimposed on an IFF antenna (ABK-7, the 'ski-pole') mounted on the athwartships yard below; other 'ski-poles' (BN) are mounted on the SG/beacon platform itself, on a shorter athwartships yard. The SK air search radar is itself surmounted by a panel carrying a BL-5 IFF. Given the crowding of the antennas, and their attendant blind arcs, it was necessary to provide multiple antennas to achieve full all-round coverage: note the second SG at the after end of the funnel. Ship-to-ship radio is represented by a TBS antenna mounted facing downwards at the fore end of the foremast platform carrying SP.*

OPERATIONAL REQUIREMENTS

Development after World War II can best be understood in terms of the major naval concerns of the time. In 1945 the US Navy was designed to fight three kinds of war: (a) fast carrier strike warfare, against an enemy fleet or his shore installations; (b) amphibious strike warfare; and (c) ASW. The advent of the atomic bomb suggested that in future air and missile attack on strategic targets might become an important naval

Postwar, radars ceased to proliferate, but more and more radio antennas were needed for improved air control over greater distances. The Essex *class carrier* Boxer, *in February 1955, gives some indication of the extent of the problem. The AT-150/SRC was a short dipole for ship-to-air communication; the longer whips (66046 and 66047) were standard long-range types for longer-wave communication. As in wartime, there were two air search antennas (SPS-6B and a surviving wartime SC-5); SPS-8 functionally replaced the wartime SP or SM. AS-177/UPX was an IFF antenna, in effect replacing the BN or BK series. SG-6B was a dual-purpose antenna, for surface or zenith search, depending on which of its two elements was activated. Finally, there was a considerable ESM suit; URD-2 and -4 for direction-finding, as well as the specialized AS-570/SLR radar direction-finder and the wartime 66132 'top hat' intercept antenna. In the notations on the photograph 'WR' means 'Wire Rope' antenna, for longer-wave radio. Note that as yet there are no Carrier-Controlled Approach (CCA) radars in evidence, but the ship is severely limited in antenna space by her original wartime design.*

function, and after 1945 the Soviet Union became the most probable enemy. At the very least that implied increased bad-weather operations, in the Arctic and in the Norwegian Sea. Another important factor in naval calculations at this time was the advent of air-launched antiship missiles by means of which a fast bomber could strike at the carriers from long range, perhaps in weather so bad that the carrier fighters could not operate. The Germans had actually used guided antiship missiles as early as 1943, and the Soviets had captured that technology.

By far the bulk of radar development in the postwar Fleet went into the Fast Carrier Tast Force, in particular into its all-weather defense. Guided missiles were pushed as a bad-weather answer to bomber attack, while Carrier-Controlled Approach was expected to reduce the periods of weather when fighters would be grounded. To some extent the all-weather fighter program made practical by CCA used a technology parallel to that of the ship-launched missiles, *ie* beam-riding.

Above all the requirement was for early warning of impending raids. A 1948 Supporting Plan for Fleet Air Defense Policy observed that search radars then in service 'were designed to detect a 200-knot airborne target, with a reflecting area of 50 square meters at a range of 75 miles, operating at an altitude of 15,000 feet. The airborne target of 1953 is expected to provide a reflecting area of only one square meter and will operate at speeds in excess of 500 knots at altitudes up to 90,000 feet. This increased speed alone necessitates the extension of the detection range to 300 miles'.

This was only the beginning. By the late 1950s attacking aircraft might well be supersonic, with even lower cross-sections. Even the supposed 1953 threat appeared to demand a 35ft antenna, presumably SPS-2, and this set would not fill a requirement for hemispheric search.

There was a long series of attempts to realize hemispheric coverage so as to close the hole directly *over* a ship: SPS-3 and its abortive successors. The interim solution was zenith scanning (SPS-4, SG-6), but hemispheric search became really practical only with the advent of electronically scanned (FRESCAN) sys-

tem in the late 1950s (SPS-26/39/42/52, SPS-48). There were also new height-finders (SPS-8/30) for fighter, and later missile, control.

Beside SPS-2, there were under development SPS-3 (hemispheric search) and a medium-range L-band search radar with a higher data rate, SPS-6. Part of the incentive for the move to L-band was a Navy agreement to vacate the VHF band assigned to television, 174–216mc/s.

The increase in data rate from the 13 seconds of SK to the 4 of SPS-6 was balanced by an increase in PRF from 60 to 600, but even so the L-band sets were ill-equipped to handle the small radar cross-sections of jets. This accelerated the development of P-band radars such as SPS-17, -28, -29, -37, and -43, and the cautious exploration of somewhat higher frequencies (SPS-31/40 and now SPS-49). Meanwhile, the L-band attempt at long range, SPS-12, was never fully developed, although it did enter service in some numbers.

THE AIR SEARCH PROBLEM

Throughout the postwar period there were four great goals in air search radars: (a) great range (two-dimensional); (b) high data rate; (c) accurate height-finding for missile and fighter control; and (d) hemis-

Missile operations also brought a demand for effective very long range search combined with effective height-finding at maximum range and data rate. The tentative solution, under development from about 1946 onwards, was the massive SPS-2 stacked-beam radar, seen here aboard the missile cruiser Little Rock *in April 1960. Her Talos missiles (controlled by the two SPG-49 directors clearly visible aft, as well as by a pair of SPW-2 dishes above and below them) greatly outranged the Terriers of her predecessors, and so required better information. Operational tests soon showed that the frequency-scanned SPS-39 amidships did not provide data at sufficient ranges, and ultimately the surviving Talos ships received the far better SPS-48. SPS-2 itself was dropped as too complex and too unreliable; a contemporary of the British Type 984, it appears to have been far less successful, a sort of dinosaur among US naval radars.*

Missiles brought great complexity, as well as the urgent need for long-range, high-precision radars. Boston, *the first US missile cruiser, is shown here some years after her completion, in July 1960. Missile guidance itself was accomplished by the two SPQ-5 directors immediately forward of her launchers, but they in turn were designated onto a target via a Target Designation System Mark 7, based in turn on a high-data-rate precision search radar, CXRX (a modified SPS-8), on her mainmast, below her TACAN. Abaft it, on a short tower, she carried her long-range air search radar, an SPS-29, the best then available. Forward were back-up and air control radars: SPS-8 and, just visible, SPS-6B, plus an independent IFF working with the forward height-finder (SPS-8). Only forward of that lattice radar mast was she comparable to a World War II cruiser.*

pheric coverage. Great range meant a relatively low PRF coupled with high average projected power. Late in the 1950s *pulse-compression* was introduced; it is common to all of the current long-range air search sets (SPS-37/43 and -40, and presumably also the new SPS-49). The only serious attempts at combining high data rate with very long range were SPS-32/33, SCANFAR, which attempted to achieve a data rate of about one second to track very fast targets at extreme range.

Height-finding at long range was at first the province of stacked-beam sets such as SPS-2; but one of the great lessons of the postwar period was that beyond a point altitude data was of little value – especially to a semi-active homing missile. This observation, and the complexity of the stacked beam sets, led to their abandonment. In the end only the pencil-beam types (*eg* SPS-30) were left.

Moreover, fire control radar would not match the likely threat. In November 1948 Captain M E Murphy, of the Bureau of Ordnance, told the General Board that 'Radar, to be effective, must be capable of acquiring a target in sufficient time so that the weapon can open fire at its maximum range. It must start tracking in sufficient time so that the solution can be made, and the weapon fired, and the projectile or missile get out there during the time of the flight of the projectile before the target comes to within the maximum range of the weapon. In other words, the faster the target the greater the range the radar must have. This is axiomatic. Now, more range can be attained in a radar by either increasing the antenna size or increasing the transmitter power, or both. Since the range varies directly as the antenna size and only as a fourth root of the transmitter power, it is naturally easier to increase the antenna size than transmitter power to get the required range, but we all know there are practical limitations to the increase of antenna sizes on board ship.

'. . . We have here [indicating first chart] depicted various targets at 200 knots, 600 knots, 1600 knots and 3500 knots with targets approximately the size of the present day fighter. This one, the 3500-knot, is a V-2 type guided missile which is about the same size. These ranges on the top illustrate the ranges that these targets must be acquired and radar tracking started in order to have the burst at the maximum effective range of the weapon, which, for the purposes of this display, is considered 10,000 yards. Naturally in the case of guided missiles it would be larger. For instance, to get a 1600-knot target in sufficient time, it must be acquired at 50,000 yards. We have used as a standard here our Mark 25 Mod 2 radar, which is the latest in the fleet, having a five-foot antenna dish and 50 kilowatts power. If we want to obtain the requisite range in the radar to do the job and keep the transmitter peak power output constant at 50kW, you see the antenna size will have to

The newly completed missile cruiser Galveston, *photographed in January 1959, displays a combination of long- and medium-range air search radars to support her dual role of missile defence and air control. The original Characteristics drafted in 1955 called for SPS-2 and a hemispheric-scan radar (SPS-39), with CXRX (as in the* Bostons) *as back-up for the latter; however, the CNO ordered procurement of CXRX stopped in April 1956 'in view of its excessive size and weight and its limited range capability'. By that time the new long-range SPS-17 had been added; it is visible on the foremast.* Galveston, *the first of the Talos cruisers, was also to have had SPS-8A (with SPS-8B as an ultimate replacement), and SPS-12 for medium-range air search. SPS-2 was dispensed with as not yet available; the later Talos cruisers were to have had only SPS-39 and an improved version of the massive SPS-2. In fact SPS-12 was not needed, and the ship operated with the three radars shown, fore to aft: SPS-17, SPS-39 and SPS-8B. The small antenna forward served an SPS-10 for surface search, and the ship retained her former 5in and 6in fire controls with their World War II-era radars. The Terrier conversions of* Cleveland *class cruisers were provided with similar search radars; however, they had only two guidance radars aft, rather than the four (one set of SPG-49/SPW-2) per launcher rail of a Talos ship. By 1963, SPS-30 had displaced the SPS-8B aft, and all of these cruisers ultimately had the very long range SPS-37A/43A forward. There must have been topweight problems, since by 1967 both* Oklahoma City *and* Little Rock *had dispensed with their FRESCANs, although* Galveston *went into reserve with hers in place, in 1970. Terrier ships had their height-finders amidships and their FRESCANs aft, and* Topeka *differed from her half-sisters in having a tripod foremast. She was laid up in 1969, her two sisters in 1973; all retained their FRESCANs to the end. None of these ships ever had SPS-12; all ended up with SPS-29 and then -37A/43A.*

go up to high values, to 20 feet in diameter for a 3500-knot type. That is not acceptable. It is too large. Ten-foot for 1600-knot target is rather large. Most ships could not take it. Also, if we hold the antenna size constant at a five-foot diameter and increase the peak power, we go up to phenomenally large powers in kW. At the extreme the figure is 12,800kW, which is beyond present capabilities. Even 800kW is beyond our grasp. However, in the AN/SPG Mark 49 radar which the Bureau of Ordnance has under development, and with an 8-foot antenna and 3000kW, which is foreseeable, we can meet the requirements laid down by these targets at various peaks. This is not a dish type antenna. It is a square antenna lens type.

'This [indicating second chart] further shows the type of radar required to handle targets of various sizes at 1600 knots. We have the two-engine jet fighter here depicted, the one-engine jet fighter, guided missiles of the Bat size, and a small guided missile, all supersonic speed. Here again is the Mark 25 Mod 2 radar. If we hold the antenna size constant at five feet and increse the power, when we get down to a small guided missile we require 6 megawatts of power which is well into the future and not foreseeable. If we hold the power constant for the Mark 25 Mod 2 radar, these antenna sizes are large – 16½ feet in diameter, to take this target. Again we have shown in red the Mark 49 radar and its capabilities as regards these targets. It roughly will acquire a 2-engine jet fighter at 250,000 yards and a single jet fighter at 120,000 yards and so on, handling the small guided missile rather adequately.

'This AN/SPG 49 radar is under development in the Bureau of Ordnance under contract with Sperry Gyroscope Company and it was initiated in 1946, designed to produce an experimental radar for controlling long range guns and anti-aircraft guided missiles against supersonic targets. This radar will require power and antenna size greater than any radar now in existence.

The missile cruiser Providence *shows a typical radar suit of the mid-1960s in this July 1965 photograph: a long-range P-band air search radar, SPS-29, forward* (**54**)*, a long-range pencil-beam air search set, SPS-8B, amidships* (**64**)*, and a FRESCAN, SPS-42* (**69**)*, aft. SPS-8B added a new high-gain dish antenna to the existing SPS-8 and was the direct forebear of SPS-30. Note that the FRESCAN carries its own IFF array, UPA-22, as does the air search radar* (**68**)*, but SPS-8B does not. The small vertical cylinder just abaft the forward Mk 37 director* (**50**) *is part of the ULQ-6 deceptive ECM system, an AS-1153/SLR receiver/transmitter. There are also the usual triplet of ESM radomes on the foremast (AS-571 to port, AS-616 to starboard, AS-570 below and to port) and the 'derby'* (**58, 59**) *and 'sword'* (**60, 61**)*. The other antennas were standard communications types, such as the short dipoles (on U-shaped mast extensions) of AT-150/SRC* (**25, 27, 31**)*, the stub of AS-1018* (**29, 33**)*, twin 66047 whips* (**34***; single versions are also visible) and 66046 whips* (*eg* **15, 18**)*. TACAN* (**53***; SRN-6) and URD-4* (**52**) *surmount the foremast, and* **70** *is a missile telemetry set.* Providence *also shows two SPQ-5 missile control radars aft.*

The dimension is 8-foot antenna, 3 megawatts power. The total rotating structure will weigh about 17,000 pounds. That compares with the 32,000 pounds of the Mark 37 director. It has a clearance circle of 15 feet and a height clearance of 17 feet. Power consumption is about 40kW. I would like to bring out, however, that the below deck components are considerably heavier than existing equipment. They weigh about 10,000 pounds and require 1500 cubic feet of space. This radar is scheduled for experimental test in 1950 and, with the present budgetary outlay, should be ready for installation in the Fleet about 1955. But on a crash basis, with plenty of money, it probably could be made available to the Fleet in 1952.

'We have a picture showing the Mark 49 radar. This is not a photograph [indicating picture]. It is an artist's conception of the Mark 49 radar and how it will look aboard a destroyer. It is superimposed on a conventional photograph. The below deck components which weigh some five tons are naturally not shown. Presumably they can be put on board a destroyer if other things can make room for it. Naturally, going on the destroyer, the 49 could go on larger ships. I might add that this radar was designed specifically with the guided missile ship in mind. It will have the proper acquisition and tracking characteristics and the proper characteristics for beam rider guidance, which the Bumble-bee type of guided missile [Talos and Terrier] will have. It may be that on a guided missile ship an auxiliary type of radar will be required in conjunction with this, a secondary radar to pick up the missile after it has been launched, and swing it over into the beam of the Mark 49, which presumably is on target. That is the so-called "beam capture" problem in the guided missile business, which has not yet been solved. We have emphasized this Mark 49 radar in the presentation because it has the most apparent impact on ship design and is representative of the trend that weapon control radar is taking . . .'

AN/SPG-49 was envisaged at this time as the natural successor to Mk 25, a future radar director for guns *and* missiles. In fact it became the tracker of the Talos system. With a small dish for a 'capture beam' added it became SPQ-5 (XN-1), the prototype Terrier beam-rider projector. The ancestors of the present generation of missile control radars also included a simple semi-active homing dish, descended from AN/SPG-48, which in 1948 was envisaged merely as the basis for GUNAR and for a 'lightweight director' smaller than GFCS Mk 56. These radars were mounted aboard carriers and their close consorts, which from 1955 on were increasingly missile cruisers and, later, frigates. As we have seen, two other means of early warning had been devised late in World War II: picket ships, both destroyers and submarines, and airborne radars, AEW and the airborne CIC. The radars of the latter fall outside the scope of this book, but we can see a progression from the simple two-dimensional APS-20 of 'Cadillac' to the system aboard current E-2Cs, which is capable of height-finding and which can look down even over land.

Interest in Arctic operations shows mainly in discarded projects: SPS-7, the ice-search radar; the Arctic Picket, which was ultimately built instead as a general-purpose icebreaker (USS *Glacier*); and a series of projects for Arctic submarine radar-pickets. In 1946–48 there were extensive Arctic fleet exercises as well.

Amphibious warfare attracted considerable attention in the two or three years after 1945. Plans were drawn for a 20kt amphibious force (which is only now being realized), but little was needed in the way of specialized radars. The only exceptions were two mortar-locating systems, of which prototypes only were built.

ASW was another matter. In 1945 the Soviets captured considerable German submarine technology, and in the absence of good intelligence it had to be assumed that they were making use of it. The major radar contributions were likely to be submarine detection from the air and periscope detection from shipboard. The former turned out to be quite workable using the anti-clutter radars originally designed for AEW: thus 'Cadillac' and the Neptune ASW patrol bomber shared the APS-20 search set. Although attempts at periscope detection (SPS-15/19/20) proved abortive, the major postwar surface search radar, SPS-10, proved to have some considerable value in this role, at least in low sea states. In fact periscope and snorkel detection tests were a major feature of the Operational Evaluation of SPS-10 in the fall of 1954.

In addition there was great interest in the passive detection of submarine search radars, which is why postwar US destroyers and escorts have such extensive suits of 'ECM' (actually ESM) radomes. On the other hand, these ships do not show the HF/DF antennas so prominent on contemporary British escorts.

STRATEGIC CONSIDERATIONS

From the late 1940s on, there was yet another major preoccupation: strategic warfare, both offensive and defensive. From the US side the major developments were (a) carrier strategic attack; (b) missiles, initially Regulus; and (c) NORAD, the vast North American air defense system. The carrier radar problem was largely one of self-defense in increasingly hostile environments, *ie* during their approach to potential launch positions in areas such as the Mediterranean and the Norwegian Sea. From the mid-1950s on, the Soviets began to deploy large numbers of medium bombers with stand-off missiles, the threat foreseen by US experts as early as 1946. The responses have already been examined: long-range ship and AEW radars, the latter including, from 1960, a carrier-based airborne CIC in the form of the Grumman E-1B; antiaircraft

BY COURTESY OF A D BAKER III

▲ *The cruiser* Helena *was one of three* Baltimores *(the others were* St Paul *and* Los Angeles*) refitted extensively as fleet flagships in 1958–59. Conversion entailed replacement of the pole foremast by a heavy pylon carrying, at first, SPS-29 with SPS-10 and TACAN on a lattice foremast; communication facilities were also enlarged, with receiving whips on turret-tops and a prominent broadband antenna in the bows. Later (as here) the much larger SPS-37A was fitted. Both* Helena *and* Los Angeles *had additional modifications for Regulus surface-to-surface missile guidance: SPQ-2 (a modified SP) in place of the usual SPS-8 on the mainmast, and SPS-12 in place of the after main battery director. It was logical for both to face aft, since the missile was launched from the fantail. All three ships had the usual ECM radomes bracketed to the second funnel, but in 1960* Helena *and* Los Angeles *also had a small house there for an AN/ALT-6 jammer. Both were decommissioned in 1963; their half-sister,* St Paul, *served in Vietnam.*

missiles fired by ships and by fleet interceptors; and of course developments in ECM which cannot be discussed here.

After World War II another US form of sea-based strategic attack began to develop: the missile. Regulus was an airbreathing cruise missile with a range of about 600nm. It was mounted in submarines and heavy cruisers and, at least tentatively, on some fleet carriers. From a radar point of view it is interesting because of the specialized tracking radars it required, SPQ-2 in surface ships and the BPQ radars in submarines. By way of contrast, the successor missile, Polaris, was ballistic and so required no radar tracking at launch.

However, by far the heaviest radar impact of strategic warfare was defensive. From 1948 through the early 1960s the United States armed against a massive Soviet bomber strike. Very large nets of air search radars, including numbers of a land equivalent of the massive SPS-2, were spread across the North American continent, and the Navy provided radar coverage on the sea flanks, in the form of radar-picket destroyer-escorts (DER) and Liberty Ships (YAGR, later AGR), as well as airborne CIC aircraft flying early warning missions. The radars generally corresponded with those of the

BY COURTESY OF A D BAKER III

During the 1950s a total of sixteen Liberty ships (four each year of FY55–58) were withdrawn from reserve and converted to ocean picket ships to support Continental Air Defense. Conversion generally entailed the fitting of radars and a CIC; like the DERs, they received the most powerful sets available. USS Picket, *shown here in August 1958, had a unique enlarged SPA-25 antenna for her SPS-12 air search radar. Indeed, the special enlarged antenna of SPS-17A was designed specifically for the radar-picket ships. Although it is not visible here, there was generally an SPS-8 height-finder aft. Many pickets had quite complete radar suits; for example, in 1960,* Guardian *had SPS-17A forward, TACAN amidships, SPS-12 on her after kingposts and SPS-8 aft. About when these ships were being completed, there were proposals for armed pickets, perhaps with Talos missiles and long-range sonars (PBGs), but they were never implemented. Since the pickets had no military value other than as radar-pickets, they were quickly discarded when the radar barrier forces were dissolved (April–September 1965).*

carrier task force, for example SPS-6 air search and SPS-8 height-finding in the early DERs. A major consequence of the air defense mission was the development of NTDS.

By 1965 the air threat was no longer taken very seriously, and the sea-based radar forces were broken up; many of the DERs were employed in Vietnam on coastal picket duty interdicting Viet Cong arms traffic. Regulus disappeared from service, and with it its specialized radars. What remained were the primary missions of 1946: carrier strike, now often considered in a tactical rather than a strategic context; amphibious strike; and ASW. A new element was the proliferation of ship-launched antiship missiles, even among minor powers, a point emphasized by the experience of the October 1973 Middle East war. More and more effort began to go into antimissile systems, especially those designed to defeat low-flying ('sea-skimming') missiles – hence the advent of SPS-58. A missile capability became more and more important, in the form of point defense systems (Sea Sparrow and its successors), and we see a version of Tartar in the new FFGs. In both cases new lightweight fire control radars were required: for example, in the FFGs there is no three-dimensional radar, only a search radar and an improved missile guidance radar (STIR).

THE 1980s

On a larger scale, we are beginning to see systems designed specifically to defeat saturation missile attacks on carrier task forces: Aegis and Phoenix, supported by an improved AEW/flying CIC airplane, the E-2C. Future development will probably be in the direction of smaller and cheaper electronic scanning systems analogous to the Aegis SPY-1, as missiles proliferate and as saturation attacks on less valuable targets become increasingly practical. In 1981 attention is concentrated on the radar suit to be fitted to the new-design destroyer, tentatively designated DDGX. This radar and its associated combat system may well drive up the size, and thus to some extent the cost, of the DDGX. One possibility is a modified SPY-1, redesigned to take advantage of new technology and thus considerably lighter and less expensive. There are two competitors: Sperry and Hughes, the latter with a FLEXAR (Flexible Adaptive Array Radar) derived from the AWG-9 which controls the Phoenix missile in the F-14 fighter. It recalls a 1974 proposal for a destroyer armed with a shipborne version of Phoenix, a weapon proposal revived from time to time for carrier self-defence. Electronic scanning is also being proposed for smaller ships. For example, at the end of 1980 Congress authorized a study of a *Perry* class upgrade incorporating a fixed-array replacement for the present Mk 92 fire control system.

It seems probable that antiaircraft warfare, in which radar performance is central, will come to dominate

▲▶ *The frigate Reasoner (shown on 24 February 1977) displays the complexity, particularly in ECM and in her communications, of the modern surface warship. Ironically, her design was 'work-studied" and initially it was hoped that virtually all such functions could be accommodated in a wrap-around 'billboard' antenna atop her 'mack'. That structure, which is largely empty, is now merely surrounded by a few AS-1018 broadband dipoles (such as* **18**), *and the ship still requires the usual array of whips (such as* **13**, **14** *and* **15**; *NT-66046 type) as well as a satellite communications system* (**25**; *AS-3018/WSC-1) and a wire fan* (**12**; *for 2–6mc/s). There are also HF satellite receivers, AS-2815/SSR-1* (**10**, **11** *and two others, forward, which are not visible here). Finally,* **7** *is an NT-66047 whip forward. The ship's primary sensors are her SPS-40* (**27**) *and SPS-10B* (**26**) *radars and her ESM arrays. The latter include AS-571A/SLR* (**32**), *AS-616A/SLR* (**33**), *AS-899B/SLR (not visible, on the forward of the three ESM dome platforms around the mast), AS-1174/SLR* (**35**), *AS-1175/SLR* (**36**), *and AT-924/SR* (**38** *plus another not visible here) and one other antenna not shown; all are part of a WLR-1 system. The usual 'derby' and 'sword' warning antennas are not present, however. Active ECM defense is by the usual ULQ-6C system* (**40**, **41**; *AN/SLA-15s), and the Mk 68 director (with SPG-53 radar) is visible* (**45**). *There are three antennas for her LORAN systems: AS-2283/SRN-12* (**42**), *AS-2822/SRN-15 (not shown) and AS-1982/SRN-42* (**44**; *for SPN-40). Items* **47** *and* **48** *are infrared beacons (SAT-2 system), and* **28**, **29**, **30** *and* **31** *are the usual IFF transponders (AS-177/UPX).*

naval warfare. That is, the submarine threat is increasingly an antiaircraft one as submarines come to be armed with submerged-launch missiles such as the US Harpoon and the Soviet SSN-7. Current US efforts to upgrade fleet air defense are threefold. First, there is the new Target Acquisition System Mk 23, which is to be used to control the short-range Sea Sparrow. Connected to a computer, it automatically detects an increasing threat and makes the decision to engage it. The TAS will be deployed aboard units not fitted with longer-range systems of the Standard missile type. The latter will benefit from the introduction of Integrated Automatic Detection and Tracking (IADT) systems, which link together the radars of a ship through AN/SYS-1 systems. In missile cruisers, the equivalent systems are designated AN/SYS-2 (CG/New Threat Upgrade). These systems combine, coordinate and correlate radar inputs to form a single track file, taking account of the different characteristics of the different shipboard radars. Radar detection is automated, with the SYS controlling the detection threshold to overcome clutter without losing barely defined targets. Tracks formed within the AN/SYS system are automatically passed to the ship's NTDS and Weapon Direction systems, as well as to other ships via the standard NTDS data links. In effect, the IADT system, which imposes very little cost in weight, power or volume, considerably improves the performance of radars already in service. This is vital, since the US naval radar inventory is extremely large. Thus improvement to the fleet will generally come through modernization rather than through mass replacement. Improvements will be both in reliability and in automation, through provision of ADT features in existing radar types. The reliability improvement program is exemplified by the introduction of SPS-67, essentially a re-engineered SPS-10 surface search set.

Yet another important current and future development is the spread of effective ESM, *ie* effective methods of detecting enemy warships and aircraft by means of their emissions. A sufficiently directional radar can even provide targeting information for anti-ship missiles – if good ESM equipment is available. Perhaps, then, navies will try to move radars off ships, perhaps into helicopters such as LAMPS (Light Airborne Multipurpose System), aircraft such as the E-2, even into towed balloons or kites – if the systems can be made light enough.

Two more remote locations can be envisaged with current and near-term technology: space and the land. The Soviet Union is already credited with active radar satellites suitable for limited targeting of ships on the surface of the sea, and it is not difficult to envisage a corresponding US system employing short-wave, perhaps imaging, radar. There would be major problems, to be sure: unlike his Soviet counterpart, a US commander is concerned largely with vehicles *not* on the surface, *ie* aircraft and submarines. Most ships a satellite sees would be neutral or friendly and, therefore IFF would present serious difficulties. Moreover, merely to analyze the data produced by a system of real-time ocean surveillance satellites might be a massive, possibly less than worthwhile, task.

Similar problems would afflict land-based radars. Large low-frequency over-the-horizon (OTH) systems have been used for some time to warn of approaching ballistic missiles, and it is not hard to imagine their capabilities extended to the detection of air or even surface targets. Such radars might be linked to a fleet by radio (*eg* via satellite), permitting a relatively low level of fleet radar emissions.

A relatively unrecognized aspect of such remote operations is the requirement for very precise navigation. The OTH operator may know that 100 'Badgers' are flying on course *x* at speed *y* relative to *him*, but that does a carrier commander little good unless he, too, knows his position relative to the OTH radar. Satellites present similar requirements: in each case position-fixing and IFF interact.

In 1981 such systems are relatively distant. For the near term there is a well-defined 'product line' of US naval surveillance radars, the results of attempts to achieve diversity competing with attempts to reduce costs by standardization. The standard air search radars (two-dimensional) are SPS-40 and -49, with the older 200mc series (SPS-37/43), too inaccurate in bearing, fading away. Even SPS-40 is often criticized on that score, but its superior performance against antiradar missiles will probably keep it in service for a long time to come. These radars combine with three-dimensional types: SPS-52 for lightweight Tartar installations and SPS-48 for heavier ships. Then there is the multifunction array radar category, embracing, in 1981, only

Recent Soviet naval radar practice is exemplified by the 'Kara' class missile cruiser Ochakov, *photographed in 1977. Her fore and aft SAN-3 batteries are controlled by 'Head Lights' radar directors, each consisting of a pair of target-tracking dishes with a pair of missile trackers above, for effective command guidance. The primary air search radar is the large 'Top Sail' amidships, but it is backed by a three-dimensional (vee-beam height-finding) 'Head Net-C' on the shorter lattice mast forward. The large radomes on the side of the 'Top Sail' mast are for ECM ('Side Globe'). Other systems visible here include the twin 76mm guns, controlled by 'Owl Screech' radar directors (one of which, pointing forward, is visible at the base of the forward deckhouse) and the Gatling close-in defense guns, controlled in pairs by the 'Bass Tilt' radar visible against the large louvered section in the uptakes. Less evident is the silo for SAN-4 point-defense missiles, abeam the 'Top Sail' mast, and controlled by a 'Pop Group' radar director to its right. There are also small navigational radars, one of which is visible on the after side of the 'Top Sail' mast.*

SPY-1, but potentially including other contenders for DDGX.

In the surface search category there are SPS-10/67, the old standby, and SPS-55, an attempt to bring such systems out of the band associated with missile guidance. The existence of SPS-67 is an admission that mass replacement of existing surface search radar systems is unlikely at this time.

Finally there are the point defense radars, SPS-58/65 and the Target Acquisition System Mk 23. SPS-58 was an emergency development after the sinking of the Israeli destroyer *Eilat*, but Mk 23 was considerably more sophisticated. Just how each will fare over the next decade is problematical. There is no indication of any new radar start, in this or any other area, although there has been pressure for a new two-dimensional utility air search system.

It does seem likely that, as computer costs fall and digital electronics becomes more and more affordable, FRESCANs such as SPS-48 will become less and less attractive compared to pure phased arrays. In particular, their inherent lack of range resolution translates into an inability to resolve small targets (such as missiles) in phenomena such as range clutter. Thus one might suspect that the near-term future lies with phase-to-phase scanning arrays. Some of the new radar proposals (such as those for DDGX) call for small single-face arrays rotating relatively rapidly to achieve something approaching the performance of full four-face systems.

The new Sovremmeny *shows a typical Soviet back-to-back radar installation, the 'Top Steer' FRESCAN radar mounted with the pulse-compression (two-dimensional) 'Strut Pair'. The helical delay line of the former is visible in the original print. The smaller directors are reportedly the illuminators of a new semiactive missile system, SAN-11; externally they are quite similar to the 'Bass Tilts' which normally control Gatling guns such as the one at left. The use of three separate surface search radars (at the masthead) is unusual: they are synchronized, to avoid blind spots. As for the larger radars, the radome is similar to that observed aboard 'Nanuchka' class corvettes, the contents of which have not been published. The typical dish director presumably controls the ship's guns, which had not yet been installed at the time of this photograph; it closely resembles 'Owl Screech'.*

It should, however, be emphasized that the size of the investment in existing radars is such that all systems currently in use (with the probable exception of the SPS-37 and -43 series) are likely to continue in use for some considerable time. Both SPS-48 and -52 are subject to large-scale improvement programs involving new antennas with lower sidelobes and new and more reliable transmitters and receivers. The advent of integrated combat systems such as SYS-1 and -2 should also greatly increase their value over the next decade or two. Moreover, the very slow advent of SPS-49 is a demonstration of how long it takes for new programs to take effect. SPS-49 was, after all, first designed about 1962. Almost two decades later there is a program for new and improved surface surveillance radars (SSURADS), but no new designations have yet been announced.

POSTWAR TRENDS IN WESTERN EUROPE

British naval radars dominated the immediate postwar European export market. The sets in question were wartime types: for example, the new Swedish light cruiser *Tre Kronor* had, in the early 1950s, Type 281 or 960 on her foremast and Type 277 and 293 on her mainmast. New Swedish destroyers had British-style radar suits, and many existing European warships were equipped with British systems. However, from about 1955 on French and, more importantly, Dutch radars began to offer effective competition. In recent years British naval radar exports have been confined largely to warships of British origin, or at least of British design.

The principal British commercial naval radar manufacturers are Decca (navigational radars), Kelvin-Hughes (surface search), Marconi and Plessey. Marconi in particular has long operated in partnership with the Admiralty; it is responsible for the production of such systems as Type 909, 965 and 1022. Plessey has been more interested in the export market, with its AWS series of S-band air search radars; however, Plessey now produces Type 994 and a 294 which is a modernized Type 293 for export.

The first Marconi naval export radars were an SNW series, superficially similar to wartime US air search sets. Subsequently the company began an entirely new radar series for naval installation, its 800 series of naval (801 to 839), coast defense (840 to 849), and land-based (850 and higher) radars, prefixed by 'S' for surveillance or 'ST' for surveillance and tracking. Marconi advertisements imply strongly that some of the naval sets have been adopted by the Royal Navy, but security prohibits the publication of details. The data in Part 2 of this book therefore indicate the current level of British naval radar technology but not the details of particular systems. Most of the S-series radars appear to be intended for fast patrol craft, Marconi having abandoned the frigate and destroyer market to firms such as Plessey, at least for the moment.

For its part, Plessey has concentrated on a series of S-band air search radars for ships of corvette or larger size, with considerable commercial success. The antennas generally roughly resemble the Dutch LW/DA series; examples appear in such ships as British-built or -designed light escorts (mainly Vosper frigates) and in refitted former British destroyers such as the Iranian *Artemiz*. In addition, the Royal Navy has adopted a modified version of the AWS-4 as its Type 994, succeeding its former standard target indication radar, the Type 993 'quarter-cheese'. There are five major variants.

West European ideas make an interesting counterpoint to the US concepts which dominate this volume. In particular, France and the Netherlands built up large naval industries as part of their postwar industrial recovery. Their systems are to be seen in many lesser fleets, and some of them show points well worth noting.

More recently Germany, Italy, Japan, and Sweden have begun to manufacture naval radars, Italy in particular having enjoyed considerable commercial success. However, too little has been published about the products of these nations to make connected accounts useful. Hopefully the detailed data presented in Part 2 on US, British, Dutch, French and Italian systems will prove useful to those who have access to detailed data on other systems.

One can recognize distinct stages of radar technology. The UK began to produce equipment at a 1935 level (hence the long wavelengths of the CH and 79 systems) but although the British were able to devise far more advanced systems, severe wartime material and

The Nigerian corvette Dorina *illustrates current tendencies towards commercial radar development. She is fitted with a Plessey AWS-1 air/surface search radar aft, a Dutch M22 fire control set forward, in its characteristic radome, and a Decca TM-626 navigational radar.*

VOSPER THORNYCROFT

labor shortages and a postwar financial squeeze dictated that one reminder of prewar technology, Type 960, remain active two decades after the end of World War II. In similar vein the US Navy began to deploy radars at a 1940 level, which shows in the CXAM/SC/SR sequence – and which survives, in a sense, in the big P-band air search sets. Both the Netherlands and France emerged from World War II determined to modernize, *ie* to emphasize high technology. The Dutch in particular could build upon a prewar background in advanced computing fire control systems for AA guns, typified by the Hazemeyer system which was applied to 40mm guns. Their postwar Hollaandse Signaalapparaten (HSA) company was extremely successful – its systems even drove the British from markets such as Australia. The French Thomson-CSF organization was not quite as successful, but it has certainly done well.

Both countries entered radar technology at a 1945–50 level. The French began with P- and S-band search radars, types quite accessible to the US technology of 1945, and indeed not nearly so sophisticated as SX or even SM. For their part the Dutch appreciated very early the charms of L-band, and produced a series of distinctive high-gain antennas in L- and S-band, as well as an extremely successful series of radar fire control systems. In both countries, too, radar developers benefited from considerable operational experience: both navies were already familiar with modern radars, thanks to the supply of British and US systems during World War II and the early postwar period. For example, in 1945 the new French battleship *Richelieu* had both British and US air search radars (Type 281 and

GIORGIO ARRA

▲ *Photographed in 1972, the carrier* Clemenceau *displays the range of French naval search radars. The general arrangement of the mast resembles that of American converted* Essex *class carriers. It is topped by TACAN; below is DRBV-50 for surface search, then an ESM level (including the US URD-4 UHF D/F antenna), then the ellipse of the DRBV-23B stabilized air search radar, similar in concept to the US SPS-6 and -12. The massive mattress is a DRBV-20C for long-range air search (as in the US SPS-37A/43A), and fore and aft of the island are DRBI-10C height-finders, using the Robinson-type scan of SPS-8. A small navigational radar antenna is just visible above the bridge (DRBN-31), and the large radome covers a CCA radar, probably comparable to the US SPN-35.*

The West German Navy has been a Dutch radar customer; the destroyer Schleswig-Holstein *is shown in September 1975, after a refit in which her original LW-02 was replaced by the newer, elliptical LW-04. This class was the first to employ the new system. The original DA-02 was later replaced by an elliptical DA-05. The small conical fire control radar antennas belong to SGR-105s.*

BY COURTESY OF NORMAN POLMAR

GIORGIO ARRA

The Indian Leander *class frigate* Udaygiri, *shown here in June 1977, is equipped with standard Signaal radars: SGR-105s on her British-type directors, DA-05 forward (in place of the usual British 'cheese') and LW-04 aft.*

SA). No truly postwar equipment entered service in either navy until the completion of the first generation of postwar indigenous warships: *De Grasse, Chateaurenault* and *Surcouf* in France, and *De Zeven Provincien* and *Holland* in the Netherlands. In each case these developments of 1954–55 set the pattern for the next decade, after which the current radar generation began to appear.

In US terms the first generation of West European air search radars comes midway between SPS-6/8/10/12 and later P-band systems such as SPS-17/28/29/37. Both France and theNetherlands invested heavily in L- and even in S-band air search radars, largely eschewing the 75cm and 1.5m equipment favored by the US Navy; in the case of the Royal Netherlands Navy this may be very much a matter of rough water considerations. The French tried P-band sets but apparently much preferred the L-band of early US postwar systems. In both navies great effort later went into complex three-dimensional systems suitable for use with antiaircraft missiles, such as the French DRBI-23 and the Dutch SPS-01 (MTTR). However, neither has produced a simple, lightweight FRESCAN comparable to the US SPS-39/52, and indeed both imported that system from the United States.

Italy entered the field of naval radar manufacture rather later than did France and the Netherlands, but nonetheless has succeeded in exporting a wide range of search and firecontrol radars. There are two principal

GIORGIO ARRA

The Australian destroyer Vendetta, *shown here in 1977, emerged from refit in May 1973 with the characteristic egg-shaped radomes of a Signaal M22 fire control radar system, and with LW-02 aft. Given its considerable precision, and the Track-While-Scan capability of the M22, she needed no separate target indication radar, that role being filled by the search antenna of the M22 system. Conventional British ESM and radio antennas were retained, the typical 'candlesticks' being visible on her lattice mainmast.*

AMBROSE GREENWAY

The Danish frigate Herluf Trolle *was reportedly equipped with Selenia (Italian) export radars: RAN-14X (on the foremast); the after radar has been identified as SF-1, but that bears no resemblance to any existing Selenia series. Danish designations have been reported recently as CWS-2 and -3 (possibly not in that order), which may reflect replacement of radar electronics by a Danish or Swedish product.*

manufacturers, Selenia and SMA (Segnalamento Marittimo ed Aereo of Florence), Selenia being the more important of the two. Italian naval radars did not appear in any quantity until the 1960s; until then the Italian Navy used a mixture of US sets, some of which had been built under license. Perhaps the first major installation of an indigenous Italian system was the large Selenia Argos 3000 installed aboard the converted missile cruiser *Giuseppe Garibaldi* for long-range warning. It appears to have been designed for land use, and had a peak power of 5MW. The other radars on the cruiser were of US origin: SPS-6 for search, SPS-39 for height-finding and SPG-55 for Terrier control. Meanwhile, however, Selenia was developing its own series of search radars.

8. Radio

Radio is the older brother of radar, its precursor and partner in the shipboard electronic forest. As such it provides contrasting (and hence illuminating) applications of the same physics on which radar depends. There are of course important differences: radios operate on a far wider spread of wavelengths; they often depend for their operation on their environment (*eg* on the metal of a ship or the ionized layers of the atmosphere) to a far greater extent than does a radar; and the mathematics of radio is intimately affected by the one-way character of radio transmission. Moreover, that directionality, so highly valued in radar, is useless in a radio intended to communicate with a receiver the precise location of which is either unknown or rapidly varying.

Radio operation can be divided into three categories: ship-to-shore, ship-to-ship, and ship-to-aircraft. Typically ships which are not operating in close company do not communicate directly with each other; each will send its batches of messages to shore communications terminals, and powerful shore stations will broadcast messages for the fleet, containing messages for individual ships. The latter will 'copy' those signals applicable to them, rejecting or noting the others. On the other hand, rapid ship-to-ship communication is necessary for ships operating in tactical combination, and the same applies to naval aircraft. Amphibious operations

The carrier Oriskany *(April 1975) shows a variety of communications antennas, as well as a highly-directional double helix* (**106**) *which in this case was used for meteorological data (AN/SMQ-6A). It was employed primarily by carriers, although a few appeared aboard fleet flagship cruisers. UHF Phasor-90s* (**59**, **60**, **63**, **64**) *were used to communicate with aircraft; they are circularly-polarized conical antennas. Also evident are a pair of HF whips* (**20**, **21**) *for longer range communication. Radars shown here are the standard SPS-30 pencil-beam height-finder* (**90**) *and the ageing Mk 25 for 5in fire control* (**117**).

extend the ship-to-shore mode to include such short-range functions as control of fire support; and given modern, very extended formations, it may also be necessary for the commander of a task force to communicate directly with a ship well over the horizon.

The development of NTDS added a requirement, for at least some ships, to be able to pass signals from one computer to another over a high-capacity digital link. At first this installation was often designated HICAPCOM, or High Capacity Communications, but the digital link in question is now called Link 11, in a series of numbered NATO and US communications links. Others include Link 4A, for ship-to-air, Link 10 (British-Dutch simplified digital link), and Link 4 (ship-to-ship via teleprinter, for non-NTDS ships).

These different modes of communication carry with them very different requirements for radio range. In one kind of communication, very long range is extremely important: the fleet talks to Washington from Okinawa, for example. However, long range is just what is *not* wanted for tactical communication. Long range might mean that a carrier pilot at Leyte would hear commands intended for *another* carrier pilot at Pearl Harbor, or that a Japanese intercept officer in the Home Islands might hear a pilot being talked down over Hawaii. Proper frequency choice prevented both types of embarrassment, but even so, radio intercepts were a useful source of intelligence in both World Wars. In the past few years an increasing desire by Presidents to control crisis operations has blurred these distinctions. Many of the tactical commands in the *Mayaguez* incident, for example, came directly from Washington.

BANDS AND BEAMS

Like radar, radio can best be visualized in bands with distinctive properties (see Chapter 2). These are largely a matter of how the waves propagate. Very generally, a radio antenna sends out a combination of 'ground' and 'sky' waves. The latter are reflected from the upper atmosphere at some wavelengths, a property responsible for most forms of long-distance radio communication. 'Ground' waves propagate along the surface and through the space directly above it ('space' wave). Just as in the case of radar, there is interference between direct and indirect transmission of the space wave. In all cases the space wave is effective out to the radio horizon, but at low frequencies (18–300kc) the ground wave can travel considerable distances, depending upon the conductivity of the ground. For example, the sea has about 5000 times the conductivity of soil. LORAN uses this effect. As the frequency rises, less and less of the signal remains in the ground, and the ground wave loses its value for long-range communication. This is much the same conduction effect as that responsible for the use of coaxial cables in radar.

Long-range operation is thus possible in three bands: low frequency (LF), with its ground wave (up to 300kc); medium/high frequency (MF/HF, up to 30mc), which reflects off the ionosphere; and ultra-high frequency (UHF, 300–3000mc), which reflects off the troposhere. Those frequency regions allowing for reflection of sky waves have, in the past decade, been exploited for over-the-horizon (OTH) radar: the ionosphere or troposphere is used to send signals beyond the horizon and back. Losses each way are enormous, which is one reason why OTH is a very subtle and delicate system.

Long-range communication is less so. The primary subtlety is the variation in height and reflectivity of the ionosphere and troposphere with season, time of day, sunspot activity, etc. Just where the sky wave will touch down depends on the height of the reflecting layer; typically in practice there are several layers, and signals may undergo multiple transmissions, some parts of one signal arriving just as parts of another (via a different path) do. Even the maximum frequency the ionosphere will reflect can vary, and an exotic form of meteorology thus becomes an essential element of long-range communications. One other point worth making is that, depending upon antenna orientation, waves will be emitted at a range of angles implying some *skip distance*, some region in which they will not touch ground.

There is one other important means of long-range communication: relay via satellite. In satellite systems sky waves at frequencies which do *not* reflect from the upper atmosphere are directed up at a satellite. Such transmission is simplified by the short wavelengths involved, *ie* by the ease with which they can be formed into tight beams. The satellite can then act either as a passive reflector or as a transporter; the latter mode, now universal, makes ultimate reception far easier. An important virtue of satellite, as opposed to conventional, radio communication is its potential covertness. The beam up can be so well defined that it is almost impossible for anyone not directly in its path to detect it, and hence to detect its originator. The broadcast from the satellite gives no hint, except a cryptographic one, of the location of the originator of the message.

Thus a really widespread satellite communications net can, at least in theory, satisfy the requirements both for communications security and for control from Washington. Satellites permit very long range using signals carrying so much information that millions of simultaneous messages can be addressed without confusion. On the other hand, the beam from ship or shore *up* to the satellite can defy interception. Encryption is required to provide the messages down with a measure of security; but once more, the high data capacity of the satellite channel allows for a high degree of encoding, which can be provided automatically by computer. Whether this capacity, which can permit a President to bypass the chain of command to order air strikes directly, is a good development from the Navy's point of view is another question entirely.

As in the case of radar, radio development shows a gradual increase in radio frequency; however, radio differs from radar in that such an increase was not always desirable from the point of view of range requirements. On the other hand, every increase in operating frequency increased the number of available channels, given a constant width (in frequency) per channel – and channel width determines the rate at which information can be transmitted. Voice radio requires far wider channels than Morse code, and the narrower the channel the worse the transmission. Thus early US experiments with HF voice radio were generally unsuccessful, whereas the development of VHF radio made voice fighter control practical. Once again, as in the case of radar, it proved far easier to provide high power at lower frequencies: high-power UHF and VHF transmitters were products of the post-1945 period.

US NAVY INSTALLATIONS

The first US naval radios were Marconi sets tested aboard the armored cruiser *New York* and the battleship *Massachusetts* in November 1899, ranges as great as 46nm being achieved. In 1902 the first Navy sets were installed: 1kW Slaby-Arco (German) spark-gap instruments, operating over the 200–400m (750–1500kc/s) band. Such radios transmitted by producing high voltage sparks, mixtures of many frequencies, which could be received as continuous signals, punctuated by Morse code. In effect the wavelength transmitted was determined by the antenna rather than the transmitter, the former peeling away from the complex signal that frequency (or, rather, range of frequencies) which it sent most efficiently.

That this was not a particularly efficient system was realized quite early. Many inventors worked towards a more controlled generation of high-frequency waves. In the United States 'arc' equipment was used as the basis for radio telephones (1907); this system was actually used during the Great White Fleet cruise, but its hasty procurement and lack of trained personnel made it unreliable, and the Navy dropped voice radio for about ten years. At this time typical Navy wavelengths varied from 600–1000m (300–500kc/s) for ships to as much as 2700m (111kc/s) for shore stations – which shows that although half-wave dipoles mattered for transmission, they did not matter half so much for reception. The point is better made by the size of table radios designed to receive commercial stations operating at about 1mc/s, *ie* about 300m.

From the end of World War I, naval frequency allocations were more and more affected by the needs of civilian radio users. Thus immediately after the war the Navy (and civilian users) reserved the MF and LF bands (*ie* those below 600kc/s) for long-range communication (mostly via the ground wave); the 600–1250kc/s (240–500m) band would be used for short-range intrafleet (tactical) communications. However, commercial broadcasting in the 550–1500kc/s band grew to the point where those frequencies were

The command ship Wright, *an ex-carrier, was designed almost completely for optimum radio performance, as part of the National Command Post Afloat Program (FY62). Her sister-ship,* Arlington *(ex-*Saipan*, FY63), was to have been similarly converted, but instead became a radio relay ship (AGMR), similar externally but lacking the headquarters facility of this ship. Fiberglass masts permitted unusually high antennas: two big steerable log-periodic arrays (forward and amidships) on the centerline (with spreaders below for conical-monopoles) and three somewhat shorter poles to port supporting a flat-top. There were also a pair of conical-monopoles, shorter than either of the other systems, on the centreline. Whips were supplied along the edges of the former flight deck. The radar suit included a small radome for SPN-8 (presumably for helicopters bringing national leaders aboard in a crisis) and SPS-6B, plus the usual ECM radomes (retained from* Wright*'s light carrier days). There was also a small satellite-communications dish, not readily visible in this August 1963 photograph. In 1968* Wright *was refitted with a tall solid mast carrying a big tropospheric-scatter dish on her starboard side abeam No 2 funnel; the former ECM mast was removed and its radomes transferred to the new structure. Somewhat later the big flat-top, with its three tall masts, was removed.* Wright *was deactivated in 1970 as the NECPA program terminated.*

reserved for civilian use. Meanwhile the Naval Research Laboratory convinced the Navy that higher frequency signals, transmitted via sky wave, would suffice for long-range communication. This solution had been rejected in view of the irregular behavior of the ionospheric layers, but extensive NRL research during the 1920s permitted the construction of tables for prediction, *eg* of the maximum usable frequency (MUF). HF equipment was used experimentally during 1924–25, its performance during a fleet cruise to the South Pacific (1925) convincing the CNO and the Bureau of Engineering of its virtues.

HF thus became the principal fleet radio band of the prewar period. The earliest HF transmitter, TV (1923), operated in the 2–3mc/s (100–150m) band, but within a few years transmitters operating at up to 18mc/s (17m) were in use. For example, XE, a 2–18mc/s transmitter, was developed for installation in the confined spaces of submarines (1928). HF penetrated water well enough for communication to be maintained with submarines at periscope depth. At the upper end of the HF spectrum wavelengths were short enough for free-standing half- or quarter-wave dipoles (whips) to be effective, and the latter began to displace some of the more conventional wire antennas during World War II as the need for radios increased enormously.

Lower frequencies were used for tactical purposes. Examples, for capital ship installation, included TP (150W, 75–600kc) and TU (2kW, 195–565kc), both first ordered in 1923 – and both still in service in 1941. A 1926 Fleet Radio Plan for TU and the high-frequency TV was put forward for use in cruisers. In 1931 the heavy cruiser *Louisville* (CA-28) was completed with TAF (1kW, 4, 12 and 16mc/s bands), TAQ (2kW, 175–600kc/s), main transmitters and TAD (100W, 2–3mc/s) and TAJ (300W, 195–600kc/s) emergency sets. In the 1926 plan additional XA sets (4, 8 and 12mc/s) were reserved for fleet flagships, *ie* for very long range transmission.

Ten years later the standard cruiser installation included two LF/MF (175–600kc/s) transmitters, one of 2kW for shore, strategic and scouting, and one of 500W for shore and tactical communication. One 1kW set of medium frequency (300–2000kc/s) served for aircraft. There were three main HF sets, two of 500W on the 2–18.1mc/s band for aircraft and strategic/tactical use and one of 1kW in the 4–26mc/s band for strategic/tactical use only. There were also emergency sets on both 175–600kc/s and 2–18.1mc/s bands (500kW), and a pair of the new TBS on the bridge.

Other classes were similarly equipped. The table indicates the upward trend in cruiser frequencies. Main sets *only* are indicated; note that the 115–156mc/s band was introduced for communication with aircraft. The 1957 CLG is a force flagship.

US Cruiser Radar Frequencies

	CA (1941)	CA (1944)	CAG (1950)	CLG (1957)
175–600kc	2	–	–	2†
300–2000kc	1	1*	2*	–
2–18.1mc	2	3	3	–
2–8mc	–	–	–	4
6–18mc	–	–	–	2
4–26mc	1	–	1	–
8–30mc	–	–	–	1
115–156mc	–	2	1	2
Short range				
TBS (60–80mc)	2	2	–	–
225–400mc	–	–	11	10
2–15mc	–	–	–	2
2–7mc	–	–	–	2

*One set is 300kc–18.1mc
†300–600kc

VHF AND UHF

The advance in frequencies for aircraft communication should be noted. With the advent of fleet HF, similar sets, many of them voice, began to appear in large numbers aboard naval aircraft. However, airborne HF suffered from several defects, its long range allowing too great a scope for interference between signals, and multiple-path transmission causing garbling. There were also complaints of insufficient channel capacity. VHF radio came into use during World War II as a reliable voice-transmitter for air-to-ship operations – indeed, the development of VHF radio was accounted a primary factor in the success of naval fighter-direction.

The other major wartime VHF system was the short-range voice radio, TBS. This designation was originally only the result of the position of the set in the standard US Navy 'T' series, but the happy coincidence of 'TBS' with 'talk between ships' made these letters a general designation for short-range tactical voice radios. This low power (40W, 60–80mc/s, 3.75–5m) transceiver had as its major virtue short range, so that ship-to-ship communication within a task group could be considered secure both from direction-finding and from interception for intelligence. Coding became unnecessary. Unfortunately, under some unusual circumstances VHF transmission was not quite limited to the line of sight; late in the war the Navy began to switch to even more closely horizon-limited UHF systems. At present part of the VHF band (roughly 30–160mc/s, 2–10m) is retained for amphibious operations and short-range land-mobile systems.

The other reason for postwar adoption of UHF was a need for even more channels. Some experimental equipment was produced as early as 1936; it could be made extremely directional, and in one 1938 experiment was used for bridge-to-bridge communication. In 1944 production equipment (RDZ/TDZ, 200W,

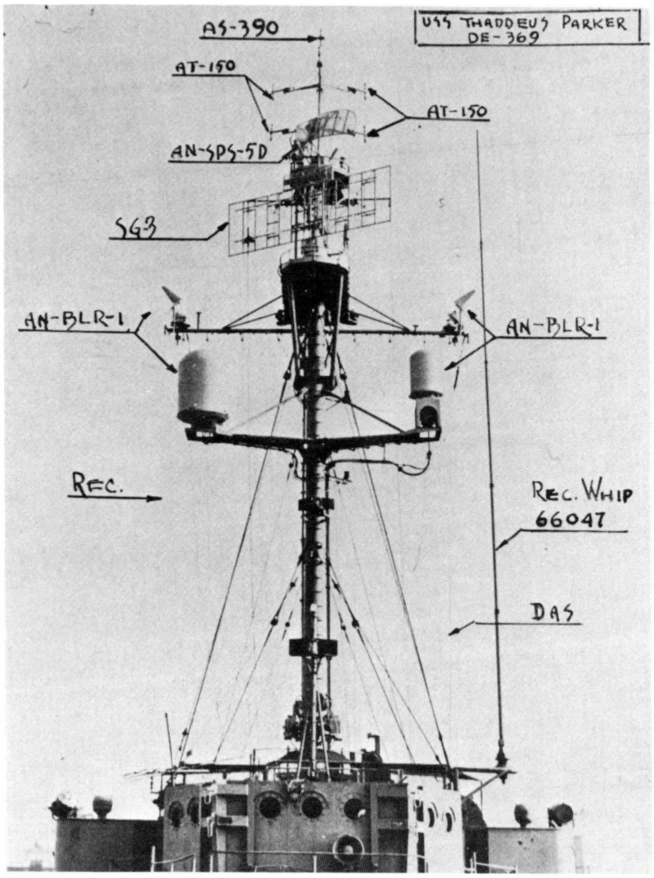

The destroyer-escort Thaddeus Parker, *photographed in the early 1950s, shows a typical small-ship radio system of its time, concentrating on short-range ship-to-ship and ship-to-air dipoles (AT-150 and AS-390), with a single whip and some wire antennas for long-range reception and transmission. The radar marked 'SG-3' is actually an SC-3; DAS is a LORAN receiving antenna.*

225–400mc/s) appeared, and now the Navy was using the P-band both for long-range air search *and* for short-range communication. At the same time an inter-departmental committee assigned the 225–400mc/s band to the Navy, all other UHF bands going to the Army and the Air Force.

Postwar economies stretched out the transition from VHF to UHF tactical radio, so that as late as the mid-1950s many escorts had only the former. The situation was worse in Allied navies and air forces, where VHF equipment continued in service for many years. Joint operations with these forces, then, required the retention of VHF radio aboard US warships even after the conversion to UHF had, in principle, been completed.

However, there was one important factor militating against UHF. As the threat of nuclear attack on ships at sea grew, naval formations began to spread out, so that one hit could not destroy all. Moreover, higher aircraft speeds required that picket ships stand farther out, often beyond the line of sight. Such stand-offs were required just as the density of signal traffic rose sharply as a consequence of NTDS – hence the use of HF for the latter, and the continuing wide display of HF (whip) antennas. Unfortunately, HF signals have so long a range as to be easy to intercept, *ie* to home on. However, in the 1950s and 1960s there was no other over-the-horizon system practicable.

Now there are two. One is the satellite, operating at microwave frequencies and accepting from the sender a beam so narrow as to be almost impossible to intercept. The other is tropospheric-scatter, at UHF frequencies such as P-band. This is nothing new. Early CXAM radars had a keying feature permitting operators to communicate via the troposphere, and in 1940 *Yorktown* was able to use an equivalent of OTH radar (her CXAM directed up at an angle) to observe the Californian coast 450 miles away. As in the case of HF, NRL collected sufficient scattering data (from 1954 to 1959) to make tropospheric communication practical. In 1955 a surplus SK-3 sufficed to pick up clear telephone conversations at 250 miles (USS *Achernar*), and a similar dish aboard USS *Thuban* permitted ship-to-ship P-band communication (40kW) at up to 630 miles (in February 1956). Similar results were obtained at X-band, but the smaller X-band antennas could not absorb enough of the incoming signal. Moreover, rain could cause serious problems at X-band. Hence the Navy adopted P-band UHF tropospheric communication.

These signals propagate in so well-defined a direction that there is little point in trying for ship-to-ship conversation over any substantial range. Hence it appears that the fleet uses tropospheric scatter mainly for communication with the shore – generally, now, via directional log-periodic arrays such as those aboard the now-defunct communications relay ship (AGMR) and the *Blue Ridge* class LCCs. The command ships *Northampton* and *Wright* sported big dishes.

RADIO ANTENNAS

Just as in radar, the simplest radio antenna is a dipole half a wave long, producing an omnidirectional signal. Longer wires produce more sharply directional emissions; shorter ones emit less efficiently. Unlike radar, much radio equipment operates at such long wavelengths that a half-wave dipole would be uncomfortably large; this problem accounts for much of the ingenuity displayed in radio antenna design. An important element in that ingenuity is the fact that wavelength in a conductor, especially a thick one or one specially terminated, may be far shorter than the free-space wavelength associated with the same frequency. In addition, an antenna extended outward from a conductor ('grounded') such as a ship's superstructure presents to electromagnetic waves an 'image' in the conductor. Hence a properly grounded quarter-wave dipole will transmit almost as efficiently as will the equivalent half-wave dipole. Since World War II the simple dipoles have been supplemented by a variety of array antennas; but the dipoles, most prominent as whips, still dominate.

As in radar, it is vital to keep in mind the equivalence of transmission and reception. However, it is also vital to appreciate that reception is possible with an antenna far from optimum (far from half a wave in length). For this reason the use of multiple receivers on one antenna was practical well before World War II, but the use of multiple transmitters, well separated in frequency, was not. It should be noted that, of the many current shipboard radio antennas, about half are associated with receivers – relatively few sets combine reception and transmission.

Early naval transmitters operated at frequencies of 100kc – wavelengths of 3km – or less; it will be appreciated that a half-wave wire antenna would be out of the question. Even a quarter-wave antenna would be excessive at 750m. The earliest solution was the flat-top, which takes its name from the flat horizontal element or curtain of wires, with a vertical wire leading down to the radio transmitters; one variation, in many ships, is the 'inverted-L', in which the lead runs down from one end of the horizontal or nearly horizontal element. Generally in such an antenna the combined length of vertical and horizontal elements is a quarter-wave; in other flat-tops each half, from one end of the horizontal flat-top to the down- or up-lead, plus the length of the vertical lead, is a quarter-wave. In operation, it is the vertical element which radiates: the dominant means of transmission at low frequencies is a ground wave requiring vertical polarization, hence a vertical antenna. In fact in many flat-tops multiple or folded horizontal wires are used to effect a phase cancellation which reduces undesirable high-angle radiation that might otherwise be emitted by the horizontal elements. Such sheets, or the alternative wire 'sleeves', are a hallmark of low-frequency (ground wave dominated) operation.

At higher frequencies the flat-top itself begins to radiate, so that a given flat-top antenna can cover a very broad range of frequencies. Depending upon how it is coupled, a flat-top can radiate efficiently at wavelengths four and eight times its length, using the ship as a ground plane. In addition, most antennas can often operate at higher frequencies, whole multiples (harmonics) of the basic frequency – albeit with different characteristics.

Unfortunately flat-tops are difficult to maintain on ship, their support at mast-ends in particular being

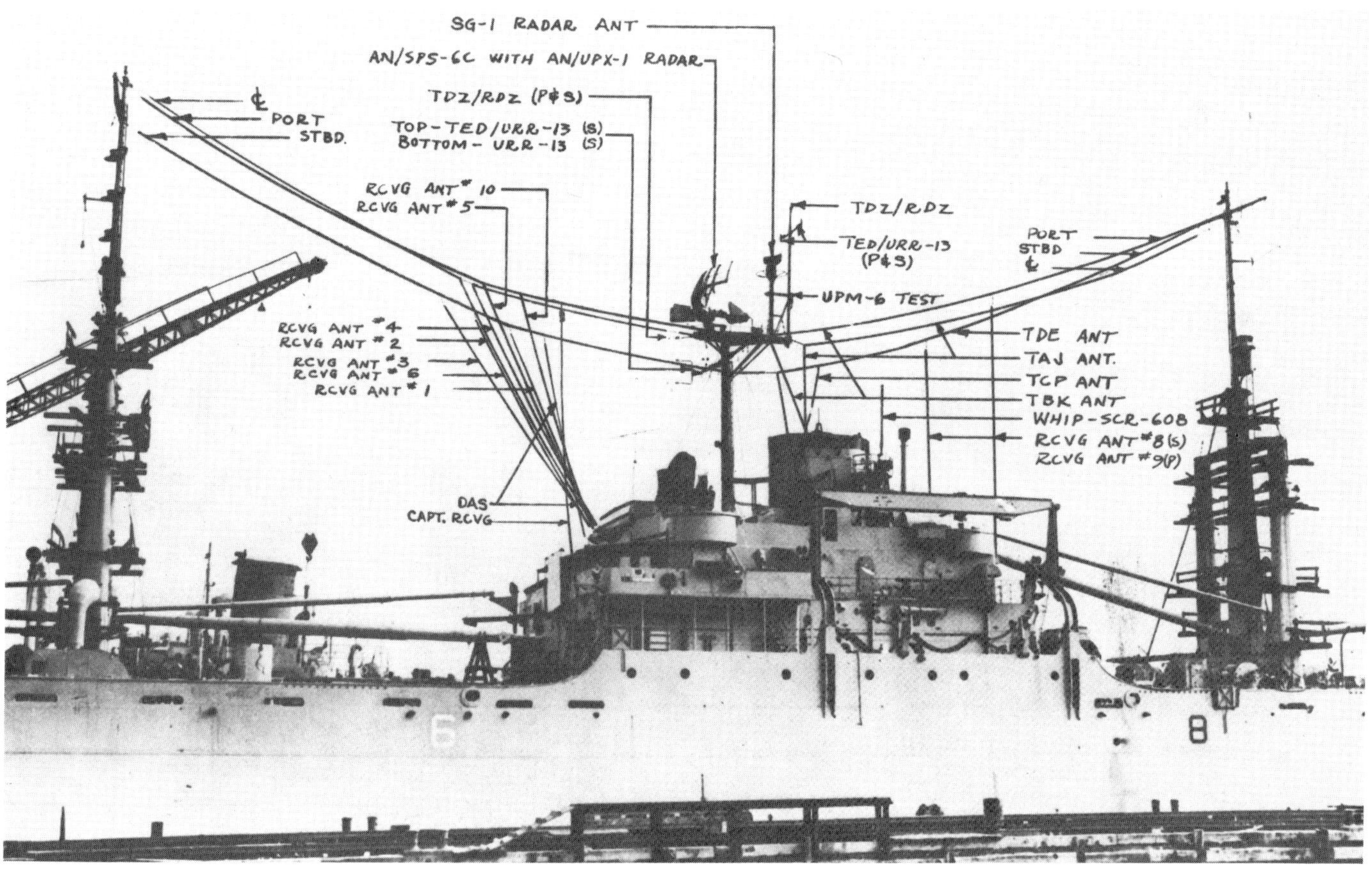

An attack transport shows a typical flat-top antenna arrangement; the transmitters are the devices with 'T' designations (eg TDE, TDZ). SCR-608 was an Army radio, to communicate with the beach, and receivers were designated in an 'R' series. This is the attack transport Bellatrix, *February 1955, shortly before going into reserve.*

subject to breakage. Hence as shipboard transmitters grew in power it became possible to substitute less efficient single wire antennas, which in any case were practical at the higher frequencies of the post-World War I period. For example, 1mc corresponds to a wavelength of 300m. Generally, wiere antennas would not be exactly a quarter-wave long; they would be electrically 'loaded' or 'tuned' to the proper equivalent length. The US abandonment of sheet antennas after World War I marks a considerable rise in standard frequencies, but also the abandonment of broad-band operation: each narrow-band transmitter would now require its own antenna. This situation led to problems solved only after the Second World War.

At still shorter frequencies a quarter-wave is very practical for shipboard installation. For example, a 40m wavelength corresponds to a frequency of 7.5mc/s. During World War II US warships began to sprout 'whips', quarter-wave free-standing dipoles mounted on their superstructures to take advantage of grounding. In fact most whips were not quite a quarter-wave long, and their electrical length was adjusted by coils – which have the disadvantage of absorbing much of the power, so that some whips were no more than about 5 per cent efficient. Their great virtue, however, was their efficiency in topside arrangement. There are three current standard types: NT66046 (28ft aluminum, 500kc–30mc), NT66053 (25ft stainless steel, quarter-wave at 9.35mc), and NT66047 (35ft).

At the same time shorter wavelengths were used for short-range transmissions, the best-known example being TBS. For this system a quarter-wave dipole was combined with an artificial 'ground plane' consisting of two crossed wires – at such wavelengths a radio wave cannot differentiate between two thin wires and an endless sheet of conducting material, and acts as if there were an 'image' dipole doubling the length of the real one. Even shorter dipoles came into use for VHF and then for UHF circuits (air-to-ship). Current examples include both the short quarter-wave dipole with artificial ground plane (AS390/SRC, 225–400mc/s) and the full half-length dipole for installation at the end of a short horizontal arm (VHF: NT66095, for 100–156mc/s; UHF: AT150/SRC, 225–400mc/s).

World War II brought about an enormous proliferation of radio systems – and massive interference, to the point where retuning one antenna on some command ships would throw off the others. Postwar research emphasized broad-band antennas which could be used simultaneously by several transmitters of very different frequencies. The first fruit of this program was the sleeve antenna, essentially a whip in a large-diameter grounding sleeve. Both specially designed sleeves *and* parts of a ship's structure could be used, and in the early 1950s the Naval Research Laboratory used the stacks of *Gearing* class destroyers as sleeves for whip antennas. A

One way to get reliable long-range communication is to scatter directional signals off the troposphere, a technique tested by the Naval Research Laboratory. One of its earliest applications was to the command cruiser Northampton, *which had a large dish (apparently a converted SK-2) on a foremast, as shown here. The small antenna which replaced the earlier SPS-2 has not been identified, but appears to have been an independent IFF.*

more conventional installation of sleeves on the command ship *Northampton* replaced fifty narrow-band (conventional) antennas with ten, five for transmission and five for reception. The broad-band sleeve principle extended into the UHF spectrum and many ships now carry a UHF sleeve antenna, AS1018/URC, which appears to be nothing more than a fat fiberglass-covered 6ft mast but actually contains an elaborate array of internal antennas consisting of a slotted dipole for overhead transmission, an upper dipole, then a series of spacers, then a lower dipole. Together these give full hemispheric coverage with circular polarization, over the 225–400mc/s band. The accompanying program to develop multicouplers for radio transmitters and receivers shows merely in topside simplification compared to a mess we can only imagine, but it occupied very considerable effort at NRL.

FANS, CONICAL-MONOPOLES AND HELICES

In the late 1950s new classes of antennas began to appear: 'fans', 'conical-monopoles' and 'log-periodic arrays'. All are broad-band types, intended to preserve their electrical properties over a wide range of frequencies. The fan consists of four wires cut for a quarter-wavelength at the lowest freqeuncy to be transmitted, all of which fan out from the transmitter. In effect they function together as a single radiating dipole of great diameter, whence the broad-band effect. The conical-monopole was originally designed in 1957 to reduce antenna heights at naval airfields, but later appeared aboard many ships. It consists of a series of wires extending out and down from a mast, meeting horizontals, and then falling inward to the base of the mast. The electrical effect of the horizontals is to make the lower elements electrically equivalent to the far longer upper ones – in effect, to lengthen the mast. Typical ranges of HF frequency of 3 or 4 to 1 (such as 2–8mc/s) are claimed. At low frequencies the entire antenna radiates in a conventional manner, although at a higher frequency only the lower elements radiate, the upper ones acting to drive radio waves outward at a low angle.

There were also 'conical-monocones', consisting of four sloping-wire radiators supported by a central mast. Four more (impedance-matching) wires met the mast lower down. These wires are often so inconspicuous on photographs as to be missed entirely, but they proved very useful in replacing sleeve antennas which obstructed essential vision. The first example was installed on the instrumentation ship *Observation Island* in 1959. Any one antenna could cover a frequency range of 5 to 1 (3 to 1 for the conical-monopole and the sleeve), but the monocone could be stacked in up to three sections. Thus an antenna on the foremast of USS *Northampton* contained upper (2–6mc/s) and lower (6–8mc/s) sections. In theory, a triple monocone could, then, cover the entire HF range from 2–30mc/s, saving consider-

BY COURTESY OF NORMAN POLMAR

A conical-monocone antenna aboard the command ship Arlington; *smaller variants of the same basic design were common aboard US missile cruisers of this period, the early 1960s. The 'cone' is formed by wires spreading from the upper part of the array, but there is also an inverted cone below it.*

able deck space.

The conical-monocone at least *looks* like a more conventional whip antenna. The log-periodic array looks like a yagi with elements of rapidly tapering length; such arrays are prominent, for example, aboard the *Blue Ridge* class amphibious command ships. The name derives from the ratio of element length and separation to the next element, which is held constant, so that the elements and their separations grow in geometric progression. By way of contrast, the elements of a yagi are all more or less the same length.

A more fundamental contrast is that *all* the dipoles of the array are active. However, they are driven alternately exactly out of phase by lines extending along the array from its apex. Dipoles near the apex are very nearly exactly out of phase, and their effects cancel out. However, further down the array, delays in the transmission line cancel out the phase difference between adjacent dipoles, and the dipole fields add. Just where in the array this 'active region' occurs depends upon the frequency. The remaining dipoles act as reflectors and directors, as in a yagi. The broad-band effect is

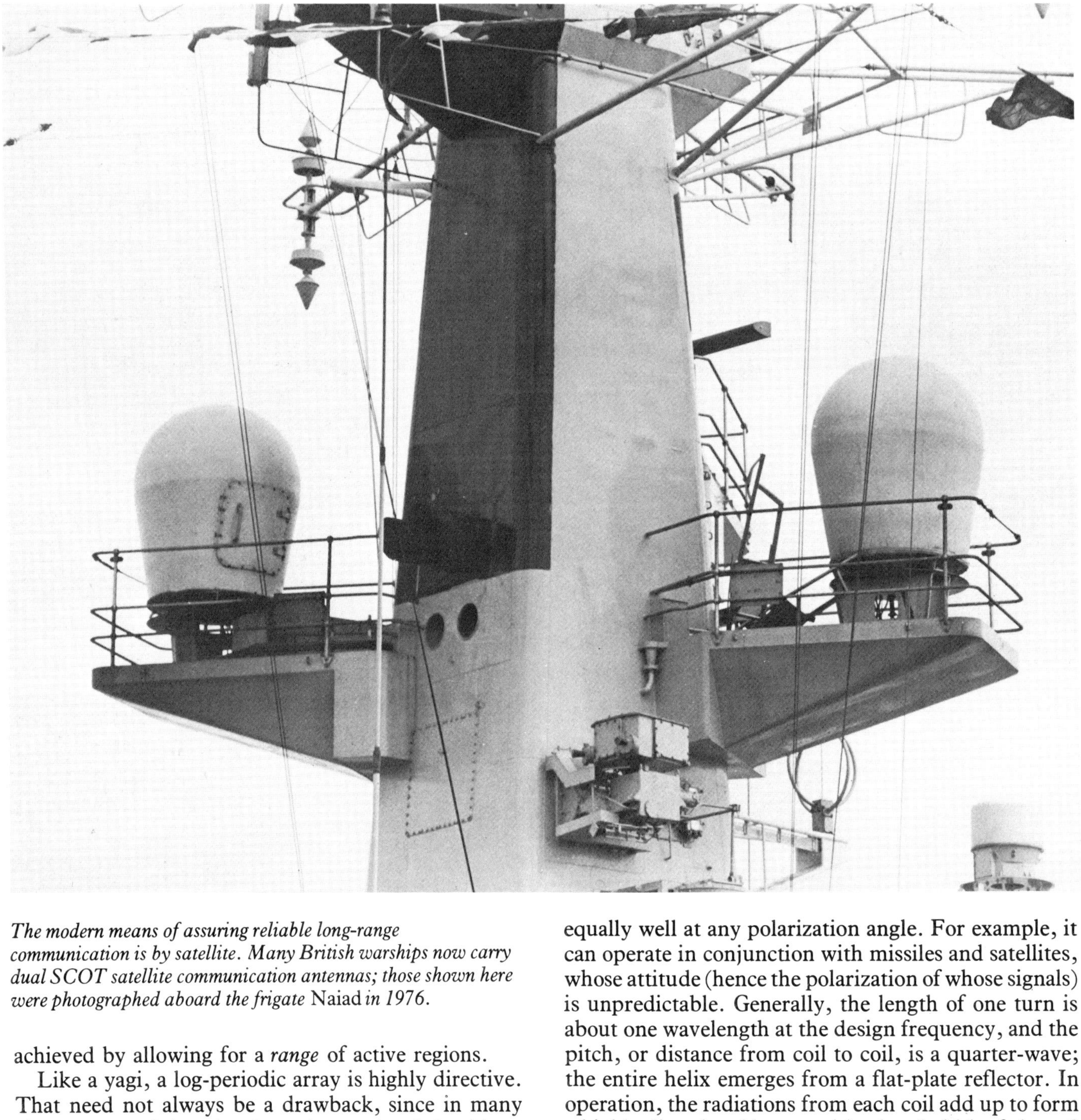

C&S TAYLOR

The modern means of assuring reliable long-range communication is by satellite. Many British warships now carry dual SCOT satellite communication antennas; those shown here were photographed aboard the frigate Naiad *in 1976.*

achieved by allowing for a *range* of active regions.

Like a yagi, a log-periodic array is highly directive. That need not always be a drawback, since in many circumstances it is quite clear where a signal should go, *eg* to San Francisco from Vietnam. In such a case it would be valuable to be able to project HF signals at a low grazing angle (for maximum range) and in a fairly concentrated directed beam.

The *helix* is another directional antenna, effective over the 100–300mc/s band; for a time in the 1960s large helical antennas were mounted on some carriers. They were considered effective for communications, radio telemetry and countermeasures, the advantage of the helix over more common antennas (such as yagis) being its ability to receive linearly polarized signals equally well at any polarization angle. For example, it can operate in conjunction with missiles and satellites, whose attitude (hence the polarization of whose signals) is unpredictable. Generally, the length of one turn is about one wavelength at the design frequency, and the pitch, or distance from coil to coil, is a quarter-wave; the entire helix emerges from a flat-plate reflector. In operation, the radiations from each coil add up to form a highly directional beam, with a bandwidth of about 2 to 1, and in effect the helix is to VHF what the log-periodic array is to HF, although the latter has a far greater bandwidth, typically 10 or even 20 to 1.

With the rise of NTDS the need came about for a compact high-efficiency broad-band HF antenna. The result was the disk-cone type, which most NTDS ships display near their bows. This is a conical antenna with a disk atop it, one Navy text referring to it as a groud-plane (quarter-wave) antenna in which the ground plane has been folded down in the form of a cone and the vertical quarter-wave radiator replaced by a disk element. The result is a vertically polarized, omnidirectional antenna with a frequency range of about 8 to 1; the disk in effect prevents high-angle vertical radiation, and thus improves antenna performance as compared to conventional dipoles. Disks are used to terminate other types of radio antennas (to adjust effective electrical length, or to form flat-tops), so that the disk-cone is in effect one end of a long spectrum of fat dipoles with flat tops.

For more than a decade this inventory of ship-to-ship, ship-to-air, and ship-to-shore antennas has been supplemented by a variety of small upward-looking satellite links, such as the dishes of AN/SSC-2 (6ft dish, 1965, 5kW, on *Canberra* and *Midway*); AN/SSC-3 (1966, 55kW, on *Providence*, *Wright* and others); and the current numerous AN/WSC-2 (4 and 8ft dishes). AN/WSC-1, which is also on many ships, employs a small mattress reflector instead. All of the dishes operate in the X-band; however, it has proven possible to use satellites as repeaters of UHF (P-band) signals, which may explain the appearance of WSC-1 antennas.

Finally there are the special tropospheric-scatter antennas of the command ships *Wright* and *Northampton,* which operated as strategic links while both ships served as National Command Posts Afloat, and were effective over distances of 50–200nm. Presumably these short ranges account for the short life of the tropospheric links. The recent NRL history refers to 'multi-helix type, directive steerable antennas' rather than to the large dishes actually seen on these ships, perhaps because the former system was the one originally envisaged.

9. Electronic Warfare

Both radio and radar present vulnerabilities which are in effect the consequences of their advantages. Thus radio permits a commander to communicate with subordinates at a great distance, but he cannot – at least often cannot – avoid at the same time communicating inadvertently to enemies who may be listening. At the very least he may give away his position to direction-finders (D/F); at the worst an enemy intercepting his broadcasts may break his code and so discover his intentions. Even if codes remain unbroken, the mere rise and fall of traffic volume may give away operational plans. Radio messages can also be jammed, and an enemy can insert deceptive ones.

Radar presents similar opportunities. The same radar signal which seeks an enemy also alerts him to the existence and at least the direction of its transmitter – and at a far greater range, since a radar intercept receiver detects the one-way radar signal, whereas the radar set must listen for the much weaker signal reflected back from a small target. Signals can be jammed, or the radar can be deceived, either by the insertion of spurious signals or by the use of radar decoys. All of these techniques have seen, and are in, wide use. In fact countermeasures – more generally, electronic warfare – developed almost as rapidly as did electronics, so that the catalog of antiradar devices was nearly complete, in outline, as early as 1945.

Electronic weapons fall into three categories: (a) Electronic Support Measures (ESM); (b) jammers; and (c) deception devices. Electronic Support Measures are passive sensors of signals. Such 'intercept equipment' detects an enemy through his own emissions, and can provide intelligence for more active countermeasures – indeed, for the application of nonelectronic weapons such as missiles. Moreover, the existence of intercept equipment tends to discourage an enemy from free use of his radars and radios. Radio direction-finding falls into this category as does, by extension, means of decrypting enemy communications. ESM has an important intelligence function, since more active countermeasures require a detailed knowledge of enemy equipment for their success. Jammers destroy the information content of enemy signals, radio or radar, by covering them in meaningless noise. Deception devices on the other hand, produce misleading signals. They include such electronic items as blip-enhancers and deception transponders, and also radar-reflecting decoys such as chaff.

Against the arsenal of electronic countermeasures (ECM), radio and radar can be provided with a matching arsenal of counter-countermeasures, or ECCM. ECM is most effective when its user is well aware of the characteristics of the target radio or radar, and when the target operator does not know the particular electronic weapon to be employed against him. Therefore in peacetime all major powers make great efforts to collect electronic intelligence without at the same time allowing their rivals to witness the use of their own equipment in wartime modes, and particularly without allowing potential enemies to observe the use of active ECM gear. For example, many radars have alternative peacetime and wartime frequency bands.

THE GENESIS OF ELECTRONIC WARFARE

Navies became interested in radio countermeasures almost as soon as they adopted radio for command and control. Thus the US Navy experimented – unsuccessfully – with jamming during its 1903 summer maneuvers. During the Russo-Japanese War both sides practised radio warfare, intercepting and on occasion trying to jam enemy signals. Jamming was as yet very difficult, as sufficient power was hard to come by, but intercepts proved relatively simple and quite useful. This state of affairs continued a decade later: electronic intelligence became a primary source of British information on German naval movements. The Germans themselves learned to practise deception by continuing to broadcast from port after their fleet had sailed.

One of the ironies of the Battle of Jutland was that this deception succeeded even though the British intercept organization was aware of German practices and even of the fact that the German fleet was at sea. By 1916 the British were so used to the fruits of radio intelligence that Admiral Jellicoe asked 'Where is the call-sign [radio station] of the German flagship?' rather than 'Where is the German flagship?' The British radio intelligence organization knew that the second question was the proper one, but did not volunteer this information; consequently Jellicoe was not nearly so well informed as he might have been. Although he sailed on the basis of radio intelligence, Jellicoe lost some of his faith in its accuracy when he inadvertently asked the wrong question before sailing. Later he failed to intercept the German fleet the day after the battle largely because the British signals intercept organisation had decrypted a German radio message giving an erroneous position for the German High Seas Fleet – apparently a piece of quite unintentional deception.

The key roles of code-breaking and D/F were little publicized, but all navies took the lesson to heart. Most major warships of the interwar period, for example, were equipped with radio direction-finders, loops of wire which could be rotated to discover the direction from which signals came. These were generally publicly described as navigational aids, but they were expected to be most useful in tracking down emitting enemy warships. Navies also built systems of fixed land-based D/F stations, such as the US Navy system in the Pacific, as a means of tracking enemy fleets. The best proof of the significance accorded this early form of ESM is the great reliance on radio silence early in World War II. Perhaps another indication of the significance of ESM was the US adoption of a deliberately short range radio, TBS, which could be used freely only because it was unlikely to reach out to enemy ESM receivers.

The US Navy had a primitive radio direction-finder at sea aboard the collier *Lebanon* as early as 1906, although it proved unsatisfactory because, among other things, it was on a fixed mount, so that the ship had to be swung to take a bearing. However, in 1916 a rotating-loop D/F, SE-75, of the type later familiar, was adapted to naval use by the Philadelphia Navy Yard. Combined with a vertical antenna, it became Model SE-995, effective between 300 and 2300kc/s. Twenty SE-75s were fitted to battleships and cruisers in 1916, but they proved unpopular. However, SE-995s were mounted on all destroyers in 1917–18 and proved useful in locating submarines, foreshadowing the use of HF/DF 25 years later, as well as in navigation – in effecting concentrations of ships at sea. A somewhat more limited form of SE-995, designated DA (250–600kc/s) was standardized after World War I. Other postwar types included DB (45–600kc/s, 1922), DK (1930), DM (1934), DN, DP (100–1500kc/s, 1934), DAE (240–2000kc/s), DAK (250–1500kc/s), and DAP (290–550kc/s, for small craft). All were externally similar loop antennas, and they were very nearly universal. A feature not immediately obvious was a sense antenna near the direction-finder, in the form of a vertical wire, which permitted the loop to resolve its 180° directional ambiguity.

The Allies employed both radio and radar warfare in World War II. The most famous examples of the former are probably the code-breaking successes 'Magic' and 'Enigma'. Their effects, which are only slowly coming to light, were very great. For example, the Germans ran their end of the Battle of the Atlantic from a base in France, which required a great volume of radio traffic – in both directions. Many writers have suggested that such a mode of operation was foolish, even suicidal, but the motive was to achieve a concentration of U-boats – which could communicate with each other only with some difficulty – sufficient to overwhelm convoy escorts. Moreover, the individual U-boats had only the most limited ability to detect shipping around them, even when surfaced, but the U-boat command could utilize broken Allied convoy codes and reconnaissance aircraft – and the reports of every individual U-boat – in order to plan convoy battles.

HF/DF

One of the great Allied technical successes, HF/DF ('huff-duff'), exploited the German dependence upon radio communications. Unlike the longer wave D/F gear, HF/DF generally used a fixed antenna consisting of four Adcock aerials, the combination of whose outputs indicated the direction of an incoming signal. A spinning rotor detected the signal and permitted display even of very short signals on a cathode ray tube (CRT), which would remain bright even after the signal had ceased.

The Royal Navy worked on a more conventional HF/DF even before the outbreak of war, and a prototype (FH1) was installed on the destroyer *Hesperus* on 12 March 1940. Its coil was rotated manually until a signal was heard. It could not be entirely satisfactory because incoming signals might be so short that the D/F loop would not be pointed at them during its slow cycle. An improved FH2 was developed in July 1941 and installed in August, but its development was not pushed further.

Instead, the fixed-aerial type was evolved, independently in Britain and in the United States. The British series was descended from ideas that Sir Robert Watson-Watt, later the inventor of radar, had evolved in 1926 in an attempt to D/F electrical storms automatically. At the beginning of 1940 his papers were discovered in the archives of the Signal School, and the first FH3 with fixed antenna and CRT display was mounted aboard the ex-US Coast Guard cutter *Culver* in October 1941. She had previously mounted the prototype FH2, which had suffered from the short duration of U-boat transmissions, the Germans being well aware of the vulnerability inherent in their system. Later there was the more sensitive FH4. Many British and some US escorts carried FH4, FH3, or its US equivalent, DAR (modified FH3), at the masthead in place of air search radar – a clear indication of its value. Indeed, Jurgen Rohwer has suggested that HF/DF was considerably *more* valuable than surface search radar in ASW, because the U-boat was so small a radar target (and so close to the surface of the sea) that the effective range of radar against a surfaced U-boat might actually be far less than the range at which the U-boat's lookouts could see an escort ship coming. Shipborne radar was far more useful for night and low-visibility station-keeping and maneuvering. On the other hand, HF/DF allowed escorts to run down a U-boat transmitting the signals on which other members of a wolf-pack could home.

Meanwhile, the US Navy developed its own fixed-aerial/CRT HF/DF system, based on work begun by

HF/DF was so important in the Battle of the Atlantic that many US escorts were completed with it in place of their air search radars. This is Francis M Robinson; *she has an SL surface search radar. Later most such escorts received a short mainmast for their HF/DFs.*

the French Navy in 1938. By May 1940 two French scientists, Deloraine and Busignies, had developed experimental models, which they brought to the United States in December 1940. By 1 April 1941 they had completed a prototype shore-based system, and later they developed a shipboard HF/DF designated DAQ. Later an improved version, DAU (1.5–22mc/s rather than 1.5–21 or 1–20 for DAR), appeared. US warships often carried their HF/DF on an auxiliary mast, *eg* a mainmast in destroyers and DEs.

These devices were effective from about 1 to 20mc/s, at which transmission is a combination of sky and ground waves. The former refract so freely as not really to be worthy of D/F; in effect HF/DF worked on the ground wave, at distances of up to 200 miles – an entirely satisfactory range for tactical D/F, in which escorts might home on a submarine over ranges of a few miles. Late in the war NRL developed a much smaller unit, which spun continuously within a radome. This DBA led to a series of postwar VHF and UHF direction-finders – AN/URD-2 (VHF) and AN/URD-4 (UHF) – which frequently surmount the masts of US warships.

MISSILE JAMMING

The other great wartime example of radio countermeasures was missile jamming. The Germans first flew antiship guided missiles in 1942, and in August 1943 they sank one corvette and damaged another in the Bay of Biscay using Hs 293 powered missiles. At the same time the Germans used a guided bomb (FX 1400) to sink the Italian battleship *Roma*. The latter weapons were the equivalent of modern 'smart bombs': they could be dropped beyond antiaircraft range and were essentially immune to fire. The only available countermeasures were electronic – which, in fact, proved so effective that the Germans had to abandon this form of attack after March 1944.

The guided bomb used a series of signals in the 48–50mc/s band, which were not particularly directional and so could be intercepted relatively easily. Jamming was another matter: the receiver on the bomb was quite directional, and it pointed up towards the bomber. At first the Allies were not certain of the character of the guidance signals. Two destroyer escorts, *Frederick C Davis* and *Herbert E Jones*, were hurriedly converted, with 15–300mc/s search receivers and NRL-built 10–35mc/s jammers, plus recording equipment. They reported to the Eighth Fleet on 15 October 1943. Both observed that guidance transmissions were actually in the 48–50mc/s band, and their crews rebuilt the jamming gear accordingly. Even after the equipment had been rebuilt, *Herbert C Jones* on one occasion waited long enough to record guidance signals before beginning to jam, so that more efficient jammers could be constructed. Two of the latter were available in time for Anzio in January 1944. In February all convoys from Oran to Bizerte had at least two ships equipped with jammers, and after that no ship in any protected convoy was hit by guided bombs. At Anzio, where the two original jammer ships protected the invasion, the enemy made 75 attacks and hit only five ships, two of which were not in the immediate area of the jammers. The Germans attempted guided bomb attacks both at Normandy and in southern France, but most missed; a postwar report recounted 'several examples . . . of glide bombs, headed directly for ships, which broke off under jamming and crashed into the sea'.

By the time the war ended in the Mediterranean there were eight jamming ('J') ships, four destroyers and four minesweepers, using six sets of gear built in the United States and two built by technicians in the Eighth Fleet. The jamming antenna was a whip projecting from bridge or funnel at a 45° angle. Ultimately the jamming and intercept/recording functions were separated, the minesweeper *Sustain* regularly accompanying convoys to record German signals during attacks. Additional ships were fitted out for jamming at Normandy, and fifty more were provided for the invasion of southern France shortly thereafter. Similarly, a possible Japanese missile threat, which failed to materialize, occasioned the conversion of three more destroyer escorts.

A 1945 account notes that 'Jamming was not easy. The receiving antenna on the bomb was so loaded as to discriminate sharply against signals from surface transmitters . . . Not only must the jammer be operated very nearly exactly on the carrier frequency of the control transmitter, but also the jammer modulation must be set to within two percent of one of the control modulating frequencies. The time of flight of the bomb was usually less than a minute; jammer operators were trained to intercept and jam within 15 seconds after the signal appeared.'

However, according to the US history of wartime countermeasures, 'Intelligence reports received after V-E Day contained claims by the Germans that Allied jamming had not been the major cause for the disappearance of the enemy's guided missile attacks. They claimed that the real reasons were (1) that Allied air superiority prevented the parent aircraft from flying close enough to the targets and (2) that lucky hits by one of the attacking strategic air forces had destroyed all the aircraft equipped with the necessary installations and these were found very difficult to substitute. It is well known that the strategic situation of the enemy air arm deteriorated continuously during 1944, and it is clear that all of the above factors combined to make the

AUTHOR

As antiship missiles proliferate, the virtues of quiet operation grow ever greater. More and more, standard tactics envisage missile targeting on the basis of intercepted enemy emissions. That in turn requires highly directional intercept systems, perhaps typified by this searchlight-like antenna aboard a British frigate, 1976.

danger of guided missiles less serious than it originally appeared. One result traceable to the use of countermeasures was that enemy scientists responsible for the development work became prejudiced against any type of radio control which was not very carefully protected against Allied jamming. The enemy development of new weapons was thereby slowed down considerably and the equipment designed became very complicated and difficult to build. By the end of WW II, no newly developed guided missile had seen operational use against Allied ships.'

Radio and radar deception was also extremely important in the European landings. For example, a special force equipped with countermeasures sets and deception devices such as barrage balloons and corner reflectors simulated a large invasion force and succeeded in capturing islands off the Italian coast. Again according to the official history, 'A few boats, carrying countermeasures sets (intentionally made not completely effective) and deception devices of all descriptions, were able to simulate the existence of a force comprising battleships and other large vessels. As they approached the islands, surrender was requested from the enemy and was obtained solely on the basis of a bluff supported by the indications appearing on the enemy's radar scope. This tactic, originally used for the capture of islands off the coast of Naples, was repeated to simulate the approach of a landing force at a point north of the real beachhead at Salerno. This particular deception is credited with delaying by several days the operational employment of one of the enemy's crack divisions against our invading forces . . . A substantial effort was also carried on to protect the main invading force at Salerno. Here the purpose was to use countermeasures to hide the exact spot of landing during the night before D-Day, even though it could not hide the fact that an operation was impending . . . One evidence of effectiveness was that several batteries of German 88's were firing very wildly, and in fact shells were fired over the convoys (protected by radar jammers) and landed on the opposite shore where there were some German troops . . .' Both radio and radar countermeasures were used at Normandy, the jammers there being carried aboard landing craft, with monitoring installations aboard some of the larger ships.

When radar appeared, practitioners of shipboard ESM assumed that it, too, would be subject to intercept; many commanders early in World War II counselled radar silence on this ground. For example, HMS *Hood* approached the *Bismarck* with her radar turned off. In fact the short pulses of radar are not so very easy to intercept, but even so, Japanese aircraft were able to use the US submarine air warning radar (SD) as a homing aid – which was a prime consideration in the switch to a shorter wavelength in SV, the next air search set. A parallel piece of irony was the German's use of British bomber tail-warning radar as a homing aid for night-fighters. By the middle of the war, jamming was being practiced on a large scale, usually for the protection of streams of Allied bombers over Europe. However, the Germans also made good use of ECM, for example to protect the battlecruisers *Scharnhorst* and *Gneisenau* during their dash up the English Channel in 1942.

US SHIPBOARD ECM EQUIPMENT

Germany and Japan produced radars, both air and seaborne, and the Allies produced jammers against all of these threats. At first most effort went into airborne equipment, which was sometimes adapted to shipboard use, but by 1944 a standard series of ECM devices for use aboard ship had appeared.

The shipboard ECM system might be divided into defensive and offensive parts. The defensive portion consisted, first, of sets of more or less omnidirectional intercept receiving antennas mounted on yardarms and other projections, feeding specialized broad-band receivers and pulse analyzers by means of which the intercepted transmissions might be identified, and, second, of radar direction-finders. The offensive portion consisted of directional jammers as well as chaff-filled shells, which could create deceptive apparent targets for enemy fire-control radars. A complete ECM system also included monitor antennas, which could observe the behavior of the enemy radar during jamming in order to indicate the success of the jammer.

All ships from DEs up received intercept equipment. In October 1944 the standard allowance consisted of one RDO and one SPR-2 receiver plus a DBM radar direction-finder, and submarines received one SPR-1 and one SPR-2, to cover the ranges, respectively, of 40–3300 and 1000–9000mc/s. RDO was credited with a range of 40–1000mc/s. At this time there was no satisfactory submarine radar direction-finder. An experimental model, XCV (later designated DBU), effective over the range between 100 and 1000mc/s, was fixed to the conning tower in the form of two arrow-shaped dipoles, the first being installed aboard the submarine *Pike* in July 1944. Direction-finding required that the submarine be swung – DBU was basically a warning device. A second intercept receiver, XCY (2000–3500mc/s), consisted of a pair of antennas fixed to the periscope shears and was tested aboard the submarines *Dentuda* and *Skate*.

These systems were often used to detect enemy air and naval units. For example, intercept operators aboard the battleship *South Dakota* were able to locate Japanese radar-equipped aircraft from two to twelve minutes before the search radars could. Submarines were equipped with radar intercept gear primarily as a means of air warning. However, the submarines also learned to locate and then to hunt down enemy ships using their radars. One outstanding example was the submarine *Batfish*, which intercepted and then homed on the radar emissions of three Japanese submarines in succession, sinking them all.

Their external manifestations were a series of antennas. Unlike radar and radio antennas, intercept antennas had to be capable of receiving relatively weak signals over a wide range of frequencies; typically several antennas were provided to cover the entire range. In submarines, however, there was very little topside space for even one additional antenna, let alone several, and so these vessels were equipped with one or two short whips for their SPR-1 search receivers. Surface ships could afford more elaborate installations. The initial SPR-1 system, for example, employed a combination of the AS-56/SPR-1, a relatively thick dipole, and the AS-57/SPR-1 double cone, the latter looking like a very fat dipole or a small drum. Both were mounted in various positions, including special ECM masts on destroyers and carrier bridge structures. Their replacements were a series of cone (upper end of frequency spectrum) and stub (thick dipole, lower end)

BY COURTESY OF A D BAKER III

From 1944 onwards, many US destroyers, particularly (but not only) Fletchers, *were provided with ECM installations on a special mainmast aft. This example, aboard the* Anthony *(December 1944) carries two AS-56/SPR-1 dipoles on its yardarm, and two double-cone AS-57/SPR-1 antennas below. The 'wagon wheel' spokes are the ground planes provided for two removable stub dipoles (AS-37/SPT-4), although in the photograph it appears that longer, thinner dipoles have been installed in their place.*

antennas mounted atop splayed-out dipoles forming a ground plane, and resembling the spokes of a partial wagon wheel. These antennas are still in use at present. The cone antenna, effective at 300–3000mc/s, is mounted within a plastic cup, and the combination is frequently called the 'derby'; it is designated CAGW-66131 in the old Navy system, in which '66' referred to an antenna. Its companion, the 'sword', is designated CAGW-66132 (40–300mc/s). In standard installations these antennas are paired port and starboard, with the stub and cone pointed up at a 45° angle so as to intercept both horizontally and vertically polarized signals.

Both derby and sword appear to have been introduced with the SPR-2 receiver, but they survived to become part of most later US ESM systems; they are also in service in other navies, for example the Royal Navy. Their main defect was that they did not extent into X-band. SPR-3 was developed in 1945 to cover the 100–12,000mc/s range, using a combination of the 66131 antenna and a dual waveguide, AS-45A/APR-6. SPR-4 was redesignated BLR-1 (for submarines). Postwar countermeasures receivers were designated in an SLR or WLR series instead of the earlier SPR series.

The standard wartime radar direction-finder was DBM, a radome-enclosed spinning antenna. Generally two different antenna assemblies were used to cover

The heavy cruiser Indianapolis, *at Mare Island in July 1945 after her last refit and shortly before her loss, displays the full range of US Navy ECM devices of her day. Her foremast carries a TDY, covering both low and high bands, back to back (circled); there were several alternative reflector configurations. The 'sword' antenna on her centerline (visible just above the Mk 33 director) acted to sample enemy reactions to jamming. There was also an S-band jamming system (TDY), consisting of two deck-edge jammers in radomes (one of them visible at the break of the forecastle) and a sampling antenna on the foremast (circled, just forward of and below the Mk 34 main-battery director, with its new radar, probably Mk 8 Mod 3). Her yardarm carried a pair of 'ski-pole' IFF antennas and the antenna of her TBS tactical radio, soon to be displaced in the fleet by UHF types with shorter dipole antennas. The mainmast was devoted to ESM and active radar; it was topped by a DAQ high-frequency direction-finder (HF/DF), and a yardarm (circled) carried a pair of DBM radar direction-finders. The mainmast also shows a zenith-search radar (canvas-covered, just beneath the HF/DF).*

different frequency ranges; wartime installations consisted of one antenna for the 200–1000 or 90–1400mc/s range and another for the 1000–5000mc/s range. Each consisted of a horizontal and a vertical dipole mounted back to back, to provide for both possible polarizations; the output of the antenna showed on a CRT whose sweep was synchronized with antenna rotation. Frequency was also indicated. DBM and DBM-1 can be distinguished from later (postwar) radome-enclosed ESM devices by their relatively squat radomes. In the mid-1950s the DBMs were replaced by a series of SLRs which had been developed from about 1948 onwards.

The offensive side of wartime naval ECM began with adaptations of the heavy jamming devices developed originally to protect the bombing offensive over Europe. SPT-1 (30W, 95–210mc/s) was similar to APT-1 and often called DINA, the Direct Noise Amplifier; it was originally developed to counter the German 125mc/s *Freya* early warning radar. SPT-2 (5W,

500–700mc/s), similar to APT-2 and often called 'Carpet-I–Ia', was developed to counter the German *Wurzburg* gunlaying radar. SPT-3 (10W, 85–135mc/s) was similar to APT-3 and SPT-4 (15W, 200–550mc/s) to APQ-2; the latter was often referred to as 'Rug'. SPT-5 (20W, 475–585mc/s), often designated 'Carpet III', was similar to APQ-9; SPT-6 (10W, 350–1400mc/s), called either 'Robe' or 'Carpet-IV', was similar to APT-5; and SPT-7 (1000–2500mc/s), named 'Mat' but apparently not used, featured an APQ-21 transmitter.

These were relatively low-powered transmitters, especially when one notes that a radar might well produce a peak power of hundreds of kilowatts. However, the returning radar signal would not be nearly that strong; it is best to compare jammer power with *average* radar power. For example, averaged over its cycle, SK (200kW, sixty 5-microsecond pulses per second) put out only 60W. On the other hand, simply because it operated only intermittently, a radar power tube could produce very high peak power, whereas a continuously operating jammer tube could not do as much nearly as easily. That is why most of these early- and mid-war systems operated at the 10 or 20W level. They were generally referred to by names like 'Carpet' and 'Rug' because of the 'grass' or 'carpet' they were able to impose on an enemy radar's A-scope by virtue of their emissions.

Wartime development tended towards higher and higher powers, in part because a basic ECCM development was higher radar power for 'burn-through'. This led to TDY, the late-war standard jammer, and also to postwar very high powered systems such as SLT-1. Another ECCM technique was frequency-switching: a low-powered jammer might mask one radar frequency, but it would have to be tuned so precisely that a sudden frequency switch might destroy its effect. One wartime development, Pimpernel (SPQ-1), was designed to sense the radar frequency and then automatically tune its L-band 5W transmitter.

TDY was the first of the US Navy's high-powered jammers, and it required a series of different antennas to cover an expanding frequency range due to enemy radar developments. It was highly directional, and monitors incorporated in the antenna could be used for very accurate radar direction-finding, given the rough fix provided by DBM. The antenna was a radiating dipole backed by a screen of horizontal dipole reflectors the opening angle of which could be varied. The earliest installations used a single radiator to cover the 175–400mc/s band, but as Japanese radars developed, a second dipole (350–770mc/s) was added back-to-back with the first, TDY being able to rotate to bring the proper one to bear. Still later the lower band was enlarged to permit operation at frequencies down to 115mc/s, and four optional antennas, for installation two at a time, were developed: CAKZ-66AKJ (90–175mc/s), CAKZ-66AKL (146–275mc/s), CAKZ-66AKM (275-500mc/s) and CAKZ-AJY (500-850mc/s). A TDY installation also included a pair of CAGW-66132 sword antennas; they had to be shielded from the TDY itself, as they would be used while it operated. Typical positions were at the ends of the superstructure, one forward of the pilot house and one on the aft superstructure. TDY output at low frequencies was about 200kW.

By late in World War II, the Japanese had developed microwave radars, and the TDY* jammer was developed to match, with a series of four magnetrons to cover the range between 2700 and 3300mc/s. The antenna system consisted of a pair of radome-enclosed transmitters and a similar but lighter monitoring receiver. Two transmitters were required because the S-band jammer had to be located near the deck line so as to reduce the range of nulls (fades) in its pattern, a situation similar to that encountered by an S-band search radar. At lower frequencies, by way of comparison, a high location for the jammer increased range, and the fat lobes produced by interference with the sea surface were welcome. Each transmitter employed a reflector mounted above its horn feed to produce a 25° × 33° beam, with circular polarization to counter both vertically and horizontally polarized receivers. Jamming power was about 80W, and the receiver was generally mounted on a mast or high in a superstructure, out of the jamming beams.

At the end of World War II still higher powers were in prospect; the Navy had under development a 1kW S-band system, 'Elephant', or X-MBT. It was tested aboard the frigate *Asheville* and led to high-power jammers such as SLT-2 and -3.

INSTALLATION

Wartime installation policy developed relatively slowly. However, by the late fall of 1944 at least SPT-1 and SPT-4/APQ-2 had been fitted aboard 7 battleships, 3 cruisers, 12 carriers, 44 destroyers, 6 headquarters ships (AGC), 3 YMSs, 3 LCI(G)s, 45 LCI(L)s and 2 destroyer-minesweepers. In October–November 1944 the decision was made to use countermeasures equipment on a fleet-wide basis, combining radar and countermeasures operation. For example, a task force would have radar and ESM pickets stationed 20–40 miles ahead of the force, either on both bows or in the direction of the most portable enemy attack ('threat axis' in modern terms). Screening destroyers were to be disposed 'in such a manner that they can search for enemy radar signals, barrage- or spot-jam when ordered, or

*In these notes TDY and TDY-1 have been used interchangeably; they were very similar. TDYs modified for S-band operation were designated TDYa. Most ships carried both sets of antennas, with arrangements to switch frequencies (*eg* TDY-1 to -1a) in about an hour.

Radar-picket destroyers carried a full ECM suit, as shown in this early postwar view of Frank Knox. *The small thimble-like radome on the after yardarm is the 'derby', CAGW 66132 (300–3000mc/s); it replaced AS-57. Its complement is the 'sword', CAGW 66131 (40–300mc/s), replacing AS-56, not visible here. All four antennas were warning types, nearly omnidirectional. Effective jamming required direction-finding, provided by the DBM radar direction-finder (the dome on the lower tripod platform). The actual jamming was done by the large TDY, also on that platform. This combination survived through most of the early postwar years, and 'derby' and 'sword' remain in service in 1981, though they serve vastly different ESM/ECM systems.*

deceive the enemy by means of Gulls, Kites, or shell Window [chaff]'. Units closer to the task force center would also jam 'in order to create maximum confusion on the enemy's radar presentation and lend additional protection to the interior units if the enemy should penetrate the screen'. All ships with ESM equipment were assigned intercept and jamming duties in one or more standard ESM bands: 'Able' (40–105mc/s), 'Baker' (75–300mc/s), 'Charlie' (300–1000mc/s) and 'Dog' (1000–3400mc/s). Shell-loaded chaff was an emergency measure in the event that jammers were not available, permitting some considerable level of force screening.

This kind of tactical doctrine was supported by a fleet installation policy announced on 31 January 1945: all headquarters ships, cruisers, carriers, battleships and *Fletcher*, *Sumner* and *Gearing* class destroyers were to have TDY, an S-band jammer and suitable supplies of window-loaded shells covering all known enemy fire control frequencies for all ships of destroyer and larger size.

These systems were highly valued, and they proved very useful in the latter part of the war, against both sea- and airborne radars. For example, on 19 October 1944 several Japanese radar-equipped torpedo planes operated simultaneously against the Fast Carrier Task Force for the first time. According to the official ECM history,

'During the early morning hours, Japanese spotter aircraft watched fleet movements with the aid of radar. The USS *Leutze*, DD 481, intercepted and jammed an Air Mark VI signal, and the Japanese reaction was similar to the first instance off Formosa. That night, as a destroyer task group approached the shores of Leyte Gulf, Japanese Bettys and Vals, serving as night torpedo planes, approached for a radar-assisted attack. All available AN/SPT-4's in the force were tuned to the frequency range 152 to 157mc, and jamming was commenced. In the face of this barrage no torpedo attack was made. The enemy aircraft approached no closer than 10 miles and one by one turned off their radars, orbited, and finally returned to base. In the succeeding days, similar attacks, either at night or under conditions of low visibility, were made with the aid of the Air Mark VI. The Japanese, at first thoroughly confused by the surprise of jamming, began to experiment with anti-jamming techniques. On October 21 the USS *Robinson*, while jamming an Air Mark VI, noted that the enemy signal when jammed would turn off and reappear slightly later on a frequency some 2½mc higher. The operator kept pace with the shifting of the enemy radar frequency from 152 to 160mc in four successive steps. Tiring of this fruitless battle of wits, the Japanese attacker went home discouraged. Other attacks were noted in which the enemy operator, in the face of jamming, would turn off his equipment, wait for the jamming to cease, and then turn it on quickly for a free look. Even as the enemy in this particular locality was becoming educated to anti-jam measures, US operators were simultaneously becoming more skilled in their offensive ECM operations.'

One of the surprises of this period was that the Japanese did not try to home on jamming signals; however, such homing is a distinctive feature of modern tactics, extending to the use of special antiradar homing missiles – which were, moreover, foreshadowed by some experimental World War II weapons.

During the final Pacific landings, it became standard practice to employ ten or more LCI(G)s equipped with pre-tuned SPT-4 jammers as cover for the transport area; they could not accommodate elaborate ECM staffs, but their jammers could counter the Japanese airborne radars. More elaborate equipment, including TDY, was carried aboard the amphibious headquarters ships (AGC). At Okinawa, protection was particularly elaborate: intercept was arranged for the entire band from 40 to 3500mc/s, although only the airborne bands were to be jammed, with emphasis on the 150-160mc/s Air Mk VI radar. There were six AGCs and thirty LCI(G)-RCMs with 'Rug' and DINA jammers, as well as fleet units. In fact most enemy air attacks were made without the use of airborne radars, and the official history speculates that the Japanese may have avoided the use of their radars because of earlier failures.

POSTWAR DEVELOPMENT

At the end of the war, then, the position of ECM within the US Navy was well established, both for offensive and for defensive purposes. Moreover, ESM was well advanced as a means of securing intelligence concerning enemy force dispositions, a practice symbolized, unfortunately, by the experiences of the *Pueblo* and the *Liberty*.

The first US postwar ECM developments were improved submarine self-protection devices. The emergence of the snorkel suggested that future air–submarine engagements would involve 'Cadillac' or its equivalent against snorkelers. In 1948 NRL developed BPA-1, a new radar direction-finder for submarines which could cover the spectrum from 2300 to 10,600mc/s. Two large horns combined to cover the 2300–5000mc/s band, and another, back-to-back with them, was tuned over 5000–10,600mc/s. Presumably the far shorter wavelength of the latter made a single (smaller) horn sufficiently effective. The entire system rotated at up to 700rpm inside a small radome, which the submarine could project above the surface. This system was demonstrated aboard the submarine *Irex* in 1949, and saw wide service. Later it was combined with other equipment (panoramic receivers and analysis gear) to form BLR-1, the first postwar submarine ESM system.

Surface ship systems, too, were inadequate: worst of all, they could not intercept X-band signals, such as those of submarine search sets. In the absence of new intercept gear, ASW experimental units were reduced to using the Mk 25 fire control radar as an intercept receiver, and some ships were fitted with aircraft sets as well. The first specialized postwar intercept set, SLR-2, appeared only in 1952. It was essentially a surface ship parallel to BLR-1 with a new antenna. SLR-2 occupies many of the radomes usually identified as 'ECM gear' on photographs of US warships.

SLR-2 differed from previous systems in that it incorporated the signal analysis gear previously mounted separately. In its final form it was designed to receive signals between 90 and 10,700mc/s, using two or more enclosed antennas; this frequency range explains why US warships generally required several ECM radomes of different sizes. The antenna was a new type (Alford horn) in which the horn proper remained fixed in a vertical position while a reflector angled to feed it spun at up to 300rpm, so high a speed being required in view of the short duration of the radar transmissions. This arrangement avoided transmission losses in rotary joints. Compactness was achieved by mounting horns for different frequency ranges inside each other (coaxially). For example, one antenna sufficed for 2300–10,700mc/s, using two horns (2300 – 5100mc/s and 5000–10,750mc/s). In 1952 it was hoped that the very high gain of the Alford horn antenna would compensate for its narrow coverage in azimuth, so that the SLR might be used as a search antenna, replacing not merely the DBM but also the omnidirectional intercept receivers. This has not proven possible, however.

Externally, SLR-2 and its successors were generally indicated by a pair of radome-enclosed antennas, AS-570/SLR and -571/SLR, to cover a pair of frequency ranges. AS-571 was the lower-frequency of the two, consisting of rotating reflectors with dipoles (both polarizations) mounted back-to-back. In the late 1950s, many ships also mounted a small mattress antenna, AS-616/SLR, for lower frequencies; it had dipoles and horns to receive both polarizations. In the early 1960s it was replaced by AS-616A, which was radome-enclosed. Finally there was the radome-enclosed AS-899/SLR, which was designed to receive circular polarized signals. More recent antennas include other, more exotic

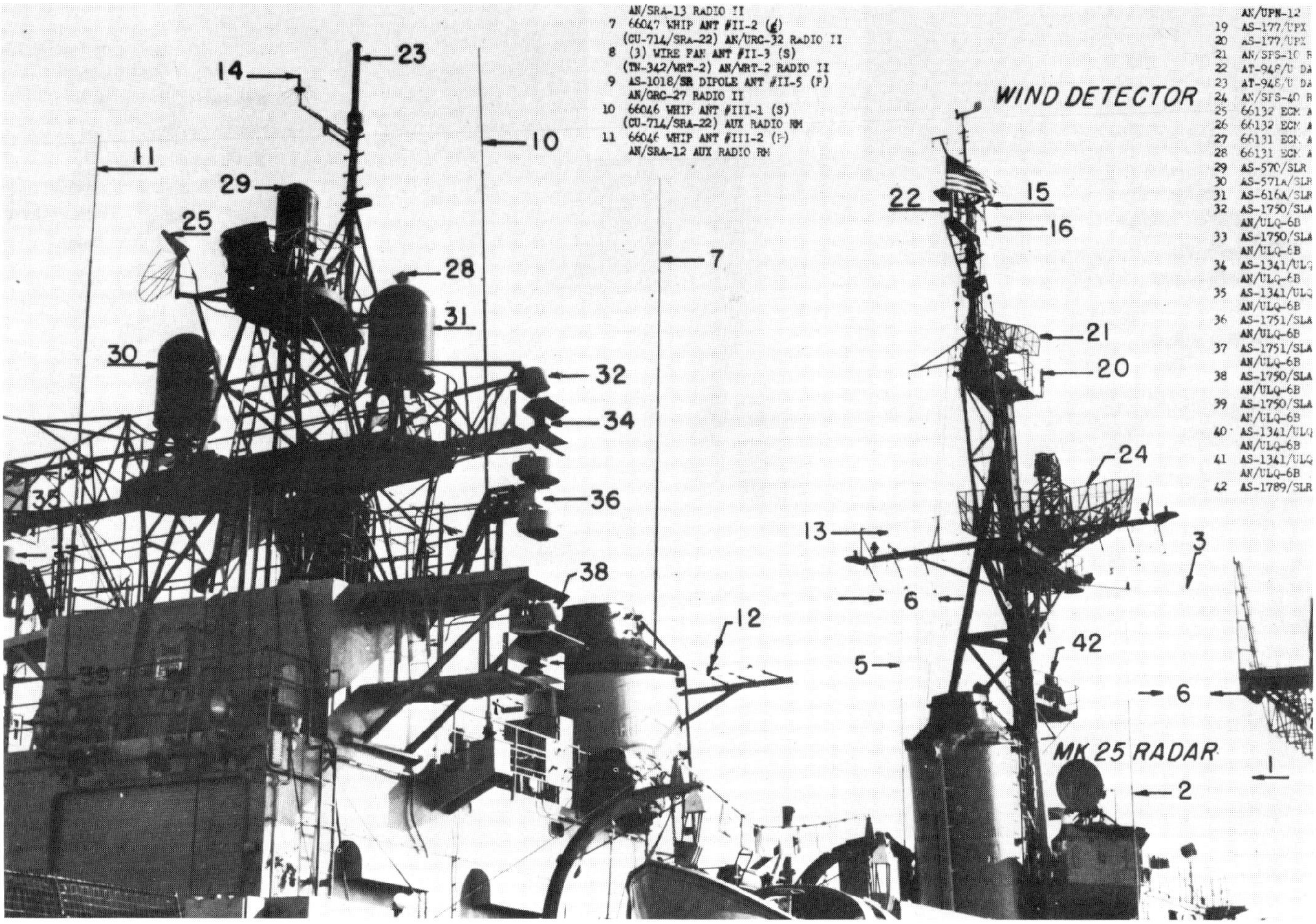

Until Vietnam US ECM systems at sea were intended primarily for warning; then most ships received active ECM as well, to lure away enemy antiship missiles. In FRAM destroyers both warning and active systems were generally mounted together atop the DASH helicopter hangar, as shown here aboard USS Hamner, *December 1965. The pure warning function is represented by the usual 'sword' antenna* (**25**) *and by a 'derby'* (**26**), *each with a similar, but less visible, antenna on the other side of the mast. Radar direction-finding was done by the antennas in the four radomes (three aft, one forward): AS-570* (**29**), *AS-571A* (**30**), *AS-616A* (**31**), *and AS-1789/SLR-12* (**42**). *This ship had the ULQ-6B deceptive ECM system, which required paired receiving and transmitting antennas. Only those to starboard are clearly visible here: AS-1750/SLA-12* (**32**) *and AS-1341/ULQ-5AX* (**34**) *for reception, AS-1751/SLA-12* (**36**) *and AS-1750/SLA-12* (**38**) *for transmission. Most of the other antennas are radio receivers, both broad- and narrow-band, plus SPS-10 and -40 search radars forward, and a Mk 25 fire control set. The ECM mast is topped by a DASH control antenna* (**23**, *AT-948/U;* **22**, *forward, is similar) and an AS-390/SRC* (**14**) *radio antenna with its characteristic spray of ground-plane vanes.* **6** *is a twin-fan wire rope antenna, as is* **12**.

types, which cannot be discussed in detail.

Operational tests of BLR-1 give some indication of postwar concepts for the use of intercept gear. In 1950 the submarine *Irex* was provided with BLR-1 incorporating a directional horn antenna. While snorkeling she was able to detect a searching aircraft using the APS-20 'Cadillac' radar, the most effective existing snorkel/periscope detection set, while the aircraft was flying at 1000ft and well beyond the radar horizon. However, this spectacular result required that the intercept receiver be set precisely to the APS-20 frequency. In a more typical frequency-scanning mode, the radar was not detected until the airplane was 2–10 miles within the radar horizon; even so, the submarine would have enough warning to spend 4–6 minutes analyzing the intercepted signal. On average, *Irex* could detect APS-20 (at 1000ft) at 45 miles, whereas the latter could not, at best, detect a snorkel outside 28 miles. The submarine's advantage increased with the sea state.

Similar results were achieved against surface ship radars: when turned to the proper frequency, *Irex* could detect SG-1B at an average of 25.1 miles, and untuned at 23.6. Submerged and snorkeling, but properly tuned, *Irex* could detect SG at 21.7 miles and SPS-6B at 34.7

The natural conclusion was that submarines faced with radar-equipped adversaries would do well to maintain radar silence, and to use intercept gear instead. Indeed, modern US submarines have no air search radars, only attack (surface search) radars for use in aiming their torpedoes.

Parallel projects provided for jammer development. The new kilowatt jammer, SLT-1, was designed to cover the 90–270 and 2460–3600mc/s ranges. Development began in 1947 but it encountered problems. At about the same time General Electric developed two other kilowatt jammers: SLT-2, to cover the range 15–1000mc/s, and SLT-3, to cover 950–10,300mc/s – using a total of eleven separate magnetrons. Few if any of these devices ever went to sea. They were extremely expensive and quite complex. In 1952 eleven SLT-1s were ordered, for installation in the new command ships (*Northampton* and the command cruiser conversion of *Hawaii* then envisaged), the battleships and some cruisers. This limited procurement was justified as a means of gaining experience. Presumably wartime equipment would be more specialized and hence far lighter. SLT-1 was described as too large for destroyers, and indeed it may never have seen service – US cruisers and capital ships of the period do not show a massive antenna not associated with any particular radar.

The few jammers which were fitted can often be recognized by their solitary positions on masts (intercept systems consist of clumps of different-sized antennas). For example, a late booklet of plans for the cruiser *Helena* shows a cluster of four SLR radomes on a bracket before her second funnel, whereas SLQ-7, the active ECM device, occupies a solitary radome bracketed on the mainmast. Here 'Q' denotes multipurpose, and may mean that SLQ-7 would analyze the signal approaching it and then react to function as a deception transponder. Indeed, the modern trend is away from pure jamming and towards methods of deceiving the approaching radar.

During the 1950s NRL developed a rapid-scan microwave intercept receiver, sweeping across frequency bands at about 1000mc/s each second; its 90-second scan through nine bands (50–10,750mc/s) was considered quite impressive in 1956, and it replaced SLR-2 as WLR-1. New intercept and D/F equipment was developed for the X- and K-bands, the latter coming into use for missile-homing and for some airborne applications. There was also a new ECM warning receiver, WLR-3, and in the late 1950s there was a new ULQ jammer series alongside the SLQs. Unfortunately the mechanical scanning of the WLR-1 has proven slow in the face of missiles; moreover, its spectrum does not include the J-band popular with recent antiship weapons. An updated WLR-8 provides digital (electronic) scan and a new upper limit of 18,000mc/s, and recent systems do cover the J-band. In many cases they are linked to deception repeaters in the ULQ or SLQ series.

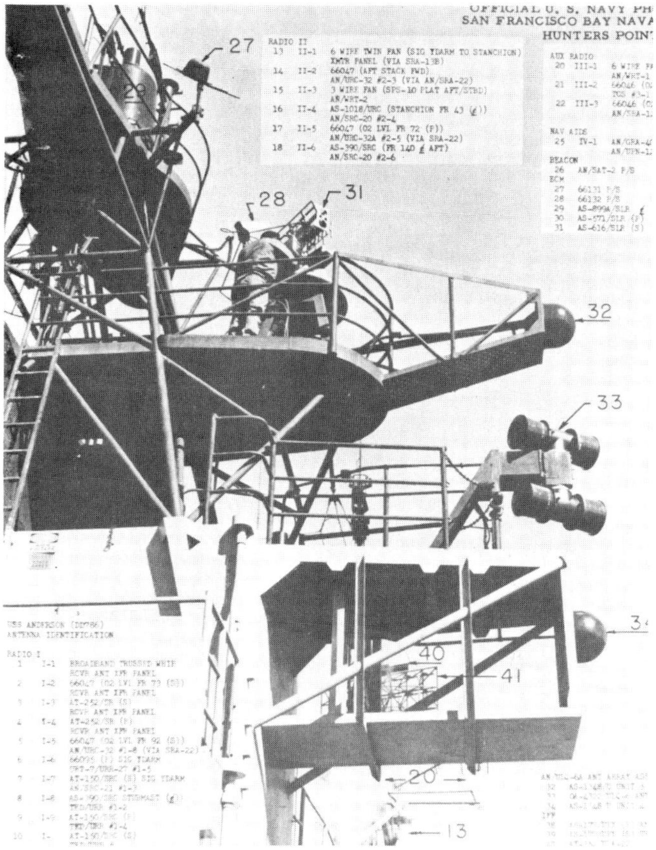

The ECM mast of the FRAM destroyer Anderson *(October 1969) shows one of her radar direction-finders* (**31**; *AS-616/SLR) uncovered, as was common in the 1950s. Also visible is an array associated with the ULQ-6 deceptive ECM system: AS-1348/U transmitter* (**32**) *and receiver* (**34**), *and OA-4322/ULQ-6* (**33**). *Antenna details varied from ship to ship, but all – including capital ships – showed the characteristic pair of small upper and lower radomes and the large trainable antenna between them. The usual fixed radar warning receivers were also present, represented here by the 'derby' of 66131* (**31**) *and the 'sword' of 66132* (**32**); *also visible is the closed radome of AS-571/SLR* (**29**).

These changes are symptomatic of a shift in the concept of ESM operations. With the advent of antiship missiles, ESM gear becomes a major sensor of the approach of enemy weapons. After a long period of development, the US Navy selected a standardized series of ESM systems in 1977: SLQ-32. There are three versions, the simplest providing minimum threat warning, and the more elaborate versions including infrared detectors and deception jammers. SLQ-32 uses fixed dielectric-lens antennas for rapid scanning and the engagement of multiple threats, the system being linked to chaff rocket launchers as well as to electronic jammers. Nearly all combatant ships, with the exception of carriers, are to receive one version or another: the first 284 are being divided among 113 low-option types for amphibious ships and frigates, 113 medium systems for missile destroyers, *Spruances* and

This is the display of the SLQ-32 EW system. In the polar display shown, the area between the two outermost circles is reserved for the display of hostile and unknown platforms, excluding missiles. The middle circle is reserved for missiles and missile-like emitters. The inner circle shows all friendlies and the own-ship heading dot. There are alternative true and relative bearing modes, with dots around the display showing emitter bearings. Special symbols, similar to those used in NTDS, are used to label the emitters, and the middle left part of the display can show bearing and engagement data for up to sixteen engaged emitters or groups; in the passive V(1) and V(2) versions this is a Threat Summary with no ECM column. The lower left portion can be used to display data about a particular emitter, labeled via the operator keyboard. There is an alternative display mode in which the entire screen displays tabular data, and into which the operator can enter data from his keyboard. In the panels below, the buttons above the main keyboard allow operator control of ECM, which can operate in either an automatic (computer-controlled) or semiautomatic mode. The operator can, for example, designate an incoming threat (generally a missile) as highest priority. Also in this panel are chaff launcher controls for up to sib SRBOC launchers. Not evident here, but essential to the success ofthe system, is a library of electronic signatures which to its computer can refer to identify particular emitters automatically. All versions of SLQ-32 can detect Band 3 emitters (missiles and their associated platforms); V(2) adds Bands 1 and 2 (navigational radars, IFF transponders, etc); V(3) adds an Active ECM (AECM) system for jamming and deception of missile guidance radars. It also has a Quick Reaction (QR) mode against pre-launch target acquisition 'pop-up' radars. V(1) is scheduled for small auxiliaries and amphibious ships, V(2) for Spruances, *DDGs, FF/FFGs and the 270ft Coast Guard cutters, and V(3) for cruisers, large auxiliaries (AOE, AOR) and major amphibious ships (LHA, LPD, LCC).*

missile frigates, and 64 high-option systems for missile cruisers and the DDG-37 class ex-missile frigates (or large missile destroyers). The carriers will receive instead the more powerful SLQ-29, consisting of WLR-8 combined with the SLQ-17 deception repeater.

These are only examples of recent practice. Modern ECM devices, particularly active devices, are rarely labeled in any way and their details are generally highly classified. Hence the discussion which follows is a matter of basic principles rather than of specific systems.

CURRENT ECM DEVICES

The threat against which most current US shipborne ECM is directed consists of Soviet radar-homing antiship missiles. These weapons are not really susceptible to noise jamming, if only because they can easily be made to home on the jammer. This threat first became important in the 1960s, and was dramatized by the success of Soviet-built 'Styx' missiles against the Israeli destroyer *Eilat*: on 21 October 1967 she became the first major casualty of surface-to-surface antiship missiles. Similar weapons were used in the Indo-Pakistani

RAYTHEON CO

SLR-32 fixed dielectric lens antenna. This systems series is scheduled to be fitted to almost all of the US Navy's surface warships, other than carriers.

War of 1971 and again in the Middle East War ('October War') of 1973.

In the face of such weapons ESM gear becomes an important source of warning information – *if* it can react quickly enough. Jamming is not nearly so useful as is deception of the missile's radar homing device; deception includes decoying as well as the production of deceptive signals. US warships off Vietnam were provided with both types of gear: radar warning systems such as WLR-1; Zuni rockets modified to scatter chaff in the path of oncoming weapons; and deception repeaters such as ULQ-6. More recently, helicopter-borne (LAMPS) ESM devices have become important. The LAMPS helicopter is also important as a means of moving a search radar away from a ship, so that she can operate in radar silence.

The missile radar is essentially a *tracker* once it has acquired its target, providing continuous corrections by tracking the radar image of the target. It is open to deception. For example, a tracker which uses some form of lobe-switching, such as conical scanning, can be deceived by signals returned out of sequence, which suggest spurious target motion. The missile tries to improve its S/N by range-gating, by looking only at returns from objects near the range it 'expects'. The 'gates' on this range are permitted to change as the target moves relative to the missile. A specialized transponder can provide return signals both much stronger than the true echoes and enough out of time to make the missile alter its estimate of relative target motion, so that it will move its 'gates' incorrectly ('gate-stealing').

These examples give some idea of the complexity of the problem. Both require for their success deception devices incorporating details of the design of the enemy missile radar: frequency, PRF, pulse shape and the character of the signal processing. The countermeasures device must be pre-programmed, using intelligence data or perhaps using the results of previous encounters. The very depth of correlation required between deceiver and deceived provides openings for counter-countermeasures. For example, the missile radar can be 'frequency-agile', skipping in frequency from pulse to pulse, using frequency as a label to distinguish real from spurious pulses. Frequency agility becomes easier, of course, at higher frequencies – which missiles must use in any case because of the necessarily limited dimensions of their radars. An analogous counter-countermeasure is the variation in the interval *between* pulses, so that the systematic return of deception pulses stands out. Once more, the deception repeater can be confused.

It becomes attractive, therefore, for the erstwhile deceiver to try to produce a decoy, a realistic radar target which the missile can be allowed to track and to home against. The earliest such device was chaff – strips of aluminum foil cut roughly to match half-wavelength to give maximum reflectance. During the Allied bombing offensive of World War II bombers crossing Europe dropped it to confuse German radars. During the Vietnam war many American warships were fitted with Zuni rockets intended to deploy clouds of chaff as a decoy against possible North Vietnamese firings of 'Styx' antiship missiles. CHAFFROC and RBOC (Rapid-Blooming Offboard Chaff) are more recent descendants. In each case a major question is just when the chaff is to be fired, but an important virtue of ESM is that it can provide warning of the approach of a missile through the missile's radar emissions – which can be 'heard' well before the missile can pick up discernible echoes.

Chaff is not the perfect decoy, however. It drifts slowly compared to an airplane, so that an MTI radar may be able to distinguish it from realistic targets. On the other hand, ships do not move very fast, so that chaff is effective at sea. A great deal has been done with more sophisticated devices, and corner reflectors have often been used to create bright radar returns suggestive of large targets.

The missile designer has available a more basic counter-countermeasure. He can try to match his homing sensor so completely to the target that no realistic decoy is possible. For example, he can use a combination of sensors, a popular choice being radar plus infrared. It is also possible, of course, to home on specific characteristic radar or other electromagnetic emissions; many radars produce some radiation even when they are in a standby mode. A somewhat more distant possibility is an optional or mixed optical-radar sensor, which holds within it an image of the target ship. However, there are already 'electro-optical' (EO) missiles which home on an image, once they have been locked-on by a human operator.

To every one of these measures there is of course at least one countermeasure. Details are often highly

GIORGIO ARRA

Sometimes the simplest countermeasures are the most effective. Chaff ('window') is nothing more than strips of aluminum foil, which produce very strong radar echoes. When combined with a deceptive ECM device, they can lure away an antiship missile. During the Vietnam War, the US Navy adapted its Zuni air-launched rocket for chaff delivery; this installation, somewhat more sophisticated than the original use of the air launcher, was photographed aboard the missile cruiser Sterett *in 1979. Many ships now have the specially designed RBOC (Rapid-Blooming Offboard Chaff) rocket system or its successor, SRBOC (Super RBOC). The counter to such deception is a missile which senses both infrared and radar signals, and in consequence a combined radar/IR decoy has been developed.*

classified, but it is not hard to see that infrared signatures can be reduced by judicious design, or confused by flare decoys, or that optics often cannot penetrate smoke. The game of antiship missile *vs* ECM is a fluid one, impossible to characterize in any comprehensive way.

COUNTER-COUNTERMEASURES

The other size of the problem is protection of one's own systems against enemy ECM. For many years, as we have noted, one of the primary Soviet threats against US carrier task forces has been mass attacks by air-to-surface missiles dropped at long range by 'Badger' bombers of the Soviet Naval Air Force. The US counter consists of air- and ship-launched missiles guided by the radars of the surface ships and of aircraft such as the E-2; and the Soviets have shown themselves well aware of the potential of ECM for disrupting our defenses. Hence the growth of Electronic Counter-Countermeasures, or ECCM, some techniques for which we have already referred to.

The most basic form of ECM is noise. If the radar power is increased sufficiently, the radar system recovers enough S/N to see through the level of noise a practical jammer can produce. In effect the radar 'burns through' the electronic haze. There are of course subtler ECM ploys and hence subtler forms of ECCM: things like MTI and its cousin pulse-doppler radar to segregate fast and slow targets; combinations of antennas of widely different characteristics to make the deceiver's job harder; and, behind it all, the radar operator, visually characterizing different returns.

An important consideration in ECM is that an enemy's techniques must often be tailored closely to the radars he wishes to beat. In particular, space aboard aircraft is limited, so that airborne jamming equipment must be carefully chosen. It follows that the presence of a *variety* of different radars, preferably operating at very different frequencies, will enhance the overall ECM resistance of a fleet. Such an argument is sometimes made to explain the value of the wide variety of NATO radars. It also explains deliberate Navy policies aimed at wide frequency diversity. For example, about 1952 policy called for 25 per cent P-band air search radars, which is why there were SRs in service as late as 1960.

A somewhat subtler example of ECCM predicated on the mismatch between jamming signal and radar pulse is the use of a Fast Time Constant (FTC), a means of discriminating against pulses *longer* than the radar pulse. FTC, which dates from World War II, militates against pulse deception devices and can also counter chaff by discriminating against echoes extending over too long a range. Other late-war ECCM electronics included an expanded sweep, which used the maximum resolution of the radar to distinguish between the chaff and target returns, and Instantaneous Automatic Volume Control (IAVC), which varied receiver gains on the basis of each returning echo. This change was delayed one pulse length, so that isolated echoes were unaffected, whereas continuous ones would be reduced. Both IAVC and FTC were also expected to reduce sea clutter. More modern radars use gain-reduction when they operate in a 'burn-through' mode. This latter technique is related to a very common modern ECCM measure, Constant False Alarm Rate, or CFAR. In a radar which decides whether a returning signal is real according to its energy, a jamming signal will be observed as an increase in the alarm rate; CFAR adjusts the detection criterion to keep the observed false alarm rate constant. In effect such a measure reduces the probability of detection of a real target – it merely permits the radar to keep operating under conditions of jamming, and is a useful adjunct to burn-through.

More generally, a radar designer can incorporate ECCM techniques in the design of his transmitter, of his antenna or of his receiver. Transmitter ECCM features seek to impose on a potential jammer or imitator a radar signal that will be easy to recognize (*ie* to receive) but difficult to duplicate. Examples of techniques include short pulses with very high power, frequency and PRF agility (jittering in the case of PRF) and wide bandwidth, so that a radar can be tuned away from a jammer. It is also possible to envisage radars designed for a low probability of interception, so that a potential jammer may be unaware of their presence. For example, pulse compression already provides the equiva lent of a very high peak power at a low effective peak power, by spreading a signal across a wider spectrum of frequencies. It is possible to imagine wider and wider spectra, and possibly even a pseudo-random rearrangement of the signals within the conventional pulse-compressed 'chirp'. Indeed, any form of coding is an ECCM feature. One might, however, note that many forms of ECCM function, too, as anticlutter mechanisms.

From an antenna point of view, the principal ECCM feature is limitation of the area of the horizon from which the antenna draws its signals. First, it is important to reduce as far as possible the side lobes through which an enemy can insert deceptive signals. A radar 'sees' a signal injected through its side lobe as coming from the direction in which it is aimed, and thus can be made to 'see' its target in quite the wrong direction. Thus side-lobe blankers or cancelers are important ECCM features. Generally, they employ simple antennas with very fat antenna patterns to sample signals coming in from directions other than those in which the radar is nominally pointed. Similarly, the narrower the radar beam, the less likely it is to admit deceptive signals when it is on target. The more frequently it observes a target, the less easily it will be decoyed away. Thus both antenna stabilization (which reduces effective or averaged antenna beamwidth) and high scan rate (for TWS operation) can be considered antenna ECCM measures.

Since the ECM operator must first measure radar characteristics and then respond very rapidly, any variation in radar operation can be an ECCM measure. For example, a variable antenna-scan rate (which is particularly easy to arrange in an electronically scanned system) can defeat some ECM measures. Polarization (*eg* circular) can be used to help distinguish the radar signal.

Receiver ECCM is potentially the most powerful tool, since it is passive and thus undetectable. For example, most of the radar data in this book is available to anyone with an effective search receiver: he can pick up PRF and pulsewidth directly, and can calculate beam angular dimensions and (from the range) peak power. He cannot, however, guess what happens to the radar signal once it has been received. Moreover, it is easier to improve radar signal processing than to alter either the antenna or the high-powered transmitter. Examples include FTC, CFAR, IAGC and MTI.

Finally there is ECCM incorporated in the overall design of a radar system. For example, monopulse is often described as an ECCM measure, since it is far more resistant to ECM than is a lobing or conical scanning tracker. Pulse-to-pulse or beam-to-beam correlation can overcome some deceptive signals. In many systems aided tracking is provided: as long as the radar tracks positively, it generates a *rate* of target motion. If the target is obscured (*eg* by jamming), the radar continues with the last available rate, hoping that the target will become observable again at its predicted position. The SPG-55, for example, has a coast mode. A system can also be designed to track a jammer, and several radars can combine to triangulate a jammer. Systems of this type are much more common on land than at sea, but it seems likely that NTDS and its successors will make passive jammer tracking much more attractive for the fleet.

Another important system ECCM technique is to use nonradar signatures to correlate with radar data. Several current fire control systems employ optical (television) sensors; the Target Acquisition System (TAS) was to have employed infrared as well. The idea is quite old – early versions of the F-4 Phantom (a fleet interceptor) had IR as well as radar search systems.

This discussion can only hint at techniques employed, as ECM resistance is a highly classified matter. It does seem worth mentioning, however, that the far greater space and weight capacity of a ship permits her to deploy a much more varied ECM or ECCM suit than can be borne even by a specialized airplane.

No discussion of hardware and ECM technique can omit the most basic ECCM element of all: the radar operator, who seems often to be able to defeat quite sophisticated forms of ECM. The ultimate goal of all ECM is to confuse the radar operator, to force him to err, to fire at false targets. The ultimate ECCM is his ability to see through that cloud of deception, to get around the jamming and the false signals by his own tactics, aided by countermeasures receivers and analyzers.

A final point worth making is that ESM devices become sensors of enemy *position*. An enemy's radars show characteristics of emission that present the possibility of definite identification and hence may be used for targeting of weapons such as over-the-horizon missiles. For many years this beaconing effect of radar and radio emissions has been well known, and in fact efforts have been made to permit the fleet to operate in some kind of radio/radar silence. Airborne radars such as that in the E-2 are a step in this direction. Others are line-of-sight satellite communication, already discussed, and relay aircraft, which present similar, though more limited, scope. Both the latter are extremely important, given the need for constant high-speed communication among ships, for example for the successful employment of NTDS. Moreover, the up-link might be not only difficult to detect but also very difficult to jam, whereas normal radio communication is not inherently jam-resistant.

PART 2
A Catalog of Naval Radar

The United States

DESIGNATIONS

The history of US naval radars is somewhat confusing because of the variety of *types* of designations employed. These reflect the organizations behind development. During World War II, the US Navy deployed two major radar series: search sets (BuShips) and fire control systems (BuOrd). The former carried over a system consisting of a *type* letter, *eg* S (search radar), and a model letter: thus SC is the *third* US search radar; and SC-5 is the *fifth* modification of SC. The Bureau of Ordnance preferred to use mark numbers, *eg* Mk 25; modifications within a mark were also numbered, *eg* Mk 25 Mod 1. However, the earliest fire control radars were also designated under the BuShips system, with the type letter F; thus FH is Mk 8. Such designations were dropped early in the war, but persist on plans drawn as late as 1945.

Three letters were generally used to extend series beyond twenty-five (the letter I was not used); thus RAA, RAB and RAC would be the twenty-sixth, -seventh, and -eighth radio receivers. There were only twenty-four search radars, so none required three letters. Lower-case letters (*eg* 'a' in SRa) indicated field changes. The prefixes 'X' and 'CX' were reserved for the Navy (NRL) and commercially produced experimental or preproduction sets, *eg* CXAM, the first US naval radar. These series included all types of electronic equipment. The Army used an entirely different class of designations, its radars being numbered in the Signal Corps Radio (SCR) series. One Army radar, SCR-720, was fitted in limited numbers to US warships at the end of the war.

The fragmented system thus described did not long outlast World War II. Even during the war it proved difficult for Navy and Army (Air Force) to coordinate airborne radar procurement. Early Navy air radars were designated in the same manner (but *not* the same series!) as the surface sets, so that ASH was the *eighth* airborne search radar. However, in February 1943 a new universal system appeared, three letters plus a number (platform-type of equipment, and function). For example, APS-4 was the forth airborne pulsed (radar) search device. New equipment designed after World War II, even when it was specific to the Navy, fitted this pattern with the prefixes 'S' for surface ships and 'B' for submarine. Generally the multiservice designations are prefixed by the letters 'AN' for Army-Navy, as in AN/SPS-6. *These two letters will generally be omitted here.*

BuShips World War II System – Prefixes

A Aircraft, used as prefix for other type letters
B* IFF Device
C* Experimental (including radios)
D Direction-finding (including HF/DF and LORAN)
E Emergency power
F* Fire control radar (later dropped in favor of BuOrd mark numbers)
FS Frequency shift keying
G Aircraft transmitting
H Sonar hoists (abortive)
I Intercept radar (aircraft only; limited to AIA, later AN/APS-6)
J Passive sonar (submarine type)
K Sonar transmitting
L Precision calibration
M Radio transceiver
N Echo-sounding
O Measuring, operator training
P Automatic transmitting and receiving (*eg* teletype)
Q Sonar (surface ship)
R Radio receiver
S* Search radar
T* Radio transmitter (including jamming transmitters and radio receivers)
U Remote control, including automatic keyers
V* Radar display
W Submarine sonar
X Experimental
Y* Radar homing beacon
Z Airborne navigational aids (superseded by AN/ARN and AN/APN series)

Multiservice System (selected items only)

Platform Letter
A Aircraft
B Submarine
C Air-portable (obsolete)
F Fixed (land-based)
S Surface ship
U General utility (more than one platform)
W Water surface and underwater

Device Letter
L Countermeasures
P Radar ('pulsed')
Q Sonar
R Radio

*Radar prefix

Function Letter
A Auxiliary equipment (*eg* antennas and displays)
B Bombing
C Communications
D Direction-finding, surveillance
G Fire control
N Navigational aid
Q Multiple or special function
R Passive detecting
S Search
T Transmitting
W Weapon control
X IFF
Y Multifunction

The SPG series continues the old BuOrd mark numbers; thus SPG-51, for example, is in the same series as Mk 26. The other series *do not continue anything*; SK, for example, was never redesignated and has no relation whatever to SPS-11, nor is there any cross-relation between, for example, the SPQ and SPS series, nor between BPS and SPS.

In the notes which follow, antenna size is width × height (width alone in some cases). Similarly, beam dimensions are width × elevation. Power is peak, *ie* pulse, power; PRF is pulse repetition frequency, pulses per second. Gain is a measure of how directional an antenna is – how much better the antenna is than an imaginary point source would be, in the direction of maximum power. It is useful here mainly as a measure of increasing sophistication. Designations in parentheses denote prototypes.

SEARCH SETS

A. WW II Series

CXAM

The prototype of US Navy long-range air search radars, derived from the experimental NRL model (XAF). Six were installed in July and August 1940, aboard *California*, *Yorktown*, *Pensacola*, *Northampton*, *Chester* and *Chicago*. The first CXAM was removed from *California* after her sinking at Pearl Harbor and installed as an Army search set on Oahu; in the summer of 1942 it was transferred to the carrier *Hornet*.

CXAM was a large 'mattress' consisting of 15 dipoles plus a second layer of dipole reflectors. It could be elevated in hopes of obtaining height data, but the beam produced, 14° × 70°, was far too broad for this feature to be useful. The 17ft × 18ft antenna operated on P-band (1.5m wavelength) with a gain of 40 (15kW pulses or 3 microseconds, PRF 1640), and scan rate was 5rpm. Accuracy and resolution were 300 yds and 400 yds, or 3°. Typically, a battleship or cruiser could be detected at 16nm, a destroyer at 12, a PBY flying at 10,000ft at 70nm and a fighter at the same altitude at 50. Total installed weight was 5000lb, of which the antenna accounted for 1200lb. The set was manufactured by RCA, based on an NRL design.

CXAM-1

Fourteen similar sets with non-elevating antennas (screen reflector) and improved servos for antenna rotation, with accuracy improved to 200yds. Delivery began late in 1941 and some remained in service as late as 1943. Initial installations were aboard the carriers *Lexington*, *Saratoga*, *Ranger*, *Enterprise* and *Wasp*; the battleships *Texas*, *Pennsylvania*, *West Virginia*, *North Carolina* and *Washington*; the light cruiser *Cincinatti*; and the large seaplane tenders *Curtiss* and *Albemarle*.

CXAM-1 aboard USS Augusta, *with radar Mk 3 below. The diagonal lines belong to official labeling.*

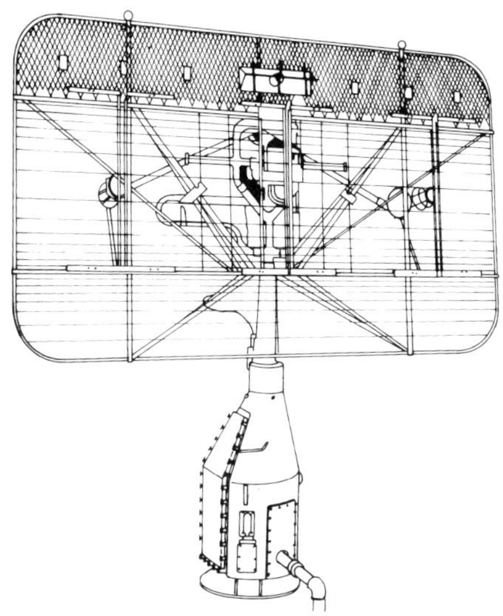

Antenna assembly for SA, SA-2 and SA-3.

SA, SA-2, SA-3 (CXBE)

One of a series of sets intended to approximate CXAM performance with a much smaller antenna, for installation in smaller ships. The antenna was 8ft 9in × 5ft, with a gain of 25; poorer directionality was reflected in a beam 30° × 52° fixed at an elevation of 5°. The set operated on P-band (1.36m), using 100kW pulses (5 microseconds, PRF 60), and scan rate was 1.25 or 5rpm. Despite increased power, range on a bomber fell to 40nm, on a fighter to 30, on a battleship to 12 and on a destroyer to 8. Accuracy was 100yds or 1° (lobe-switching) and resolution 500yds or 25°.

The antenna assembly carried both the six (3×2) horizontal dipoles of the radar proper and two sets of IFF antennas: two vertical dipoles for BL, and eight for BG. Either could be connected for operation through the same independent coaxial transmission line. A total of 400 SAs were delivered from September 1942 on, then 865 SA-2s (ending in December 1944) and 225 SA-3s. SA-2 and -3 incorporated a PPI and antijam features. The SA series was the standard air search radar for destroyer escorts and frigates and many were also fitted to destroyers. Manufactured by RCA.

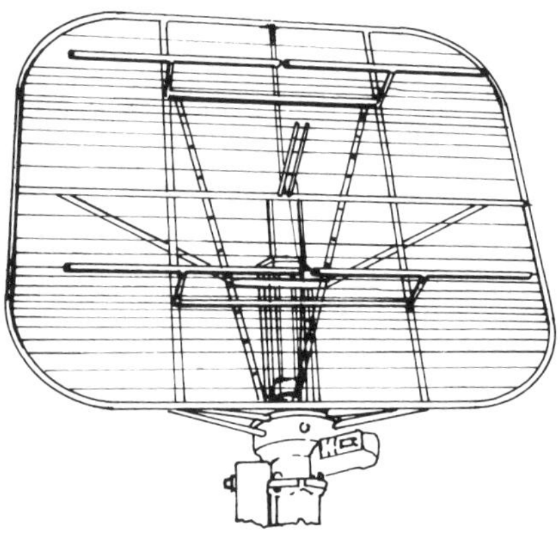

SA-1 antenna.

SA-1
A smaller SA, with a 100lb, 5ft × 6ft antenna (as compared to 500lb for SA). Wavelength rose to 1.65m, which further reduced accuracy and range, gain falling to 18, with the same peak power. Range was 25nm on a PBY, 20 on a fighter, 8 on a battleship and 5 on a destroyer. Bearing accuracy fell to 3°, using a 45° × 52° beam, the loss due to a narrower antenna. Altogether 75 were delivered, from August 1942 onwards.

SB
A portable 75cm air search set, intended to provide limited warning in an amphibious assault, until longer range and more accurate equipment could be landed. It had short range (24nm on a bomber), and gaps in coverage were filled by lobe-switching. The antenna, a converted aircraft ASA type, consisted of four yagis and was manually controlled. It used 15kW pulses (2 microseconds, PRF 1010) and produced a 26° × 33° beam. 190 were built prior to August 1943, but tests showed its inadequacy. RCA.

SC, SC-1 (XAR, CXBD)
As in the case of SA, this was an attempt to reduce the bulk of CXAM, in this instance by using a new smaller antenna, 8ft 6in × 7ft 6in, with six (3×2) dipoles. Although gain remained at 40, and power increased sharply to the 100kW level of SA, the beam, at 30° × 40°, was far more diffuse, and range fell dramatically to 30nm on a bomber, 25 on a fighter, 10 on a battleship and 3 on a destroyer.

SC air search (**1**) *and SE surface search* (**2**) *radars aboard the old destroyer* Bainbridge, *August 1943. The yardarm below carries IFF (left) and TBS tactical radio (right).*

Installation began late in 1941, and the operating forces objected strongly to the reduced capabilities of the new set. In January 1942 a new version, SC-1, appeared, with twice the power, approximately double the range, and other improvements, and all SCs were soon converted. SC used 5-microsecond pulses as compared to 2–6 in SC-1 (both had a PRF of 60) and scanned at 5rpm. SC-1 accuracy was 100yds and 5° and resolution 500yds and 10°. 400 were produced. General Electric.

SC-2, -3, -4, -5
Improved SC-1s with an antenna enlarged to 15ft × 4ft 6in with a gain of 30, using a 6×2 array of dipoles and having a beam narrowed to 20° × 50°. The integral IFF installation consisted of four vertical dipoles for BL and twelved for BG, either of which could be used. Peak power was 20kW (4-microsecond pulses, PRF 60), and minimum range was 1200yds, with a range of 80nm on a bomber at 10,000ft, 40 on a fighter at the same height and 20 on a capital ship. Altitude could be estimated to within 2000ft using a fade chart. Accuracy was 100yds (3°) and resolution 300yds (10°).

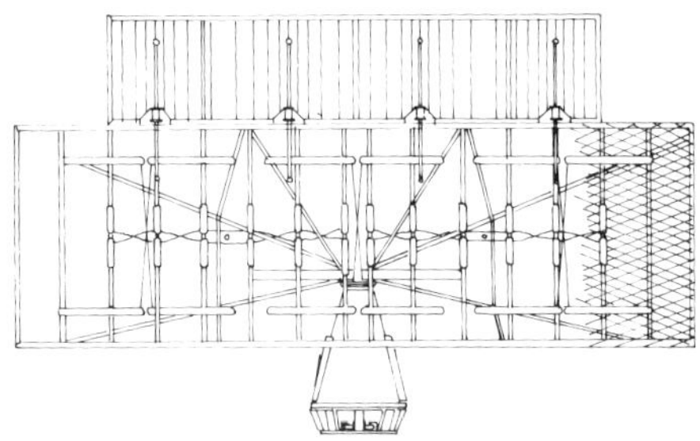

SC-4 antenna.

The SC series operated on two frequency intervals, 195–205 and 215–225 mc/s, and for each there was a slightly different antenna. A somewhat lower range of frequencies was discarded about 1942 because of interference with Mk III IFF. SC-2 incorporated a PPI, and SC-3 and -4 had minor improvements including antijam features. Production of 415 SC-2s was completed in December 1943, and was followed by 200 SC-3s and 250 SC-4s (completed December 1944). Another 100 SC-5s were produced in 1945 as a stop-gap for delays in SR production.

These were the chief destroyer radars, and as such remained in service for a long time: as late as 1963, there were still 16 SC-5s in use. General Electric.

SD, SDa, SD-1, SD-2 (XAS)
This submarine air warning radar completes the list of long-wave air search sets originally designed for US service. It was intended to be omnidirectional, on the theory that its function was to alert the submarine to the need to crash dive. At 2.65m it used the longest waves in US naval radar, but the peak power of 100kW (8.5 microseconds, PRF 60) meant little in conjunction with a poorly defined beam, and gave a range of only 20nm on a bomber at 10,000ft assuming an antenna 40ft above water. Accuracy and resolution were 1000yds and 1500yds respectively.

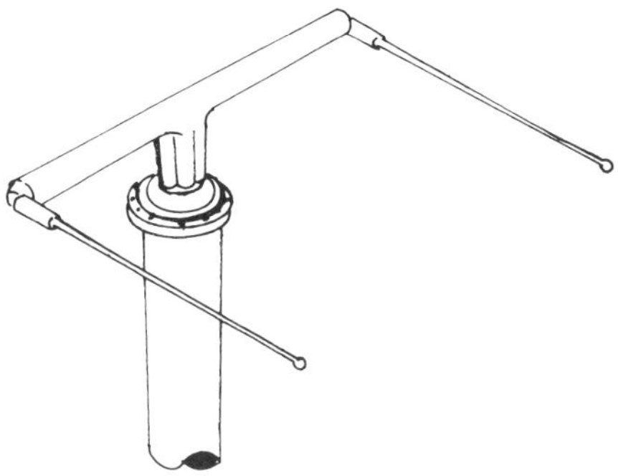

SDa U-type retractable antenna.

Sixty sets were produced between late 1941 and mid-1942. All were field-modified to SDa standard, the J-type vertically polarized antenna being replaced by a horizontally polarized U-type, and video amplification for IFF being added. These antennas were retractable. SD-1, of which 20 were built (beginning late 1942), had a fixed mast, and SD-2, of which 60 were delivered from late 1942 on, was a new production version which included provision for IFF in its indicator unit. An unusual feature of the original design was a duplexer permitting alternate operation of the submarine's radio transmitter. RCA.

SD-3

SD modified with a directional antenna featuring a 4×4 dipole array, a gain of 15 and a 48° × 55° beam, and produced for small surface craft, particularly YPs. 125 were built from the fall of 1942 on, but some were converted back into submarine sets due to the availability of better search sets for patrol craft. Assuming an antenna height of 80ft, a bomber at 10,000ft could be detected at 25nm.

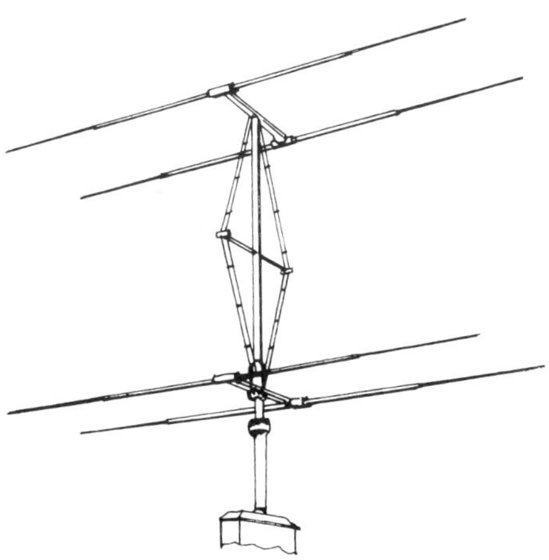

SD-3 antenna.

SD-4, -5

A further attempt to cure the problems of SD, using a linear dipole array. 104 SD-4s were delivered between July 1943 and July 1944, and production of 86 SD-5s was completed in December 1944. Power was increased to 130kW, but a range scale of 80nm seems over-optimistic. RCA.

SE

A simplified S-band surface search radar with an output of 30kW (0.25-microsecond pulses, PRF 600 and 1600), intended for small auxiliaries and at one time specified for the old destroyers. Production began in late 1942, but few were fitted, and trouble with the motor-generator, lack of a motor drive antenna training system and limited performance led to an order for its removal and replacement by SO. The 300lb antenna was a 20in segment of a 42in paraboloid (gain 600, 6° × 11° beam) housed in a horizontal cylindrical radome. Range (antenna at 100ft) was 12nm on a large ship, 8 on a destroyer and 4 on a surfaced submarine. Accuracy was 100yds plus 1 per cent of range, and 2°, and resolution was 50yds and 3°. Scan rate was ½rpm. A-scope display, with fixed range mark. Westinghouse.

SF, SF-1

Another small-ship 10cm surface search set, but far more successful. Although originally intended for such small craft as PT boats, it was so large that it had to be adopted instead for frigates and similar types, indeed for all types too small to receive SG (*eg* SCs and PCs). The first request was made by the Navy to the Radiation Laboratory on 26 December 1941, and a preproduction unit was delivered on 13 August 1942 and tested aboard the converted yacht *Gallant* (PYc-29). 1655 (600 SFs) were manufactured before production ended in October 1944.

SF shared with the other S- and X-band surface search sets the functions of navigation, station keeping, and fire control. Both A-scope and PPI were provided, and SF-1 had provision for three remote PPIs. The antenna was a 24in dipole-fed parabola with a 13° × 13° beam, rotating at 15rpm within a radome. Gain was 200 (which shows the difference between P- and S-bands) and 80kW sufficed for detection of a battleship at 12nm or a destroyer at 8. Accuracy was 75–200yds (2°), and resolution

Mast top of PC-555, *showing SF radome.*

150yds (6°), using 1-microsecond pulses and a PRF of 400. The entire system weighed only 949lb. A modified version, the 3cm SF(XI), was produced for Navy Experiments (March 1943). Six months of service aboard the test ship *Semmes* (AG-24) provided valuable experience for the development of SU. Submarine Signal Company.

SG, SG-1 SGa, SG-2 (CXGR)

The first US 10cm surface search radar, and the first Raytheon radar, tested in prototype form aboard the converted destroyer *Semmes* in June 1941. It was

SG radar antenna.

also the first microwave surface seach equipment to incorporate a PPI. The first set was installed in April 1942 aboard the cruiser *Augusta*, followed shortly by *San Juan* and *Saratoga*. SG/SG-1 became standard for destroyers and larger units, the total production of 955 (completed November 1943) including 516 SG-1s. Early problems included insufficient power and waveguide losses which limited the permissible distance from antenna to display. SGa was a field modification of SG with a higher power of 50kW, and SG-1 was the production version, manufactured from May 1943 on. SG-2S was a shore version with a much larger antenna.

The 48° × 15° parabolic waveguide-fed antenna of SG-1 projected a 5.6° × 15° beam with a gain of 550 and scanned at 4, 8 or 12rpm. With a peak power of 50kW (2-microsecond pulses, PRF 775, 800 or 825), SG could detect a 500ft high bomber at 15nm, a battleship at 22 and a destroyer at 15, and there was provision for reception of S-band beacon signals at the horizon. Accuracy was 200dys, (2°) and resolution 400yds (2–3°). Total weight was 3000lb.

SGb, SG-1b

These were SGa and SG-1 sets modified late in the war to incorporate a new PPI and an improved receiver. There were no important external changes. A postwar description of SG-1b includes a 5.3° × 15° beam, 2-microsecond pulses and a PRF variable at 800, 900 or 1000. Minimum usable range was 500yds (450yds in SG-1), while the display was marked out to 75,000yds. Scan speeds were 8, 16 and 24rpm.

SG (Zenith Search)

A zenith search reflector originally developed for SO-11 was also applied, at the end of the war, to existing SG sets. SGc was SGa with a zenith antenna, SG-1c was a modified SG-1, SGd a modified SGb, and SG-1d a modified SG-1b. The switch from surface to zenith search reflectors took about two hours. Later a specialized surface search/zenith search type, SG-6, was built. SGe referred to SG-6 with MTI.

SG-3, SG-4, SG-5 (XBT)

New versions of SG intended to have maximum commonality with the new SR series. SG-3 and -4 were to use a new band, Sw (8.6cm), and to have about 50 per cent better range than the original SGa, among the expected advantages being improved antijam performance and greater reliability. Both used a new mesh cut-paraboloid antenna with dimensions of 7ft × 1ft 6in and 3ft respectively, for antenna weights of 450 and 750lb *vs* 340 for SG; beams were to be 3° × 13° and 6.5°, and SG-4 would include provision for two-axis stabilization. Both sets were far heavier than the original SG, at 4000lb. Peak power rose to 500kW with 0.3 and 1.2-microsecond pulses and a PRF of 750, and scan rate was reduced to 2½ or 5rpm. Range on a low-flying bomber increased to 22nm, on a battleship to 30, on a destroyer to 20 and on a surfaced submarine to 12. Minimum range fell to 200yds, and accuracy and resolution improved to 100yds and 0.75°and 200yds respectively

Production of SG-3 began at Raytheon in mid-1945, and one remained in a Reserve Fleet unit as late as 1961. SG-5 was to have been a stabilized type operating on a shorter wavelength (Xb-band); it incorporated a zenith reflector.

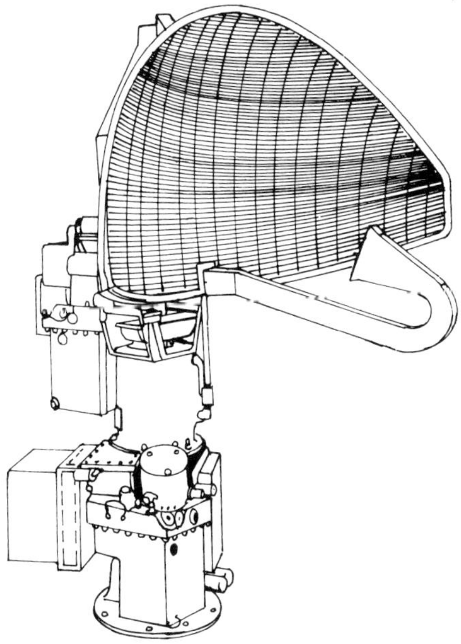

SG-3 cut-paraboloid antenna.

SG-6

The final production development of the SG series, widely used postwar. It was originally described as a lightweight version of SG-5, the radio-frequency generator being that used by SO-5/6 and the console being the new standarized type of SR-4 and -6. Two antennas were provided, one for surface search and the other for zenith watch, with a switch from one to the other. Like SG-5, it operated in the new Xb-band (6300mc/s), which would be eliminated under the 1959 radar frequency assignments – and which left its major equivalent, SPS-4, still in service. Peak power was 125kW with 0.37- and 1.3-microsecond pulses and a PRF of 625–650. The surface search antenna produced a 1.6° × 14.5° beam (zenith bucket antenna, 2.75° × cosecant squared), and the assembly could scan at 5 or 15rpm. SG-6b incorporated a modified surface search reflector, and SPS-4 was a generally similar radar operating at a lower frequency.

SG-7

The prototype carrier traffic control radar and the forerunner of SPN-6. S-band, with a narrow beam (12.5° × 1°) and a high scan rate (15–25rpm). Not very satisfactory in CCA experiments aboard the carrier *Philippine Sea* in 1948: 'at no time was it possible to see targets at ranges greater than 4–5 miles'. Only one was built.

SH

An unsuccessful S-band search and fire control set for cargo vessels, employing a 30in × 20in truncated paraboloid antenna spinning at 180rpm within a vertical cylindrical radome 51in high and weighing, altogether, 700lb. The high scan speed was to drive a PPI and the antenna could also be lobe-switched for manual train. Power was only 30kW (0.25-microsecond pulses, PRF 400–2000), which was said to be good enough for a range of 15nm on a battleship and a minimum range of 400 yards. Accuracy was 25yds (¼°) and resolution 25yds (3°)

Only 50 of 200 ordered had been completed by June 1943, when the program was dropped in favor of other systems already in production. The poor performance was a consequence of low power and the fast scan. Westinghouse.

SJ, SJa, SJ-1

An S-band surface search set for submarines developed to complement SD, the first of 160 being installed in June 1942. Early SJs had only A-scopes, but PPIs were later fitted. SJa was a field modification with greater power (50kW) for greater range (12nm on a battleship and 8 on a destroyer, *vs* 5 on a medium-sized surface ship). SJ-1 was a production version, with 300 built between

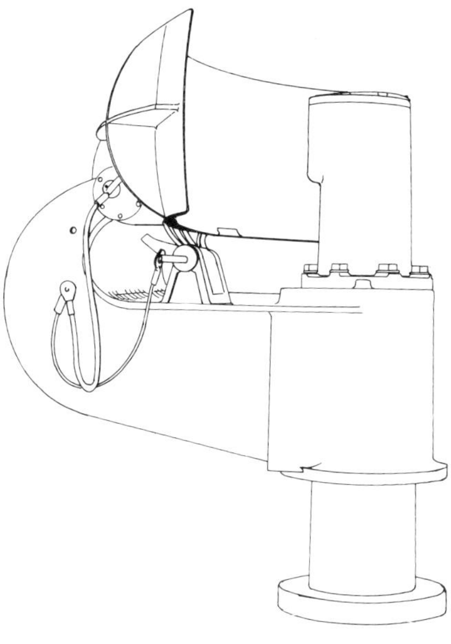

SJ antenna.

SJ-1 (**1**) *and SV* (**2**) *radar antenna, USS* Grouper.

August 1943 and February 1945. 170 SJa conversion kits were produced by April 1944. The antenna was a waveguide-fed 30in paraboloid with a gain of 300 and a 7.5° × 13° beam. It used 1-microsecond pulses and had a PRF of 1350–1650. Accuracy was 15yds, (0.3°) and resolution 50yds (5°) using lobe-switching. Westinghouse.

SK (CXFA)

The standard US wartime large-ship air search set, essentially SC-2 with a CXAM-sized antenna. The first of 250 was installed in January 1943, and production was completed in April 1944. A 1945 report on search radars noted that

SK air search antenna.

'The SK provides the longest range radar detection of aircraft targets', 100nm on a medium bomber at 10,000ft. The antenna was 17ft × 17ft reflector with a 6×6 array of dipoles. Note the effect of a longer (P-band) wavelength: so large an antenna had a gain of only 90, with a 20° × 17° beam. Peak power was 200kW (5-microsecond pulses, PRF 60) and scan rate 4.5rpm. Accuracy was 100yds (1°) and resolution 900yds (10°). The entire system weighed 4900lb, of which 2300 was taken by the antenna. SK-1M (later AN/MPS-24) was a Marine Corps land-based version. General Electric.

SK-2 air search radar, 1944, with SG above at right. The long dipole is for IFF.

SK-2, -3

SK with a new 17ft dish antenna, intended to reduce side lobes of SK; 75 were built between April and December 1944. The net effect was to spread the main beam in azimuth (to 23°). SK-2 also differed from SK in using SC-3 components. Late in the war both SK and SK-2 were considered somewhat deficient in that they could not detect high-flying aircraft, due to the low elevation of their beams. SK-3 incorporated a parasitic antenna to improve its high-altitude coverage at short range.

SL, SLa, SL-1

Another S-band surface search set, designed for destroyer escorts. Physically smaller than SG, with a 42in paraboloid with waveguide feed, having a gain of 580 and a 6° × 12° beam, but with a higher ('medium') power, at 150kW, with 1.5 -microsecond pulses and a PRF of 800. Scan rate was 20rpm, and range was 15nm on a bomber at 500ft, 20 on a cruiser, 13 on a destroyer and 10 on a surface submarine. Accuracy was 100yds (1°), resolution 300yds (6°) and minimum range 400yds.

SLa was a field modification incorporating a ringing circuit, an accurate range unit, a bearing cursor for target identification and an antenna true-bearing synchro-capacitor. Pulse length fell to 1.25 microseconds, and it used range scales at 4, 20 and 60nm as against 5, 25 and 60. SL-1 was the production version: 829 SLs were delivered by October 1943, and 480 SL-1s between January and July 1944. Manufactured by Westinghouse.

SL-X, a modification which showed the relation between wavelength and antenna size, was SL with a 3cm transmitter of 65kW. With the same reflector, gain rose to 5000 (2° × 4° beam) and range remained constant. All SL series sets had PPI displays.

SM, SM-1 (CXBL)

The first US fighter control radar, a 10cm dish, 96in diameter and having a 2.7° × 2.7° beam, using a 5° conical scan to improve accuracy. At 18,000lb (4400lb for the antenna) it was too heavy for anything short of a large carrier. Gain was 3800 (compare the larger dish of SK-2), using 600–700kW, 1-microsecond pulses with a PRF of 775–825 to achieve a range of 50nm on a bomber at 10,000ft and 25 on a battleship. An accuracy in elevation of 500ft

SM shown aboard the carrier Lexington *(CV-16), February 1944, together with FD (Mk 4) fire control radar, SG, SK, SL and SO-11.*

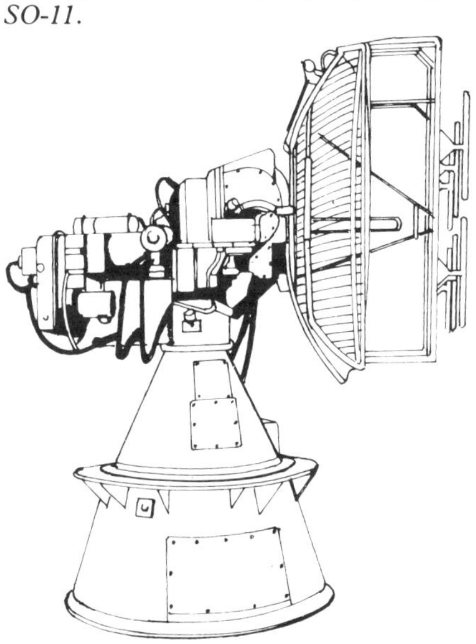

SM-1 antenna assembly.

was claimed, and accuracy was 200yds/$\frac{1}{4}$ per cent of range or $\frac{1}{2}$° with conical scanning. Resolution was 200yds, or 3°. The entire assembly could rotate at 2–6rpm, or scan a sector 7.5–60° either side to zero. Elevation was from −3° to +75°.

First requirements (spring 1942) called for reliable detection of three fighters or one medium bomber at 35nm (50nm if possible). A CXBL prototype was installed aboard the carrier *Lexington* in March 1943, and the first two production sets went to the carrier *Bunker Hill* at the beginning of September ('debugged' en route to the Pacific) and *Enterprise* in October. 23 SM sets were built through October 1944.

SP and SM-1 were lighter versions, 26 of the latter being produced for the Royal Navy, one in 1943 and the rest in 1944. SM-1 had a 6ft reflector on a light (1500lb) mount, and operated on DC rather than the AC of US capital ships; 6ft antennas were also supplied as spares for the first ten SMs. All of these sets were built by General Electric which has since been responsible for US heightfinders. These was provision for both BM (mounted in front of SM dish) and BO (framework in front of antenna) IFF.

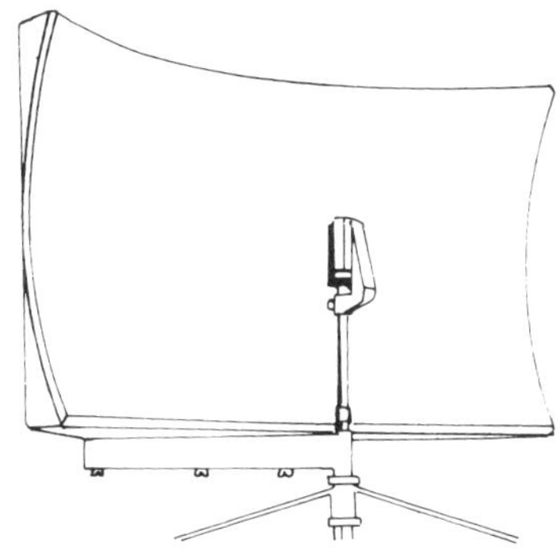

SN antenna.

SN (CXBR)

A lightweight, portable, S-band 12cm set for emergency use aboard ships and in boats. 1kW, 1-microsecond pulses, with a PRF of 1000, sufficed for detection of a battleship at 6nm or a surfaced submarine at 2. The only display was an A-scope, and the antenna, a relatively large 24in zone of a 48in paraboloid with a gain of 400 and an 8° × 20° beam, scanned manually only. Feed was a vertical dipole and total weight 361lb. 775 units were made during 1943, several proving useful aboard badly damaged carriers. SQ was a development. General Electric.

SO small-boat radar, PT-212.

SO, SOa, SO-13 (CXBX)

A large family of SO microwave (10cm and 3cm) surface search sets, about 6300 in all, was built by Raytheon. Both 3cm and 10cm versions were envisaged from the first, and in August 1942 Raytheon was asked to submit bids for eight prototypes, four on each wavelength. The CXBX S-band prototype was mounted on a PT boat at the Boston Navy Yard on 25 November 1942, and production began the next February. The original SO, for PT-boats, was a 420lb, 65kW 1-microsecond, PRF 400 set employing a 24in diameter full paraboloid rotating within a radome, the total weight being 65lb, the gain 24db, the beam 11° × 11°, and the scan rate 12rpm. Typical ranges were 15nm on a bomber at 500ft, 20 on a battleship, 10 on a destroyer and 5 on a surfaced submarine. Accuracy and resolution were 200yds and 2°, and 200yds and 5° respectively. SOa was a field kit modification and SO-13 the production model, with 0.37 or 1.0-microsecond pulses and a PRF of 650. 525 SOs had been delivered by November 1944, and 587 SO-13s were delivered between Februaary and November 1944.

SO-1, SO-8

SO-1 was similar to SO execept that it operated from a 110V rather than a 24V DC power supply and used an SG antenna; SO-8 used a smaller, 30in × 12in antenna with a gain of 125, and a 9° × 25° beam. SO-1 used 1-microsecond, 65kW pulses (PRF 400) to achieve a range of 15nm on a bomber at 500ft, 13 on a destroyer or 8 on a surfaced submarine. Accuracy and resolution were 200yds and 2°. The SO-8 antenna was housed in a radome, and its resolution (4°) was poorer in azimuth. Ranges were also lower: 12nm on the bomber, 12 on the destroyer, or 5 on the submarine. Altogether 2047 SO-1s had been delivered by November 1944, and SO-8 production totalled 1700.

SO-2, SO-9

Both of these sets were essentially identical to SO-8 except for power supply: 115V AC for SO-2 and 32V DC for SO-9. 535 SO-2s and 110 SO-9s were built by November 1944.

SO-3, SO-4 (CXBX)

These were 3cm versions of SO and SO-1. Both employed slatted cut-paraboloid antennas; SO-3's was 24in × 8in, weighed 120lb and had a 31db gain and a 3° × 9° beam scanning at 5rpm,

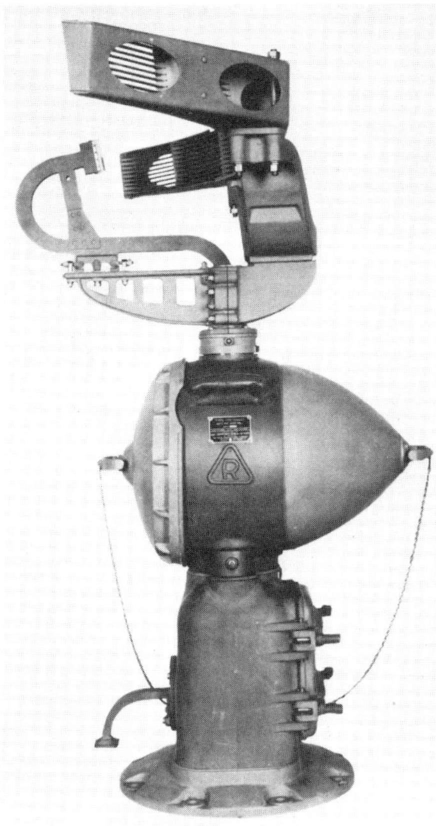

SO-4 small-boat surface search radar, 1945.

whilst SO-4 employed a 160lb, 48in × 5in antenna with a 34db gain and a 2° × 9° beam which scanned at 5rpm. SO-3 used 20kW, 1-microsecond pulses (PRF 400) to achieve a range of 10nm on a bomber at 500ft, 20 on a battleship, 13 on a destroyer or 6 on a surfaced submarine. SO-4 doubled its power (same pulse-length and PRF) to improve these figures to, respectively, 12, 22, 16 and 7nm. The greatest advantage of X-band was said to be the higher resolution possible: accuracy and resolution for SO-3 were 200yds and 1°. Similar figures were claimed for SO-4, and both had a minimum range of 200yds. Production of 350 SO-4s began in November 1944, and of 235 SO-3s somewhat earlier. Total installed weights were 750 and 1200lb respectively.

SO-5, SO-6

These were Xb-band (5cm) 120–150kW versions of SO-3 and SO-4 planned at the end of World War II. They were to have had provision for a PPI repeater and target data transmission. SO-5, designed for installation aboard PT boats and LCCs, was produced in prototype form only, two sets being built. It used a 36in × 6in cut-paraboloid 120lb antenna with a gain of 28db and a 3.6° × 22° beam rotating at 5rpm to achieve a range of 10nm on bombers at 500ft, 20 on battleships, 13 on destroyers and 6 on a surfaced submarine. Pulse length and PRF were variable at 1.33 or 0.36 microseconds and 400 or 600 respectively, to give a range resolution of 100yds and an accuracy of 50yds (4nm range scale) or 250yds (20nm scale). Bearing accuracy was 1°.

SO-6 was for auxiliaries and small patrol craft; the first of 350 on order was delivered in June 1945, although it appears that many were canceled at the end of the war. It was expected to replace SO-2 and -4 and used a 48in × 8in cut-parabola with a gain of 30db and a 2.5° × 17° beam scanning at 5rpm. Although the original design appears to have called for 0.25-microsecond pulses (PRF 600), in fact SO-6 introduced variable pulse length (0.37- or 1-microsecond) at a common PRF of 650. It was the prototype for the postwar SO-10.

SO-7

A land-based S-band version, SO-7M in a truck and SO-7N in a trailer. Some were supplied to the Soviets, and in US service both were used by the Marines. The antenna was a 3ft section of a 5ft paraboloid, producing a 4.5° × 7° beam and capable of detecting a fighter at 1000ft at 35nm. It used the 60kW transmitter (1-microsecond pulses, PRF 400) and achieved an accuracy of 100yds and 2°.

SO-10

The postwar production version of SO-6. It used 125–285kW peak power (0.37- or 1.3-microsecond pulses at PRF 650) and a 48in cut-parabolic antenna (28db, 2.7° × 17° plus 15° cosecant-squared beam) to achieve ranges up to 22nm, an accuracy of 60yds (4nm scale) or 300yds (20nm scale) and 1°, and a resolution of 75yds and 2.7°. Beam-spread upwards in SO-5, -6 and -10 was said to provide limited air search capability. Minimum range was 150yds. Relatively few were built.

SO-11

A zenith search modification of SO-13, first suggested by A Longacre of the MIT Radiation Laboratory on the basis of the shakedown experience of the carrier *Lexington;* her officers suggested a modified SL or SO. It was first tested on their ship, on 25 January 1944, and three others were placed aboard battleships. However, side lobes proved so strong that the PPI showed large echoes from nearby ships. A better reflector was developed and released for interim installation in SGs, but in 1945 it was hoped that production of SO-11s would begin in July. The special reflector was expected to cover the arc from 20° to 70° in elevation, and SO-11 was considered superior to the converted SG in view of the higher peak power of the SO transmitter (65kW, 1-microsecond pulses, PRF 400). As of January 1945 the reflector was to be a cosecant-squared bowl, cut-paraboloid, 48in × 24in, with a 12in focal length, weighing 350lb and producing a 6° × 50° beam, and it was to rotate at 12rpm while sweeping the area from 20° to 70° above the horizon. None was produced.

SO-12

The land version of the X-band SO series, in truck (-12M) and trailer (-12N) versions. A 5ft × 3ft parabolic reflector produced a $1\frac{1}{2}^\circ \times 2\frac{1}{4}^\circ$ beam with an output of 50kW, 1-microsecond pulses and a PRF of 40, and accuracy and range matched those of SO-7M/N.

SP (CXDT)

A lightweight (9000lb) successor to SM, first suggested in the fall of 1942. In fact the two sets had few components in common, due in part to the use of two pulse-lengths (5-microseconds at PRF 600 and 1-microsecond with PRF 120) in SP. There were two antennas, a 6ft unit

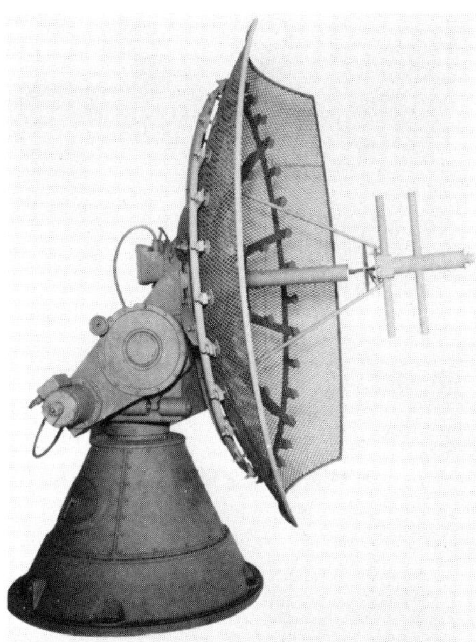

SP-1M antenna assembly (land-based version), 1944.

weighing 1700lb and with a 3.6° × 3.6° beam for destroyers and light carriers, and an 8ft unit weighing 2300lb and having a 2.7° × 2.7° beam for battleships and heavy carriers. The latter had 40 per cent more range than SM, but the former duplicated SM performance. Gain in the 6ft version was 2100, and conical scan (nutating feed) was introduced in 1945 to improve range performance by about 30 per cent. A power increase to 700–1000kW actually improved overall performance as compared to SM, although at 30nm altitude accuracy was only 1300ft. With the 8ft antenna, range on a bomber at 500ft was 35nm, on a bomber at 10,000ft 70nm, on a fighter at that altitude 40nm, on a battleship 35nm, on a destroyer 25, and on a surfaced submarine 15. The 6ft antenna reduced these figures by 30 per cent, which it was expected would be recovered with the nutating feed. Resolution was 200yds, and 0.5°, and accuracy 200yds and 1.5°, with an elevation of 1000ft at 30nm. Minimum range was 550yds. The antenna was designed to scan at 6rpm, and elevation was adjustable from −3° to +30°.

SP was simpler than SM and hence more reliable. About 300 were built, beginning in the fall of 1944, and many were installed aboard battleships, cruisers and even destroyers. SP-2 was a modified version incorporating vertical scan in height (1945); it was never service-approved. SP was later used as the basis for the SPQ-2 Regulus guidance radar.

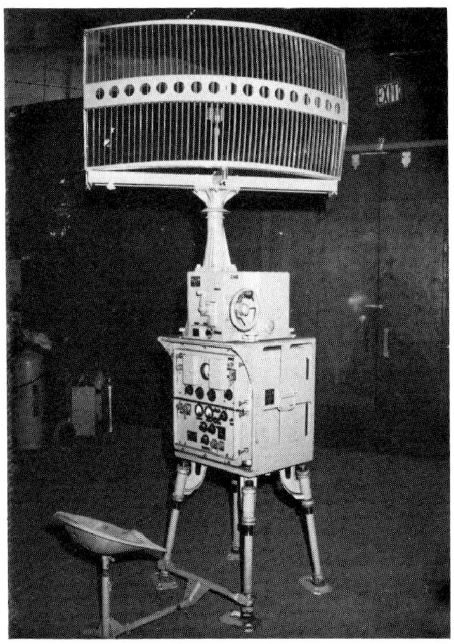

SQ emergency search radar, Naval Historical Center, 1980.

SQ (CXBR-1)
The successor to SN, incorporating a PPI and an A- or B-scope, all on the same 3in CRT. Reliable range increased to 8nm on a battleship and 2.5 on a surfaced submarine. 400 were delivered to US Navy between September 1943 and September 1944 and 300 more to the Royal Navy. Both SN and SQ were discarded at the end of the war. General Electric.

SR, SRa, SRb, SR-5 (CXCB and XBF)
The first entirely new air search set since CXAM, and the progenitor of a new series of standardized radars. SR was similar in antenna size (15ft × 6ft, gain about 30, a 6×2 array of horizontal dipoles) and wavelength to SC-2, but had a peak power of 500 rather than 200kW (variable pulse length: 20-, 4- or 1-microsecond; PRF 200 or 60), and accuracy improved to 30yds (8000yd scale) or 100yds (20nm scale) and 2° and resolution to 500yds and 10°, with a 20° × 55° beam. A 1945 report on search radars noted that 'it is the first equipment to contain the standard indicator console with PPI, A-scope, antenna control, receiver, and IFF coordinator. The construction of the SR is an improvement over the SC series in that it is ruggedly built with adequate shock mounts'. SRa was a field change incorporating an improved transmitting tube; SR-5 was to have been the production designation. SRb was a second modification with an improved receiver and a noise-figure monitor.

SR destroyer air search radar, 1945.

The pilot SR appeared in June 1944, and by the end of World War II 300 were on order, of which 100 had been delivered by March 1945. Many destroyers completed at the end of the war were fitted with SRa, which could be distinguished from SC by the rounded corners of its reflector, and 38 remained in service as late as 1961. In 1945 its claimed detection range on a bomber at 10,000ft was 110nm, on a fighter at the same altitude 75nm, and on a bomber at 500ft 25nm. About 1955 SR performance was considerably improved by the application of a new low-noise receiver, giving an 8db reduction, and a noise monitor which permitted a technician to obtain peak performance far more easily. With these modifications, a destroyer which had obtained no better than 20nm on jets began to detect such aircraft consistently at 90nm above 30,000ft, and the 8db improvement alone, which required 15 man-hours, was considered to increase the range on a jet fighter by 59 per cent. At this time it was expected that some SRs would be fitted with SPS-17 antennas, which provided 3db more gain and had smaller side lobes as well. Such sets would also obtain improved performance through the installation of the SPA-8 indicator used in SPS-17, especially for weak targets such as jets at long range. In 1961 it was claimed that the improved SRb could detect a 1m^2 target, somewhat smaller than a jet fighter, at 114nm. There was provision for integral BL, BM, BN and BG IFF. Westinghouse.

SR-1 (XBF-1)
The remainder of the SR series was an attempt to achieve frequency diversity in long-range search radars; thus SR-3 and -6 were the direct forebears of the SPS-6 series of postwar L-band sets. SR-1 was a developmental 70cm, medium-weight (3000lb estimated) air search set with an output of 600kW, 6- and 1- microsecond pulses and PRFs of 170 and 600, and an insurance against the failure of 50cm and 25cm (L-band) types. An 18ft × 6ft cut-parabola antenna with a gain of about 160 and a 10° × 55° beam was to provide detection of a medium bomber flying at 10,000ft at 110nm and of a fighter at 70. Developed at NRL. Raytheon.

SR-2
A 50cm air search set, originally intended to be compatible with DEs but ultimately far too large for that. It was the first US Navy set to require a separately rotating IFF antenna. The pilot model was tested in April 1945; 200 were ordered, but only 18 had been built by the time the contract was canceled at the end of the war. The first two went

SR-2 air search radar.

aboard the new carriers *Midway* and *Franklin D Roosevelt* during the war, and one was also installed aboard their sister-ship *Coral Sea*. The others went to the light carrier *Saipan*, to one heavy and seven light cruisers, to two destroyers, and to the first four submarine radar-pickets. All were replaced by SPS-6 on grounds of inferior performance.

Peak power was 300kW (4- and 1-microsecond pulses at PRFs of 180 and 600) which, with a 15ft × 5$\frac{1}{2}$ft cut-parabola and an 8° × 25° beam, gave a range of 110nm on a bomber at 10,000ft and 70nm on a fighter. Accuracy and resolution were 100yds and 1° and 200yds and 4° respectively. Minimum range was 600yds, and total weight was about 3000lb. RCA.

SR-3, SR-6

L-band (25cm) long-range air search radars, SR-6 being a lightweight version. Original plans called for cylindrical parabolic antennas, 17ft × 2ft and 10ft × 2ft and 750lb and 500lb respectively, the SRb mattress weighing 711lb. L-band was chosen at the end of World War II to achieve solid radar coverage with a minimum of fading. In this it was successful, but the new antennas chosen performed poorly, one postwar report speaking of 'not enough antenna', for an average range of only 20–30nm – far short of expectations.

At 500kW, and with 4- and 1-microsecond pulses, PRF 150 and 600, SR-3 was to duplicate SK performance, giving 100nm on a bomber at 10,000ft and 80 on a fighter, with a far better beam shape (3° × 65°). Production versions used slotted waveguide antennas with small angled reflectors, SR-3 actually producing a 4° × 38° beam with a gain of about 200. SR-6, 10ft × 2$\frac{1}{2}$ft and 375lb, produced an 8° × 34° beam of 2-microsecond pulses (PRF 290–310), giving fighter detection at 40nm. Accuracy and resolution of SR-3 were 100 and 200yds, and 2° and 1°.

The rotation rate of 2.5 or 5rpm was considered too low to counter jets, and both sets had short lives. They were installed as successors to SC and SK during 1957–48, and at the end of 1948 SR-3 was aboard the experimental ship *Mississippi*, the three *Midway* class carriers, two battleships, and eleven heavy and five light cruisers, the smaller SR-6 being aboard one light cruiser, one destroyer tender, 38 destroyers, one DDE and two DMs. On 25 April 1948, however, the CNO ordered the replacement of SKs to be stopped in view of deficiencies in the new sets. Existing SR-3s and SR-6s were fitted with the new parabolic SPS-6 series antennas as SR-3B (SPS-6B antenna) and SR-6A (SPS-6A antenna). Alternate modifications were SR-3A (-6A antenna), SR-3C (-6C antenna and SR-6B (-6B antenna). Other SR-3 modifications included new plumbing and a new pedestal. Westinghouse.

SR-4

Perhaps the first US counter-countermeasure radar, scanning *in frequency*, one scan per second over a range of 90mc/s. It had a nominal wavelength of 50cm, with a peak power of about 350kW (160–180 pulses per second, of 2.5–3 microseconds), and the 15ft × 5$\frac{1}{2}$ft cut-parabola was to produce an 8° × 25° beam, for performance much like that of SR-2. It is not clear whether this set was ever built. Federal Telephone and Radio.

SR-7

An early postwar attempt to combine air and surface search for destroyer escorts, using the new L- and Xb-band technology. The air search antenna, a 12ft × 4ft parabola with a 5.5° × 30° beam, was to obtain a range of 50nm on a 20m² airplane and 20 on a destroyer; the surface search section (4ft × 1ft parabola, 3° × 12° beam) was to obtain a range of 15nm on a destroyer. The entire 860lb assembly, on a stable base, scanned at 7.5–15rpm. L-band pulses were to be 500kW (1- and 4-microsecond, PRF 300) Xb-band pulses 250kW (0.4-microsecond, PRF 1800). Although the designation SPS-1 was assigned in November 1946, and SPS-1 appeared in a BuShips radar list of June 1948, it is not clear that any were ever built. RCA.

SS, SS-1 (CXEX)

The X-band replacement for SJ employing the new 1945 unit console, with a 5in PPI and a 3in combination B- and A-scope. SS-1 had aided tracking. Production of 300 began in June 1945, and this set was installed aboard all active submarines by 1949. Range scales included one calibrated to 80nm; accuracy was to within $\frac{1}{4}$° using lobe-switching, 15yds in range for a short-range scale. The 2ft 6in parabolic antenna had a gain of 26db and a 2.6° × 16° beam, and scanned at 0–8rpm, emitting 75–110kW, 0.5-microsecond pulses at a PRF of 600. Resolution was 2° or 150yds and range scales were 8000, 20,000, 40,000yds and 80nm, with 8000yds precision. SS-1 deliveries began in March 1950. Western Electric.

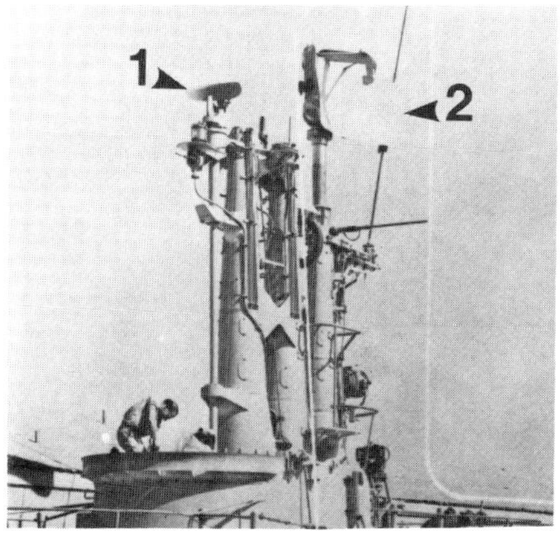

SS surface search radar (**1**) *and SV-4 height-finder for missile control* (**2**), *fitted to USS* Cusk *(Loon conversion).*

SS-2, SS-2a

A 100kW submarine search radar with a 3ft 6in antenna. Deliveries began in February 1952.

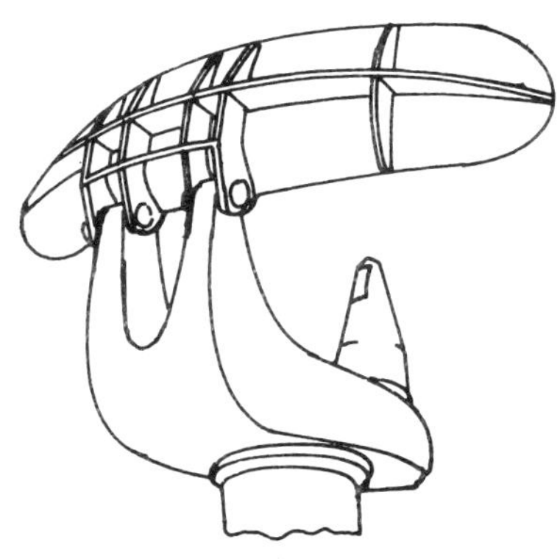

SS-2 antenna.

ST (CXGK)

An X-band submarine surface search set for use when only the periscope was exposed, employing the SJ power supply and indicator, a transmitter similar to that of the Mk 12 main battery fire control set, and a 2in × 6in slotted antenna mounted just below the lens of the night periscope. Operation was by means of a switch shutting off SJ. With the antenna only 3ft above water, a battleship could be detected at 8nm and a surfaced submarine at 3. Although only an A-scope was provided, bearings accurate to 5° (12° × 25° beam, gain 15db) could be obtained. Resolution was 20° and 75yds and accuracy 15yds. Peak power was 30kW, with 0.3-microsecond pulses and a PRF of 1500, or 600 when used with the SS radar transmitter, which had a slightly lower frequency. Production of 254 began in July 1944. Western Electric.

STa, ST-1

ST combined with SS to use the same transmitter and display; a waveguide switch would connect those components to either the SS or ST antenna. It was planned to feed a voice radio transmitter into the periscope antenna. Western Electric.

SU, SU-1, SU-2, SU-3, SU-4

The X-band successor to SF and SL, inspired by the success of tests late in 1942. It was ordered in March 1943, specifically for the DE program. The first prototype was completed at the beginning of January 1944 and tests aboard *Semmes* (AG-24) were completed by the end of May. Its shorter wavelength shows in a smaller antenna, a 24in paraboloid with a gain of 2600 and a 3.6° × 3.6° beam. The higher resolution of X-band permitted good range performance – 20nm on a battleship and 15 on a surfaced submarine – even with the low power of 15kW, using 1-microsecond pulses and a PRF of 600, for a typical antenna height of 50ft. The radome-shielded dish was line-of-sight stabilized, scanned at 6rpm and weighed 200lb. Accuracy was 30yds (on a 4nm scale) and 1°, and resolution 300yds and 4°.

SU-1 incorporated a new power supply, DC rather than AC. In 1945 contracts had been let for kits to increase bandwidth, provide either 0.5- or 1-microsecond pulses, and provide a 4000yd sweep to complement the

SU-2 surface search radar, 1945.

8000yd, 40,000yd and 80nm range scales already available. SU-2 and SU-3 employed a new truncated paraboloid antenna, without a radome; the antenna measured 54in × 24in, had a 1.8° × 3.7° beam and a gain of 36db, and weighed 200lb; they were intended for antisubmarine, rocket fire control (LSMR) and navigational purposes, and incorporated a new 0.25-microsecond pulse mode for high (100yd) resolution; and accuracy was 40yds on the 8000yd scale, other scales being 4000, 40,000 and 160,000yds. PRF was unchanged, with either pulse mode, and both employed new ECCM features. SU-3 was the DC version of SU-2; the former weighed 1900lb, the latter 1275. SU-4 added full stabilization against deck tilt, so that it could be used with greater accuracy in bad weather. As compared to SU/SU-1, Su-2 through -4 had greater accuracy, and could handle 12 instead of 3 remote PPIs; they also had a more powerful (25kW) transmitter.

Production of 522 SU/SU-1s began in January 1944, preproduction SU-2/3s in June 1945. Postwar production was slow so that as of 1949 only four preproduction SU-2s and 15 production models had been delivered. At this time SU was described as a medium-weight surface search set useful for detecting icebergs, buoys and periscopes. It was expected to replace the SF, SG, SL and SO series, and its own replacement, beginning in 1951, would be SPS-10. Submarine Signal Company.

SV, SVa, SV-1 (CXFU and CXGW)

A 500kW Sw-band air search radar, the only Navy air search set of so short a wavelength, to replace SD. Its 4ft × 2ft slotted parabolic antenna scanned at 0–6rpm (gain 1000, 5.3° × 60° beam) on a retractable mast to detect bombers flying at 2500–15,000ft at 22nm. A 300ft bomber could be spotted at 15nm, but a fighter at 10,000ft could not be detected outside 11nm. Accuracy and resolution were 15yds and 1° and 200yds and 30° respectively, using 1-microsecond (PRF 400) pulses. Range scales were 8000, 20,000, 40,000, 80,000 and 160,000yds. SV used the SJ-1 display and SVa and SV-1 used an SS display instead.

Production of 80 SVs began in January 1945 and of 220 SV-1s the following June. Both were used widely in postwar submarines.

SV-2, -2A

S-band height-finders for submarines, developed for the first submarine radar-pickets. SV-2A was modified to improve the vertical scan rate. The 2ft × 8ft antenna had a gain of about 2000 and a 5° × 2.3° beam, and scanned from –10 to +80 degrees at 20° per second while rotating at 0–6rpm. The transmitter was identical to that of SV-1

SV-3

Improved SV-1, differing only in detail, with essentially the same operating parameters. Seventeen were delivered from November 1950 through June 1951. Western Electric.

SV-4, -4A

S-band submarine missile trackers, used in Loon experiments. SV-4 was SV-2 with its antenna rotated 90° to generate a 2.3° × 5° beam.

SV-5

SV-4 plus a computer and new indicators. It became part of BPQ-1.

SV-6

Submarine height-finder for missile operations. XSV-6 was an SV-1 modified for short-range height-finding at New York Navy Yard. Peak power was 560kW and gain 560, and the 5° × 2.5° beam scanned the horizon at 2rpm while sector-scanning vertically at 2 cycles per second.

SW (CXCA)

A 75cm lightweight (281lb) medium-range portable air search set for shore use, later designated AN/TPS-2. Deliveries began late in 1943, but SW was declared obsolete in 1945. Its proposed shipborne use, similar to SA-1, did not materialize. It used an 8×4 dipole array in front of a 9ft 6in × 5ft parallel-wire reflector of 13° × 25° beam. 45kW, 4-microsecond pulses, with a PRF of 232, sufficed to detect a 10,000ft bomber at 65nm, or a 10,000-ton cargo ship at 15. General Electric.

SX (CXHR)

A powerful dual-frequency air search/height-finding system, which led to SPS-8. It consisted of two antennas mounted 90° apart: a 14ft × 4ft parabola with a 1.6° × 18° beam for air search and a 5ft × 15ft vertical parabola with a 4° × 1.2° beam for height-finding. Each used a slightly different frequency: Sw (8.6cm) for height-finding (500kW, 1-microsecond pulses at PRF 390 or 1170) and Sg (10.5cm) for search (1MW, 1-microsecond pulses, PRF 390). Although both were S-band, it was hoped that by proper beam-shaping the search antenna could avoid fades, at least out to 70nm and 20,000ft. Rather than scan in elevation, the height-finder was fed by a moving (Robinson) feed horn which allowed effective vertical scanning over a 12° vertical arc at 10 cycles per second, and the entire assembly rotated at 4rpm to provide air search via the horizontal antenna. Its fixed sector of elevation limited SX to aircraft below 20,000ft, but about 1950 this radar was modified to achieve a 31° maximum elevation. As built, SX weighed only 600lb more than the far less capable SM. Accuracy was 200yds and 1.2°; resolution was 200yds and 1.2° air or 1.7° surface targets. BuShips originally asked, in August 1943, for display of the heights of all aircraft within 50nm every 15 seconds plus high- and low-level coverage: 80nm on all aircraft below 40,000ft and below 20°.

SX radar antenna assembly.

About 45 were produced, including two prototype CXHRs in 1945, and SX served aboard carriers and the experimental ship *Mississippi*, although amphibious flagship (AGC) installations were also originally planned. The first four were installed aboard the carriers *Midway*, *Franklin D Roosevelt*, *Coral Sea* and *Tarawa* in July 1946, and by 1948 there were 25 in existence, installations including all active *Essex* class carriers – *Boxer*, *Leyte*, *Antietam*. *Princeton*, *Valley Forge* and *Philippine Sea*. Later SX would appear aboard SCB-27 *Essex* class conversions, being ultimately displaced by SPS-8. A proposed SX-1 incorporating an 8ft search dish for greater raid-handling capacity appears not to have been built, but the designation was later applied to a shore-based training version. General Electric.

SY

An Xb-band replacement for SL-1, using SS display, a stabilized antenna and an SU-2 radio generator. Does not appear to have been built.

SZ

Not assigned.

CXGQ (Project 'Henry')

A 60kW, 3.3cm high-resolution surface-search radar for small boats such as landing craft, with a total installed weight of 1240–2000lb. Resolution was obtained by short pulse-length, 0.16 microseconds at a PRF of 2000, or 0.67 microseconds at a PRF of 500, and a narrow, 2° × 7.1° beam. There were range scales for 2000, 4000 and 8000yds, and for 10, 20 and 80nm, long pulses being for the last scale only. Minimum range was only 100yds. There was also a B-scope covering any 4000yd × 30° or 4000yd × 4000yd sector of the north-stabilized PPI. The 48in × 10in horizontally slatted grid reflector scanned at 6rpm, although a later model was expected to scan at 6 or 12rpm. CXGQ could detect a bomber at 10,000ft at 15nm, a battleship at 22, a destroyer at 15 and a surfaced submarine at 12. It was designed primarily to detect ships lurking in harbors or along island shores, which suggests a PT-boat application. Accuracy was 50yds and 1° and resolution was 60yds and 2°. The MIT Radiation Laboratory built a total of 35 (completed January 1945) as a supplement to SO-3 and SU-2.

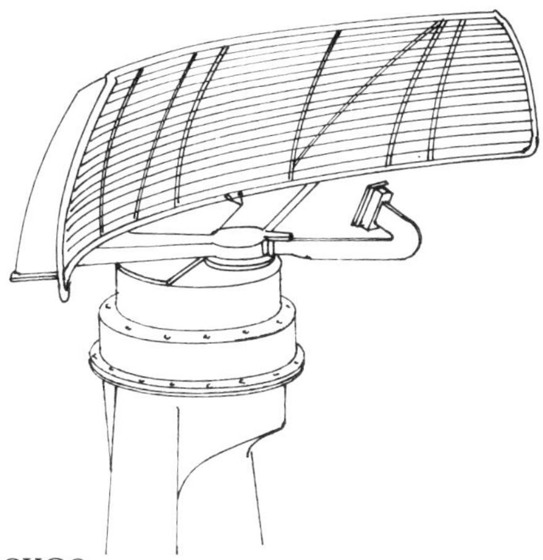
CXGQ antenna.

CXJG (Project 'Cindy')

An experimental K-band (1.25cm) high-resolution, short minimum range, surface search radar for ships down to landing craft and PT-boats. Fourteen were built, and they proved useful in Arctic ice-detection operations in 1948, but the equipment was not too reliable at such high frequencies. The 60in × 8in parabolic antenna formed a 3.2° × 30° beam, using very short, 0.17-microsecond (PRF 750) pulses of 23kW peak power, and the set was instrumented for 20 and 60nm range scales. Scan was 5–15rpm. At 305lb it was small enough to fit on a medium landing ship (LSM), and in fact it was first evaluated aboard ELSM-446 in 1947. One problem disclosed by that test was undue reflection from the topmast supporting the LSM's main surface search radar, an SO-6, and this may have led to the concept of SPS-9. SPS-7 was in effect a modernized 'Cindy'.

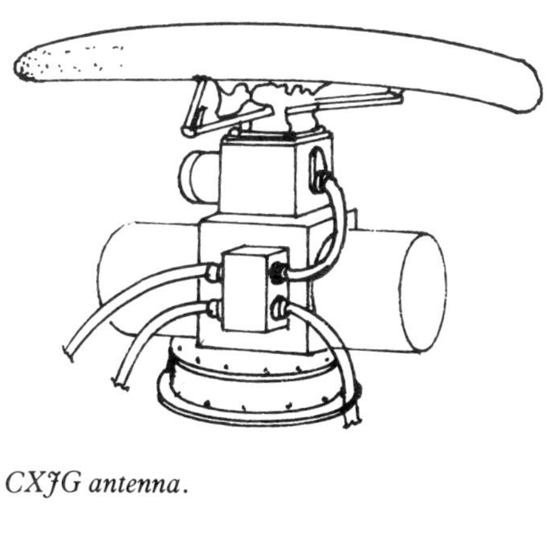
CXJG antenna.

SCR-517A

Army S-band air-to-surface vessel radar, also used in PCs and in PT-boats. The antenna was a dipole in a 29in paraboloid, with a gain of 340 and generating a 9° × 9° beam; it used 1-microsecond pulses with an output of 20–40kW and a PRF of 2000. A total of 225 were produced by Westinghouse. SCR-517B, also designated AN/ASG-1, was fitted only on bombers, but SCR-517C (225 produced by Westinghouse) was fitted to PCs and to PTs as well as to bombers. Range accuracy was 200yds and angle accuracy 5°. The entire system, which was first fitted to PC-487 in mid-1942, weighed 520lb and required 1.5kW of power.

SCR-720

Night-fighter S-band nose radar with spiral scan, used in 1945 for zenith search, mounted vertically in a big radome and using a 29in paraboloid to produce a 9° × 9° beam. Elevations from 50° to 90° were covered. It was discarded at the end of the war. Westinghouse.

B. Postwar Series (AN/SPS)

SPS-1

See SR-7 above.

SPS-2

A very ambitious L-band stacked-beam height-finding search radar, so complex that only two were built. It began as a very long range radar to detect high altitude targets such as V-2 rockets, with a required range of 300nm, then became a fleet central warning radar with some fighter control potential. However, throughout its development range was considered more important than accuracy.

L-band was chosen to give an antenna of workable size, and a stacked-beam system was chosen for economy. The antenna proper was a pair of cut-paraboloids separated horizontally by three-quarters of a wavelength. This enlarged the first side lobes but merged them with the main beam, so that slightly wider beam spacing than usual could be used. By changing the phase across each beam the minimum angle for height-finding could be reduced. Poor height-finding resolution was accepted, as any stacked beam system would become confused if two targets existed in the same range sector. A single 6.5MW magnetron, the most powerful of its time (10MW had been planned; magnetron problems contributed heavily to the unreliability of the system) sent its pulses through seven horns facing the reflector, to form 1.6° × 2.7° beams. Each beam had proportionately less power than the one below it, since at extreme range the higher angles represented targets at impossible altitudes. Returning echoes were received by all seven horns, their outputs separately amplified and compared for 'monopulse' height-finding. In principle the set as a whole could thus achieve far better data rates than could more conventional nodding-beam height-finders.

SPS-2 was also designed for both coherent and non-coherent MTI. The display systeo was as complex as the radar itself, with 26 modified VK-3a PPI repeaters and twelve specially designed AN/SPA-7 range–height indicators. The 40ft × 20ft stabilized antenna had a gain of 3000 and weighed 52,000lb, out of a system total of 97,000. Long range required both a low PRF (245) and long pulses (7 microseconds). Altitudes could be determined within ¼°, and the developers aimed at a range accuracy of 400yds. Range on a heavy bomber at high altitude was about 300nm, and on a 1m² target about 165. One SPS-2 went to the command ship *Northampton*, the other to the missile cruiser *Little Rock*.

SPS-2 inspired several smaller versions, SPS-11, -13, -34 and -44, and several stacked-beam height-finders being developed for land operations. Ultimately the naval versions were abandoned in view of their complexity, as compared to FRESCANs. General Electric.

SPS-3 (XDK)

The first attempt at a hemispheric radar and an unsuccessful one. Development began at the MIT Radiation Laboratory in 1945 and was transferred to NRL when the former was disbanded at the end of the war. Although it was widely specified for new construction and conversions in the late 1940s and early 1950s (SG-6/SPS-4 was an interim alternative), SPS-3 actually appeared only aboard the then-experimental command ship *Northampton* (ECLC-1) in 1954. This set was actually a pre-prototype, the SPS-3 (XN-1) intended for use in conjunction with the experimental Target Designation Systems Mks II and III; production versions were to be used for a combination of hemispheric search (out to 15nm), target designation to guns and missiles, surface search and air traffic control.

The original rationale had been a combination of three-dimensional target designation and 'solid' coverage against close-in small attackers such as suicide aircraft, which meant a requirement for very rapid scanning (4-second data rate). A unique dual Foster scanner antenna system was used, consisting of a conical radiation source illuminating a parabolic antenna much as the waveguide of SPS-39 illuminates its antenna, which has much the same shape. Inside the cone an inner cone spun to produce the scanning beam, the angle of emergence of the beam being determined by the relative position of this rotor, so that the beam elevated as the rotor turned. Limits on the angular variation possible with such a system required that two Foster scanners share the single antenna; they were stacked vertically, the high beam, 3.3cm, 2.9° × 2.5°, scanning from 79.8° to 39° and the low beam, 3.2cm, 2.9° × 1.7°, from −1.9° to +39°. X- rather than S-band was chosen to cut antenna weights, K-band being rejected on grounds of transmitter losses. Radio components were, unusually, placed *inside* the antenna, to minimize losses. Even so, range performance was poor, tests aboard the *Northampton* showing reliable detection of a jet fighter at 40,000ft at about 10nm, with a maximum range of 10–14 at 10,000ft, which was why SPS-3 was rejected in favor of CXRX and SPS-24/25.

Each Foster scanner scanned in elevation at 1800rpm (600 optional), while the entire assembly rotated at 15rpm (5); there was also a 'search mode' in which only the low beam was used. Both transmitters produced 0.7-microsecond pulses (PRF 3000, 150kW; alternatively 1500 per second at 250kW, scanning more slowly). In operation, for at least some time it proved impossible to run both scanners simultaneously, and only the lower beam could be used. In effect SPS-26/39 replaced the troublesome Foster scanner with a folded waveguide, while retaining much the same basic principle of hemispheric search. CXRX, a modified SPS-8, served as an interim hemispheric search system. It had better range performance.

SPS-4

A Raytheon surface/zenith search set contemporary with, and very similar to,

SG-6, but operating in the C-band (5.4cm) rather than the Xb-band (4.6cm). It used 0.37- to 1.3-microsecond pulses (PRF 625–650) and scanned at 5 or 15rpm. The 7ft × 2ft surface search antenna had a gain of 1000 (2° × 26° beam), and the 4.75ft zenith search reflector had a gain of 200 (2.2° × 50° beam). With a peak power of 200kW, reliable range on a destroyer was 20nm, on a battleship 23nm and on a fighter 20nm. Minimum range was 150yds, and range accuracy 100yds. First deliveries were made in August 1952. Later, high aircraft speeds made zenith search less useful, in view of the short range practical, and a field change eliminated the zenith search feature from many SPS-4s about 1958.

SPS-5

A pure surface search set by Raytheon, roughly equivalent to but 292lb lighter than SU and with performance comparable to that of SG-6 in surface search. It was presumably intended as the postwar counterpart to non-zenith SG; SPS-10, which actually filled this requirement, was originally to have replaced the smaller SO-6/10 series. It was considerably lighter than SPS-4 (810 *vs* 3300lb), presumably in part because of the omission of the zenith search feature. Range on a destroyer was 20nm, using pulses of 170–285kW (0.37 microseconds long, PRF 680), the fixed pulse length being a compromise between long- and short-range performance. SPS-5 had an antenna which could tilt to 65° for limited air search. During operational evaluation aboartd the minesweeper *Peregrine* and several motor torpedo boats in 1952–53 it tracked fighter sections at 10,000ft (but not 20,000ft). This feature was later eliminated by a field change.

Deliveries began in March 1952, and SPS-5 and its modifications (through SPS-5D) were widely deployed, for example aboard the old destroyer-escorts. SPS-5A incorporated a new, non-tilting antenna which produced a narrower beam, 1.5° rather than 2.2–2.6°, both beams being 15° wide in the vertical. SPS-5B was similar. By the late 1950s there were complaints that the Xb-band (6400mc/s) of this radar interferred with commercial communication links, and under the new frequency assignments of 1959 it was eliminated as a radar band. SPS-18 was designed as a C-band (5600mc/s) replacement, but

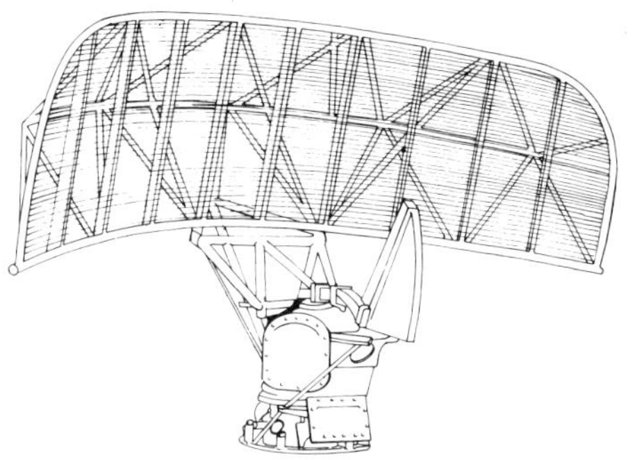

SPS-5C and -5D antenna.

instead existing SPS-5s were modified by a field change (all completed by June 1961). The resulting SPS-5C had higher power (350kW) and used longer, 0.5-microsecond pulses, its new antenna producing a 1.7° × 15° beam. In both cases the scan rate was 17rpm.

SPS-6

The first postwar air search radar, an L-band successor to SR-3 and SR-6, using a new parabolic horn-fed antenna. Its development appears to have been inspired by NRL tests of a land-based, portable L-band air search set, AN/TPS-1. The SPS-6 series was modular, using antennas of differing capabilities aboard ships of different sizes: SPS-6 (18ft × 5ft antenna, 3° × 10° beam, 80nm range on a fighter); -6A (similar antenna dimensions but 3° × 20° beam and 70nm range); -6B (3° × 30° beam, 60nm); and -6C (similar but lighter – 800 *vs* 1000lb – antenna with lower shock resistance and lower scan rate, beam as in -6B but 50nm range). In 1963 SPS-6C was credited with the detection of large conventional aircraft at high altitudes at 70–140nm, and fighters at 60–80nm, using 500kW pulses. These sets could operate in two modes: long-pulse (4 microseconds, PRF 150) for long range; or short-pulse (1 microsecond, PRF 600) for high resolution.

SPS-6 was operationally evaluated aboard the heavy cruiser *Macon* (installed September 1948). SPS-6B was installed aboard the destroyer *Winslow* at Norfolk the following December, and was also tested aboard the light carrier *Saipan* and the destroyers *Massey* and *Zellars*. SPS-6 proved far less satisfactory than -6A and -6B: it was considered useful against jets only up to 20,000ft,

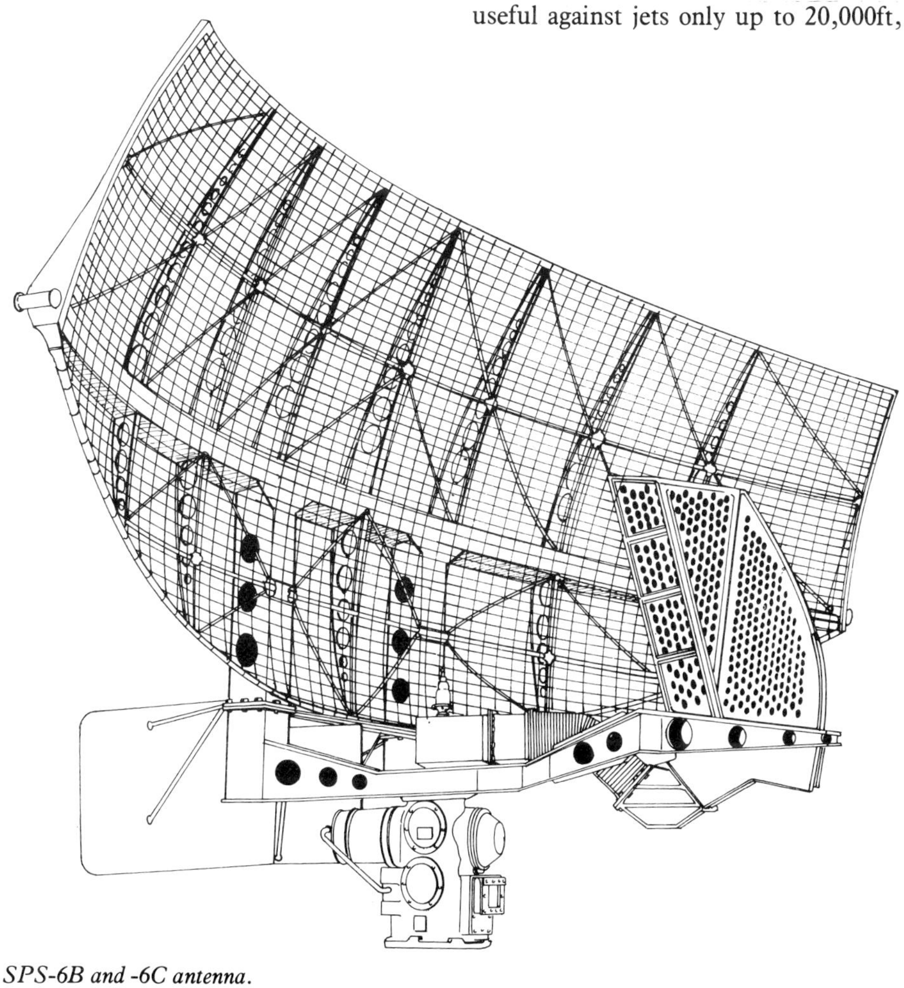

SPS-6B and -6C antenna.

and quite poor above 25,000, whereas -6A and -6B were useful as high as they were tested, *ie* up to 43,000ft, where -6B detected an F2H at 80nm, with no fading down to 70. In 1949 it was stated that, against a jet fighter closing at 400kts, earlier radars would seldom give half as much, with practically nothing above 25,000ft. SPS-6 detected two propeller fighters at 60nm at 6000ft to 30,000 feet, but lost one third of its range against a FH-1 jet fighter. SPS-6B detected a B-29 bomber at 31,000ft and 145nm, and was expected to detect the much larger B-36 at 160 or 170nm. However, effective range was limited by weather, which also limited altitude performance on some days. SPS-6 was also subject to ducting, and sometimes detected ships at 30–50nm. For the first time in a ship borne air search radar, it was affected by sea returns at long range, a problem previously encountered only in AEW systems.

In some cases paired installations of -6 and -6A were planned; they were originally specified for the postwar radar-picket (DER), the *Essex* reconstruction (SCB-27A), and for the ASW conversion of the *Independence* class light carriers (SCB-54), but production problems made such a policy impractical, and each ship received only one set. In such circumstances it was natural for the versatile SPS-6B to be selected: in all, 25 SPS-6s, 45 SPS-6As and 110 SPS-6Bs were built. First deliveries of SPS-6 were made in June 1948, -6A and -6B following in 1950 and 1952. Most sets were of the -6C through -6E models, physically indistinguishable from -6B. SPS-6D was -6C without the integral IFF, and the final version, -6E, introduced a better receiver. Replacement of the entire series by SPS-28 began about 1959, although some sets were replaced earlier by SPS-12. Westinghouse and Bendix.

SPS-7

An abortive 40kW Ku-band set for ice navigation, in effect a modernized equivalent of CXJG ('Cindy') using a shorter wavelength and higher power. The 65in antenna produced a 0.6° × 10° beam, and the 5-microsecond pulses (PRF 1500) gave an accuracy and a resolution of 5 and 80ft and 5 minutes of arc (*ie* 60ft at 1 mile). The antenna rotated at 1.75 or 6rpm. An earlier version used the CXJG antenna, pulse-width (0.17 microseconds) and pulse-rate (PRF 750). Sylvania.

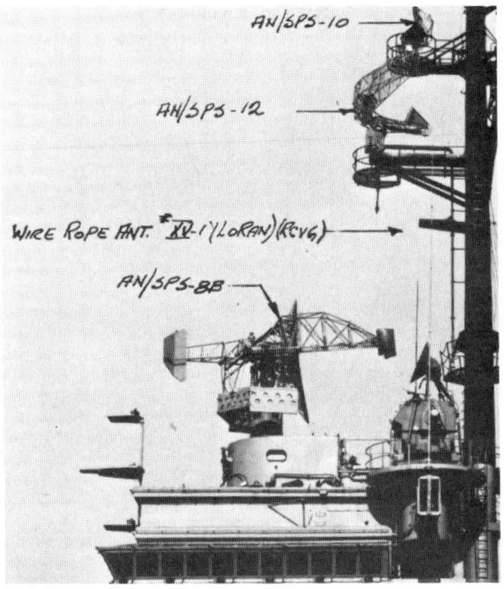

SPS-8B, -12 and -10 as fitted to the carrier Independence, *March 1959.*

BY COURTESY OF A D BAKER III

SPS-8, CXRX

A pure height-finder descended from SX, initially described as a lightweight version of that radar. It used an SX height-finding antenna but could elevate to 36°; the scan covered an 11° sector and had a 3.5° × 1.2° beam. It operated in S-band with a peak power of 650kW (1 microsecond and PRF 1000, or 2 microseconds and PRF 500) which gave a maximum range of 83nm on an airplane at 10,000ft in the short-pulse mode and of 165nm in the long-pulse mode. In a test, SPS-8 was able to detect two F2H fighters at 60nm (1-microsecond mode, 5rpm), but the more powerful SPS-8A/B could do as well at 72nm (700 pulses per second, 5rpm). Vertical scan rate was 5, 10 or 20 per second, and scan in azimuth 1, 2, 3, 5 or 10rpm. Continuous height accuracy was 500ft (relative). Deliveries began in March 1952.

SPS-8A was more powerful: 2MW was desired, but sets were rated at 1MW, using 2-microsecond pulses at a PRF of 450 of 750, and range rose to 95nm. Both -8 and -8A used the same 'orange peel' antenna, SPS-8B, tested aboard the ex-battleship *Mississippi* in January–June 1956 and first delivered in May 1959, being an interim high-gain set. The ultimate nodding-beam height-finder, ordered in 1956, was to be SPS-30. SPS-8B had a narrower beam (1.2° × 1.5°, gain 41 *vs* 37.5db) and scanned a 12° rather than an 11° sector with scan rates of 6, 12, and 16.2 per second. In 1956 the ultimate SPS-30 was envisaged as an improved high-gain antenna driven by a new 2.5MW klystron. SPS-8 and -8A were to be converted first to -8C/D by the addition of the new antenna and ultimately to SPS-30 standard by the addition of the new klystron; all they would then lack would be the new anti-ECM features. Thirty of the new high-gain antennas were to have been delivered late in 1957, but it appears that this schedule was never met.

CXRX, a modified version, was an interim hemispherical search radar using a new 11ft × 17ft antenna with a beam of 3° × 4.5° at the horizon, 12° wide at maximum elevation and of 8° maximum horizontal width, to generate a stream of short pulses (0.67 microseconds, PRF 2000, peak power 650kW) for short-range hemispherical scanning. The scan mechanism was a Warren-Lewis scanner, and CXRX rotated at 15rpm while scanning vertically at 30 scans per second. It was credited with a 20nm range on a bomber. Only two were delivered: they were mounted aboard the missile cruisers *Boston* and *Canberra* in 1955, and were replaced by the more conventional SPS-30 about 1963–64. Plans for a production version (SPS-27) never materialized.

Only a few SPS-8 series radars were exported: one to Canada for the light carrier *Bonaventure*, one -8B to Brazil for the light carrier *Minas Gerais*, one -8B to Japan for the radar-picket *Wakaba* one -8B to Spain for the light carrier *Dédalo*. A total of 77 SPS-8s and -8As were in fleet service in 1957; the Brazilian set is probably the only active one remaining. General Electric.

SPS-9

An abortive combination S- and K-band surface search/ice navigation radar, using two antennas on one stable base. Official sources differ as to the power to be supplied to the two beams: 1MW or 100kW for the S-band search beam (1.75° × 14°), 250 or 300kW for the Ku-band ice-navigation beam (0.17° × 2°). The S-band transmitter was to produce 0.25- and 0.5-microsecond pulses at a PRF of 2000, the K-band 1.5-microsecond pulses at 600. Both were to scan together at 15rpm. Note the drastic effect of much shorter wavelength, using two antennas of similar size.

SPS-9 did not last long; it was a tentative specification in an offical Radar Survey of 1949, but did not figure in a BuShips radar summary five years later. The requirement for two antennas rotating together may have resulted from experience in the operational evaluation of CXJG in 1947, when the pedestal of an SO-6 radar produced considerable interference. However, the latter was necessary in view of the effect of weather on the K-band signals. The S-band component of SPS-9 was expected to provide a 70yd minimum range, and an accuracy of 20yds and 10 minutes in bearing. The set incorporated beacon facilities and antijam features.

SPS-10 and -10B antenna

SPS-10

A C-band surface search radar originally designed to replace SO-6/10 and which has since become standard for US warships, the current version being SPS-10G. The antenna is a 10ft parabolic cylinder with a gain of 1600 and a 1.5° × 16° beam, and 190–285kW pulses (0.25 and 1.3 microseconds for, respectively, high resolution of 41 rather than 213yds, and long-range performance; PRF 625–650) suffice for destroyer detection at 15nm. Scan rate is 15rpm, and there is an alternative mode with 2.25-microsecond pulses and a PRF of 312–325 in which the radar acts as a beacon. Effective IFF beamwidth is 6° × 22°.

There was no SPS-10A; SPS-10B incorporated a 500kW transmitter and SPS-10E a new antenna (1.9° × 16°, 17rpm). SPS-10F has a new PRF (625–660, beacon 317–330). Other versions correspond to SPS-10 with detail modifications. The beacon feature was removed from all SPS-10s by a 1972 field change. Deliveries began in October 1953, and it was considered so effective in ASW operations that SPS-10 was to equip 25 per cent of all ASW vessels by the end of 1956. As evaluated aboard the destroyer *Moale*, it could track buoys in to 40yds, and could detect them at an average range of 7nm. Detection range on a continuously exposed snorkel in Sea State 1–2 averaged 9630yds (maximum reliable range 12,100yds); maximum reliable tracking range on a radar periscope 6ft above water in Sea State 1 was 16,000yds; maximum reliable tracking range on an attack periscope 3ft above water in Sea State 1 was 10,100yds; and SPS-10 showed that it could reliably detect a submarine attack periscope intermittently exposed during attack situations in Sea State 2 and below.

In 1976 SPS-10 was described as the most reliable US surface search radar, with a Mean Time Between Failure of about 150 hours although with a 5- to 6-hour Mean Time To Repair a Failure. It is now being replaced by the even more reliable (and simpler) SPS-67. Sylvania.

SPS-11

A lightweight counterpart to SPS-2, essentially the same transmitter connected to a smaller, lower gain antenna described as 25ft × 13ft and of 7 tons in a 1949 report. In 1950 it was expected that SPS-11 would become the standard carrier and cruiser air search radar, with a range of 200nm (*vs* 300nm for SPS-2). However, it was not built.

SPS-12

An L-band air search radar, in effect an improved SPS-6 with greater capability. Indeed, it was originally described as SPS-6 redesigned to withstand a 70kt (*vs* 60kt) wind load. Total weight was about 4600lb, compared to 2550 for SPS-6B, antenna weights being 550 and 990lb respectively. Although the 500kW peak power did not rise, SPS-12 could produce long, 4-microsecond pulses at a higher peak rate (PRF 300 *vs* 150) to achieve better range performance, an alternative mode being 1-microsecond pulse and a PRF of 600. The PRF could be varied as an antijam measure. Coverage was similar to that of SPS-6B: a 3.3° × 30° beam rotating at 2.5–15rpm.

First delivered in September 1953, SPS-12 replaced SPS-6 in some large units. Advanced versions were proposed several times during the 1950s: first a

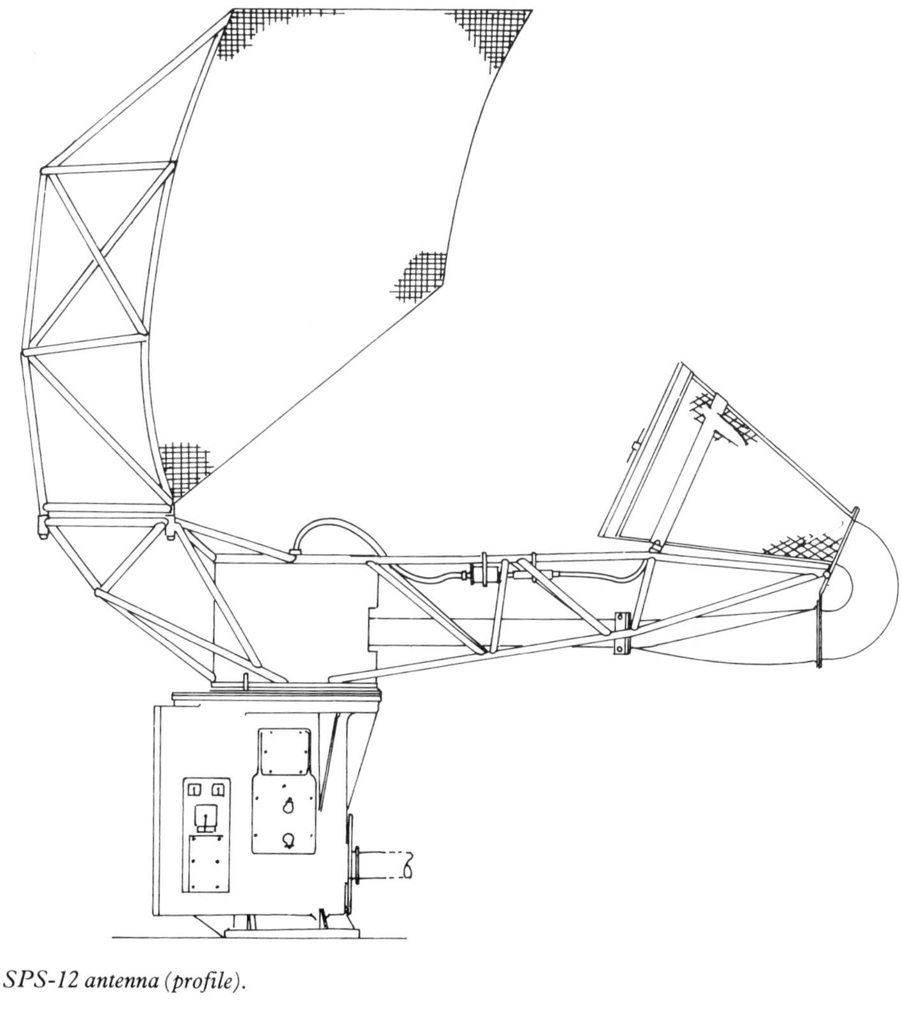

SPS-12 antenna (profile).

2MW version with a high-gain (24db, 7.3° × 24°) SPA-21 antenna; then in 1956 two 25ft × 8ft antennas with a 2.2° × 60° beam were built for the missile cruiser *Boston* and a YAGR, but they were never installed. At that time the improved SPS-12 was to have had a 24ft × 7.3ft, 3.1° × 30° antenna rather than the original 17ft × 6ft type, with PRFs of 150 and 600, for a 90 per cent probability of detection on a jet bomber at 100nm, service SPS-12 performance being only 55–65nm, or 75–90 with a new 2MW transmitter. The 25ft × 8ft antenna was to be able to achieve the latter level of performance at 700MW, and 100–135nm with a new 2MW power tube – which, like the new antenna, never materialized. The improved set, in stabilized form, was expected to combine with CXRX as a back-up for SPS-39. SPS-12A, an SPS-12 with field changes, was never produced, and -12B was assigned but later canceled. SPS-12C was a standard -12 modified by the addition of an RCA parametric amplifier in its receiver-transmitter.

RCA built a total of 139 SPS-12s in the United States; others were manufactured under license in Italy, and in 1981 this radar is far more common in Canadian and Italian than in US service. The SPS-501 radar of the newest Canadian escorts, of the *Iroquois* class, consists of a Dutch LW-03 antenna combined with an SPS-12 transmitter.

SPS-13
A stacked-beam radar intended for use aboard destroyers. Its S-band frequency was selected by scaling SPS-12 down to the requisite dimensions, ultimately a 20ft × 9.3ft antenna, which weighed 11,000lb out of a system total of 30,000lb. Power was also scaled down, to 2MW (10-microsecond pulses for long range, PRF 400). The design objectives included a range of 150nm on a conventional aircraft, or 100nm on a jet fighter with a 2m² cross-section, and a data rate of at least 10 seconds. Seven beams were required to keep the probable elevation angle error below ⅓°. Accuracy was said to be comparable to that of SPS-2, with a comparable range. However, SPS-13 was clearly too large for destroyers, and the only one built was installed aboard the missile cruiser *Canberra* in January 1959.

At that time the system was considered a potential replacement for the combination of SPS-8 and -12 aboard Terrier missile ships, and large-scale procurement for missile frigates was planned: 30 in FY60, 18 in FY61 and 5 in FY62. This program died with the success of the SPS-39 FRESCAN system. One official handbook claimed a detection range of over 200nm, and height information on targets at 95,000ft at 140nm. The seven stacked beams covered a 22° total vertical zone (individual beams 1.5° × 2.8°. Accuracy was 600yds, 0.15° in bearing, 20 minutes of arc (or 1000ft, whichever was greater) in elevation, and scan 3 or 6rpm. The antenna was to be stabilized against 30° of roll or pitch. Sperry.

SPS-14
An abortive modification of SPS-6 with automatic fault-detection. Its design was presumably a commentary on the unreliability of the former.

SPS-15
A cross-correlation version of SU-2 for periscope detection. Only one experimental model was built: 100kW, X-band, using a 1.9° × 3.8° beam (gain 1600) scanning at 6rpm.

SPS-16
A lightweight L-band air search radar designed specifically to replace the old SAs mounted aboard World War II destroyer-escorts. As tested aboard the destroyer-escort *Blair* in 1955 it could detect a pair of jet fighters at 40,000ft at 57nm (blip–scan ratio 0.5), and could maintain target indication out to 69nm. Expected performance had been 50–60nm on a 1m² target. The radar failed operational evaluation, partly because of unreliability, and was discarded. It was sometimes described as a small version of SPS-6, with similar gain; it had a 4° × 45° beam and used a 500kW transmitter and 2.4-microsecond pulses (PRF 400). Scan was 5 and 15rpm. Westinghouse.

SPS-17
The first of a new generation of P-band long-range air search radars, delivered from July 1957 on. Pulses of 750kW, or 1.5MW in some versions, sufficed for a range of 200nm on an F9F jet fighter at 40,000ft, presumably in view of sea reflection effects. By way of comparison, the smaller SPS-28 had a range of only about 100nm on the same target. PRF was 300, with 10-microsecond pulses for long range, and scan rate was 5–15rpm. The 17.5ft × 8.5ft antenna had a gain of only 63 (note the effect of a long wavelength) and produced an 18.5° × 27° beam. These dimensions were originally chosen to permit destroyer installation, but in fact SPS-17 was too heavy for that.

SPS-17A was a longer-range (300nm) version incorporating the 42ft antenna later used in SPS-37A and -43A. It was originally developed for the ocean radar-pickets (YAGR). A 75cm version was developed as SPS-31, the predecessor of the present SPS-40. The first prototype, using SR parts, was tested in 1952–53, and Westinghouse built the second, which was tested aboard *Willis A Lee* in February–April 1957. Twenty had been ordered in 1956, all for YAGRs; no more were contemplated, in view of SPS-29 development. A lightweight version was rejected in favor of Westinghouse's uprated SPS-28, which became SPS-29. SPS-17 was built by General Electric, and SPS-17A by ITE Circuit Breaker Company.

SPS-18
A lightweight C-band replacement for SPS-5, rejected in favor of a field change in the older radar. Advanced techniques were used to reduce size: the antenna was only 85in × 51in (98lb). Gain was about 1000, with a 1.6° × 14° beam (PRF 680, with 0.15-microsecond pulses), and as in SPS-10 there was a beacon mode (2.25-microsecond pulses, PRF 228). Vertical scan to 90° was also possible. SPS-18(XN) elevated to 69°; SPS-18(XN-2) was a pure surface search type. Scan speed was 17rpm, with vertical scan in 5–15 seconds, and peak power 180kW. The verdict of operational evaluation aboard the destroyer escort *Jack W Wilke* between March 1958 and January 1959 was that it was an excellent surface search set but entirely inadequate for air search; note that at the same time the air search feature of SPS-5 was being eliminated. Raytheon.

SPS-19
An experimental periscope detection radar, consisting of SG-6 plus a new Dalmo Victor antenna. The main problem in periscope detection was the 'noise', irrelevant returns from the random waves around the periscope. SPS-19 combated this noise by means of very high PRF (7000), fast scanning (1000rpm) and a relatively wide (3° × 3°) beam, which might average out some of the wave motion.

SPS-20

Another abortive periscope detector, corresponding to SPS-4 as SPS-19 corresponded to SG-6. The longer wavelength of the former showed in a slightly larger beam of 3.5° × 3.5°.

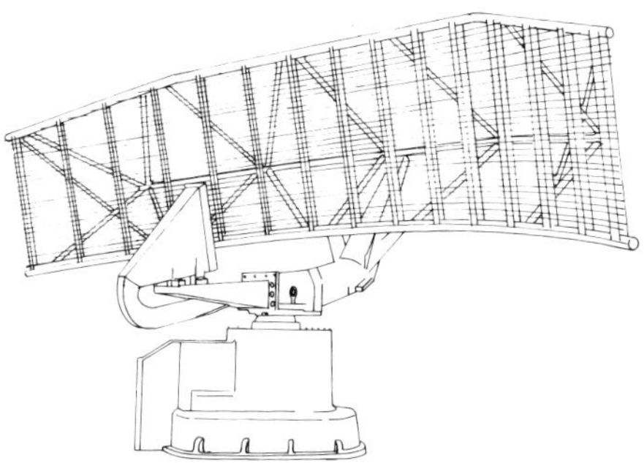

SPS-21 antenna.

SPS-21

A C-band lightweight (520lb) surface search set, for small ships or as the navigational radar of larger ones. It was first delivered in August 1954 and evaluated aboard PC-579 and PT-812 between November 1954 and April 1955. SPS-21B and -21C were modified to incorporate a variable range-marker in their displays; there were scales for 2, 4, 16, 40 and 80nm. SPS-21D was similar to the earlier models. These sets used very short pulses (0.19 microseconds PRF 1500) to give a short minimum range of 75yds. A peak power of 10kW gave a 12nm range on a destroyer, and the beam was 2° × 15°. This radar figured in Class Improvement Plans for the World War II destroyer escorts about 1960. Raytheon.

SPS-22

An abortive L-band air search radar for 'flush-decked vessels', using three radars in synchronous operation to cover up to 360° in azimuth. This is almost certainly the set of three SPS-6Cs in radomes planned for the *Forrestal*, before she was redesigned with an island and a more conventional electronic configuration. She would have had one on either beam and one right aft, and the function of the SPS-22 electronics was coordination, so that all three would feed into a common display. Westinghouse.

SPS-23

The primary Coast Guard surface search radar of the 1950s and 1960s, a militarized Radiomarine navigational set. It operated on X-band, with 0.25-microsecond pulses for the first four

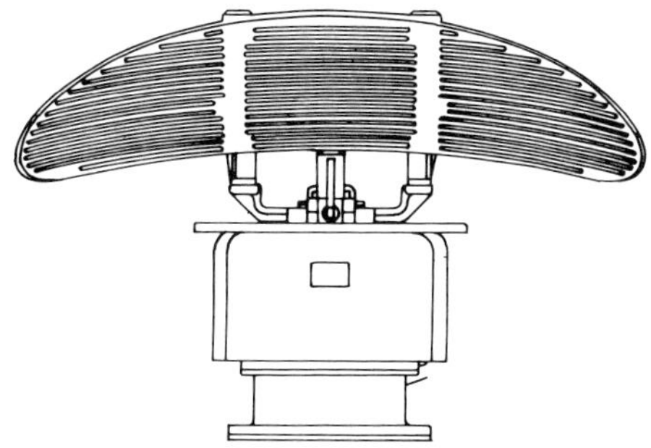

SPS-23 antenna.

range scales, 1.0 microseconds for the two long ones, a PRF of 1000 for both, and 50kW peak power, and range scales were 1, 2, 4, 8, 20 and 40nm. The $50\frac{1}{2}$ in × $14\frac{1}{8}$in, slotted, 1.9° × 20° antenna scanned at 17 or 20rpm depending upon antenna drive; it could also scan a 30°, 60° or 90° sector at 5rpm. Gain was 30db. Accuracy was 1 per cent or 25yds and 1.5°, and resolution 50yds and 1.9° at 1nm. It used STC and FTC to clarify target presentation, and could receive RAMARK beacon signals of 9310mc/s from fixed transmitters on lighthouses, lightships and harbor or other shore installations.

The original contract was let on 31 May 1951. Models differed according to antenna drive and power supply, one for small craft omitting sector and manual scan. SPS-23 and -23A required standard 208/220/240V AC power; -23 could operate on 115/230V, 60-cycle, 1 phase power; -23Y or -23XX on 115V DC; and -23Z on 32V DC power. It may have been an SPN-11 revised to meet military specifications, at what Radiomarine regarded as a disproportionate increase in cost.

SPS-24

An abortive 6cm hemispheric search radar, using the Guthrie antenna system: three stacked beams, not overlapping but 20° apart, nodding slowly at 15 nods per minute while the entire antenna rotated rapidly at 90rpm. The three 3.5° × 6° beams were not uniformly illuminated (500kW in the lowest, 300kW in the middle, and 200kW in the upper beam), to achieve greatest range at low angles, corresponding to realistic aircraft altitudes. In this way a very light reflector, sized at 6ft × 3.5ft and weighing 600lb, could duplicate the performance of the heavy SPS-3; in fact, SPS-24 was expected to provide accuracies of $\frac{1}{2}$° in bearing and 1° in elevation, out to a range of 22nm and a maximum elevation of 60°.

This system was first proposed in 1951, but it was beaten out by the FRESCANs. The light antenna weight included stabilization; below-decks weight would have been another 5000lb. SPS-24 might have gone aboard destroyer-escorts, but it failed to meet range requirements. Note that the high scan rate required very powerful pulses, since any target would meet relatively few of them. SPS-24 was also designated P-4X, in a series of developmental electronic equipment.

SPS-25

An abortive S-band lightweight hemispheric search radar, using a Kelleher scanner producing 25 nods per minute while rotating at 15rpm. Early (1954) estimates called for a 500lb antenna providing elevation coverage from 0° to 86°, with a 50nm range on a 1m² target, given 2.5MW pulses. Below-decks components were to be derived from the airborne AN/APS-20B (AEW) radar, and would have weighed only about 1500lb, compared to about 8000lb for CXRX. About a year later a table of air defense radars included a stabilized version, with an 1100lb antenna and 600lb of below-deck components, with elevation to 70° and a range of 40nm. Beamwidth would have been 4.4° × 12.5°, depending upon elevation. SPS-25 was also designated P-10X.

SPS-26

The first of the electronically-scanned three-dimensional (hemispheric) radars, using frequency scan (FRESCAN) in elevation; it solved the hemispheric search problem. The first laboratory model, using a 1MW klystron, was ready for demonstration in August 1953. It began tracking targets that September, and development of a naval set began in March 1955. The first was delivered for evaluation aboard the experimental frigate *Norfolk* in August 1957.

In tests, SPS-26 tracked pairs of F3H, F11F or F9F aircraft to the horizon at low altitudes, and to ranges of 100–110nm for all altitudes up to 45,000ft; a single F8U (Crusader) was tracked at 55,000ft out to 65nm under favorable atmospheric conditions. Maximum range and altitude limits were 160nm and 75,000ft, and MTI was incorporated in the design. For each scan of the antenna in elevation, 26 pencil beams, each 2.4° × 3°, were emitted

SPS-26 radars on land test; the one at right has an IFF attached. The original version of SPS-39 used the same antenna. HUGHES

at different frequencies, the centers of the upper 13 beams being separated by about 3°, the lower 13 by about $\frac{3}{4}$° in elevation. Normal scanning rate was 15rpm but was reduced for each of four MTI modes. Peak power was 1MW (4-microsecond pulses, PRF 495–3380 for alternative range limits).

The Operational Development Force found that SPS-26 was effective in controlling a CAP section at ranges of up to 80 or 90nm, and considered its maximum range too short to make it effective as a primary air search radar: '. . . early warning information from some other source such as the SPS-17 radar is necessary in order to realize the full potential of the SPS-26. The limited capability of the SPS-26 in providing tracking at ranges of 100 to 110nm is considered marginal for the present Talos missile and completely unacceptable for missiles with longer ranges'. In addition, SPS-26 was considered unreliable. It was developed into the SPS-39, which saw wide service. Hughes.

SPS-27
Designation canceled; it would have been the production version of the CXRX three-dimensional air search radar (see SPS-8 entry).

SPS-28
The lightweight counterpart of SPS-17 for destroyers, using the SPS-17 antenna but a less powerful transmitter, incorporating some SR components. First delivered in January 1957, it was widely deployed aboard destroyers in the late 1950s. Pulses of 250kW (4 microseconds, PRF 150) gave a reliable range of 100nm. The 17.5ft × 8.5ft mattress had a gain of about 70 and produced a 19° × 27° beam. SPS-28A, slightly heavier in appearance, used an SPS-29 antenna. In all, 55 were produced for destroyers and DERs. SPS-28 was rapidly replaced by -29, and at present the sole surviving example is the main radar of the Philippine destroyer escort *Rajah Lakandula* (ex-USS *Camp*). Westinghouse.

Foremast of USS Mansfield *(DD-728), showing SPS-29 long-range air search radar* **(20)** *with IFF* **(18)**, *SPS-10* **(19)** *and Mk 25, March 1965.*

SPS-29
A Westinghouse P-band long-range air search set (October 1958) which achieved the performance of SPS-17 on the weight of SPS-28 (which it soon replaced). Of six versions produced, all but SPS-29D used a 7×4 element mattress very similar to that of SPS-28, about 18ft × 11ft and with a 20° × 25.5° beam, producing very long, 10-microsecond, 750kW pulses 300 times per second. SPS-29D, which is standard in large Coast Guard cutters, used a corner reflector similar to a part of the big SPS-37A/43A antenna. SPS-29 had an instrumented range of 270nm. Altogether 89 were produced: 50 were in the fleet by 1 July 1959, and at that time another 32 were scheduled for delivery by May 1960.

SPS-30
The culmination of General Electric pencil-beam height-finder development, achieving very long range through a combination of very high peak power (2.5MW), long pulses, and a low PRF. The instrumented range of 240nm implies a PRF of about 300, and the reported average power level of 9kW implies pulses about 12 microseconds in length. Scanning is by means of an organ-pipe scanner, which is much faster than the Robinson feed of earlier types, the only motion being a rotation of the feed *within* the organ-pipe structure. Larger motions are reserved for shifts in the sector to be scanned. Thus SPS-30 scans any 12° vertical sector between 0° and 36° at 4, 8, 12, 20 or 40 scans per second. The 12ft × 15ft solid antenna produces a 1.5° × 1.2° beam, and antenna gain is reportedly 41db. Sector scan in azimuth is 1, 2, 3, 5 or 10prm. It was first delivered in May 1962, and it replaced the SPS-8 series; in turn, SPS-30 is being succeeded by electronically scanned radars. A total of 57 was built.

SPS-30 pencil-beam height-finder, USS Shangri-La, *November 1969.*

SPS-31
A long-range P-band (75cm) air search set, a product of the same program which resulted in SPS-17 and -29. It was the progenitor of SPS-40, which latter benefited from considerable weight-savings. The contract was awarded in June 1956, and SPS-31 was closely related to a new AEW radar intended to replace APS-20E, the APS-70 (425mc/s, 2MW, employing Airborne Moving Target Indication); tests with APS-70 mounted aboard the destroyer *Richard E Kraus* from September 1958 through

February 1959 encouraged the adoption of SPS-31 and then of SPS-40.

Although it was in effect a 75cm version of the long-wave mattresses, SPS-31 employed a horn feed, as in the shorter wave SPS-6. This was nearly the lightest (and the smallest) of all the types considered by NRL, even though it had the severe defect of covering much of the reflector area. The alternatives were parabolic cylinders fed by waveguide slot arrays or by arrays of dipoles. Peak power was 2MW, with 6-microsecond pulses and a PRF of 300. The 18ft 6in parabolic antenna had a 9.5° beam (gain 23db) and an accuracy in range of $\frac{3}{4}$ per cent plus or minus 100yds. Range scales were 5, 10, 100 and 250nm. SPS-31 appeared in February 1959 as an alternative to SPS-29. It was experimentally installed aboard the radar-picket destroyer *Furse*, 1958–62. Lockheed Electronics.

SPS-32

The fabulously expensive SCANFAR very long range P-band electronically (frequency) scanned search radar. Four panels, each 40ft × 20ft, are required for 360° coverage. Including deck equipment, the four weigh about 48.5 tons. Each has 36 panels set in 18 columns. The vertical columns connect to taps on an end-fed undulating coaxial cable 1100ft long and compressed into a total of 40ft. Note the use of a coaxial, as opposed to the waveguide of S-band FRESCANS. The individual radiators are set in plastic for better blast resistance. Each antenna has a beam pattern roughly 7° × 50°; detection ranges of up to 400nm are claimed with 1.5MW (20-microsecond pulses, PRF 200). Only two were built, for the nuclear warships *Enterprise* and *Long Beach*, and the lack of success of this system is suggested by the absence of follow-ons in later carriers. It and SPS-33 were removed in refits, 1980–81.

SPS-32 (**1**) *and -33* (**2**), *USS* Enterprise *(CVAN-65).*

SPS-33

Companion to SPS-32 in SCANFAR, a tracking radar with frequency-scanning in one dimension and phase-switching in the other, to achieve a rapidly moving pencil-beam. Each panel is about 20ft × 25ft and reportedly weighs about 125 tons, which shows the effect of requiring beam motion in two directions. Both SPS-32 and -33 were developed by Hughes Aircraft, which had begun work on frequency scanning in the early 1950s.

SPS-34

SPS-2 repackaged with a lighter antenna, intended to support the Talos system. For a time the standard missile cruiser radar system was to have included either this radar or SPS-30, but in 1960 the latter was selected, even though in theory one SPS-34 would be far more effective than two SPS-30s in range and target-handling capacity. One reason for the rejection of the larger radar was the poor reliability of its predecessor, SPS-2; it also cost $5 million, as against $750,000 for SPS-30. SPS-34 was a shipboard version of the FPS-7 radar of the USAF SAGE system. Detailed data are not available, but published descriptions of the FPS-7 mention a range of 300nm and height coverage to 150,000ft. General Electric.

SPS-35

A Raytheon small-boat X-band radar. 7kW pulses, with a PRF of 1500 for 1, 2, 4, 8 and 16nm range scales and 750 for 32nm scale (0.2 microseconds in both cases), in a 2.2° × 14–18° beam scanning at 20rpm, gave a resolution of 50yds and 2.5°. Range on a destroyer was 16nm, compared to the 20 of more powerful sets and the 12 of SPS-41 or -46. Antenna weight was 80lb. First delivered in December 1957, and widely known as Pathfinder 1500. Originally designated SPN-21.

SPS-36

A nonmagnetic short-range surface search radar for minesweepers, originally designated SPN-23. Peak power is 10kW (X-band), using a constant PRF of 1000 but a variable pulse-width of 0.1 microseconds for $\frac{1}{2}$, 2, and 4nm range scales, and 0.45 microseconds for 8 and 16. The 2° × 28° beam scans at 17rpm, and minimum range is 25yds. Antenna weight is 90lb. Principal employment is in minesweeping boats (MSB). First delivered March 1958. Edo.

SPS-37

A pulse-compression derivative of SPS-29 with 200-microsecond pulses compressed to 6 but using the same antenna. It is deployed aboard many destroyers and small missile cruisers. SPS-37A, aboard carriers and the large missile cruisers, uses the big antenna originally developed for SPS-17A. The change in antenna increases effective detection range from 233 to 300nm. First delivered in September 1960, having been operationally evaluated aboard the cruiser *Los Angeles* (installed at Long Beach in March 1960). 46 were built. Westinghouse.

SPS-38

An abortive very long range (300nm on a $1m^2$ target) air search radar, designed to emphasize frequency-diversity as a counter-countermeasure. It incorporated three distinct bands, 510–570, 570–630 and 630–690mc/s, with 15 channels per band, and could jump from one to another on a pulse-to-pulse basis, manual, automatically programmed, or randomly. The operator had a special countermeasure receiver and analyzer. Externally SPS-38 would have resembled the contemporary SPS-37A, with a 40ft × 6ft 6in antenna, producing a 3° × 15° beam, and roll-synchronized at 3–6rpm; it scanned at 5 or 15rpm. The frequency-hopping design precluded the pulse-compression features of its contemporaries, but instead very power-

SPS-38 anti-ECM long-range air search radar, showing stabilization.

ful (3MW) 8-microsecond (PRF 250) pulses were used to achieve a range resolution of 1300yds.

The specialized counter-countermeasure design was very expensive, it being estimated that in 1961 that the prototype would cost about $7 million, as compared to about $280,000 for SPS-43A. At the same time the United States abandoned the 600mc/s radar band (see also SPS-45). One SPS-38 was delivered in June 1961, but it never went to sea. The contract was awarded in early 1958.

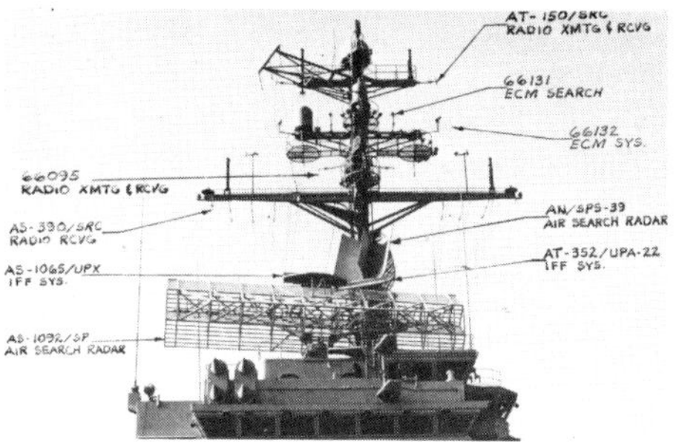

The island of the carrier America, *looking aft (December 1964), showing SPS-39 and SP antennas.* BY COURTESY OF A D BAKER III

SPS-39

The production version of SPS-26, first delivered in January 1960. Electronic stabilization replaced the mechanical system of earlier pencil-beam heightfinders. It was a medium-range S-band system, with instrumented ranges of 60 and 160nm; an effective range of about 125nm was unofficially reported, and in about 1960 this radar was credited with a range of 150nm on a 1m² target. The antenna, weighing about 3000lb, was a cylinder tilted about 15° to the vertical, with a swing circle diameter of 9.5ft and a series of 2.4° × 3° beams. Peak power was 1MW, with pulse lengths of 4 microseconds in the Long Range (160nm, PRF 925) or 2 microseconds in the Short Range (60nm, PRF 1850) mode, some SPS-39As having 3-microsecond Short Range pulses. Antenna rotation rates were 5 and 15rpm, or 11.5rpm in an MTI mode, and maximum elevation was 48° (with an alternative Limited mode elevation of 15°).

In 1963 Series III field changes introduced a new planar array antenna consisting of waveguide slot radiators, as well as a new transmitter and a new parametric amplifier in the receiver. Two alternative antennas were developed: a small one, which was to have more gain (37 *vs* 34db), lower side lobes and narrower beams (1.7° × 2.25°) than the earlier cylindrical one; and a wide one (21ft rather than 13ft swing circle) which was intended as the target designation radar for Talos and Terrier ships. The latter produced 1.1° × 2.25° beams and had a gain of 39.5db. The small antenna, which was also the type originally employed by SPS-52, was evaluated aboard the missile frigate *Sampson*; the large one, originally designated SPS-39B (only one of which was built), was evaluated aboard the missile cruiser *Galveston*, later being installed aboard *Norton Sound*.

Series III radars had three operating modes: High Data Rate (15rpm, 60nm, 2.5-microsecond pulses), High Angle Mode (160nm, 7.5rpm, 6rpm in the large antenna, 4.6-microsecond pulses), and Long Range (245nm, with the same scanning rate and 10-microsecond pulses). Maximum elevation in the high data rate and high angle modes was 42° (with an alternative limited angle of 13° in these and in the normal long-range modes), and there was a Limited Long Range mode with a 4.5° maximum elevation. Maximum height coverage was 100,000ft, which required some changes in the elevation scanning program. In addition, there was an MTI mode (5.5rpm in the large antenna, 6.5 in the small). The parametric amplifier itself was a major improvement, with a noise figure of only 3.5db, compared to 8 in the earlier SPS-39/39A.

The combination of reduced amplifier noise and increased antenna gain (which was actually a smaller improvement than that provided by the amplifier) produced spectacular improvements in range. For example, using an NRL formula, SPS-39 had an effective range of 63nm on a 1m² target. In its long-range mode, the small antenna was credited with 145nm (115 at high angle), and the large with 190 (150). In tests, the *Galveston* radar achieved an average maximum tracking range of 172nm on small, high-altitude, inbound jet fighters, using an alerted operator. The evaluation report noted that 'the planar array made the radar exceptionally free from side lobe cluttering effects in both an ECM and a non-ECM environment'. Reliability improvement was a major factor in the Series III program; tests of SPS-39 in 1960–62 showed an MTBF of only 14.3 hours, and ashore an SPS-39A (1961–62) achieved only 22.9. Aboard the *Sampson*, the MTBF of SPS-39(III) was as great as 43.2 hours, and the *Galveston* achieved 67.4

SPS-42 is an alternative model of SPS-39 adapted to interface directly with NTDS. Most SPS-39s are now being replaced by SPS-52s or, in larger ships, SPS-48s. Hughes.

SPS-40

A Lockheed Electronics pulse-compression radar operating in the 400–450mc range. It was derived from SPS-31 and first delivered in June 1961, being evaluated aboard the destroyer *Richard E Kraus* that year. Originally intended only for destroyers, it has become the standard US lightweight air search set, replacing SPS-6 and -29 and also delaying the introduction into service of SPS-49.

The original SPS-40 operated at a peak power of 200kW, with a PRF of 300, using 60-microsecond pulses which could be compressed to 0.6 microseconds in the receiver, so that system performance equated, in theory, to that of a 20MW radar with 0.6-microsecond pulses. Thus SPS-40 benefited from the excellent range resolution associated with a 0.6-microsecond pulse, 100yds or less; on the other hand, its minimum range of 6nm was imposed by the long (60-microsecond) transmitted pulse. Beam dimensions were approximately 11° × 19°. In its initial tests SPS-40 showed that it was essentially horizon-limited against single jet aircraft flying at up to 50,000ft. In the first lobe (of the fade chart), the probability of detection of a single jet fighter was estimated at 100 per cent at altitudes up to 35,000ft, and 85 per cent at 50,000ft.

SPS-40A, an improved version by Sperry, introduced a broad-band transmitter and some solid-state receiver changes. SPS-40B incorporated the new AIMS IFF, for air control of friendly

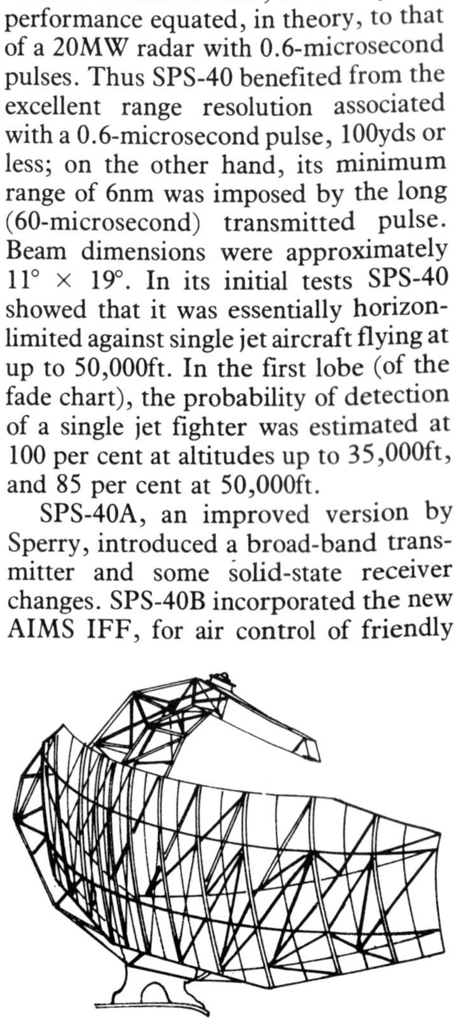

SPS-40 antenna.

aircraft, and a Low Flyer Detection Mode (LFDM). It may also include a Minimum Range Modification (MRM) tested aboard the destroyer *Cone* in 1965, which provides range and bearing data on low-flying aircraft between 100yds and 18nm, for control of friendly ASW aircraft.

Service test sets were built by Sperry, but production was awarded to Dynell Electronics, with production deliveries beginning in November 1971. By that time a total of 133 SPS-40s and -40As had been funded, and the Dynell contract included 43 new SPS-40Bs, 66 modification kits for SPS-40, and 46 for -40A. SPS-40C includes an LFDM modification, and is more reliable: the original SPS-40 and -40A were often criticized as unreliable, but modifications incorporated in the solid-state -40C and -40D were credited in 1978 with a 40 per cent improvement in reliability and maintainability. Field change kits are to ncrease Mean Time Between Failure (MTBF) from 80 to 200 hours. Other improvements currently in train include an automation module which will give SPS-40 radars an automatic detection capability, digital MTI, and an interface with AN/SYS-1. SPS-40 is one element of the DDG Upgrade program.

SPS-41
A lightweight (123lb antenna) X-band small-boat radar, first delivered March 1959. Peak power is 7–10kW with a PRF of 625 and 1600 for, respectively 0.4- and 0.1-microsecond pulses (long pulses for 8, 16 and 32nm range scales, short for 0.5, 2 and 4), using a 1.8° × 20° beam scanning at 21rpm, for a bearing accuracy of 2°. Minimum range is 25yds. Of 35 in use in 1973, one was aboard the missile cruiser *Albany* and another on an attack transport. Bendix.

SPS-42
See SPS-39.

SPS-43
An improved SPS-37, externally indistinguishable from it; in fact SPS-43 was originally designated SPS-37A, before that suffix came to denote the long antenna. The principal improvement is in ECM resistance. For example, there are 20 channels in place of the 10 of SPS-37, which probably indicates the use of longer pulses, *ie* the short base pulse is split into more frequency-shifted components. Both small and

SPS-43A long-range air search radar aboard USS Nimitz, *November 1976.*

large versions exist: SPS-43 for missile cruisers formerly rated as missile frigates, and SPS-43A for carriers and missile cruisers converted from conventional cruisers. First delivered in March 1961; 49 have been built. SPS-43 is to be replaced by SPS-49.

SPS-44
An abortive lightweight version of SPS-13, the last of the US naval stacked-beam height-finders.

SPS-45
An abortive GE-designed destroyer air search radar, intended to provide frequency diversity (600–622mc) alongside SPS-29 and SPS-40, and canceled by CNO in May 1960 when this band was abandoned to civilian use. Like SPS-38 (XN), it incorporated a frequency-hopping counter-countermeasure feature. Two were ordered for delivery in December 1961: XN-1, a roll-stabilized 1150lb type with a 7.5° × 18° beam, and XN-2, an unstabilized 1000lb type antenna with a 5° × 30° beam. Both used long pulses (10 microseconds, PRF 240) to achieve good range without very high peak power. The original contract had been awarded early in 1959.

SPS-46
A small-boat X-band radar, also used as a navigational set for larger ships: it uses a slotted array like that of SPS-55. In large scale service from February 1961. Peak power is 7kW with 0.4- and 0.2-microsecond pulses and PRFs of, respectively, 750 and 1500 for 16 and 32nm and 1, 2, 4 and 8nm scales. Beamwidth is 2.2° × 15° and scan rate 15–21rpm. Resolution is 25yds and 2°. Lavoie Laboratories.

SPS-47
A variant of SPS-39A, perhaps no more than an alternative designation.

SPS-48
A high-powered (2.2MW) FRESCAN three-dimensional radar, often described as the lineal descendant of the big stacked-beam sets. It was originally designed because SPS-39/42 could not meet the long-range requirements of the Talos missile system, and the first contract for two radars was let in June 1960. Development began in 1959, when it was described as an electronic scanning stacked-beam radar for extremely long range, a series of FRESCAN stacked beams. At the time, the chief advantage of FRESCAN was thought to be the elimination of the mechanical stabilizer base which accounted for much of the weight and complexity of other stacked-beam radars.

The end plate houses a sinuous waveguide feed. SPS-48 used multiple beams to combine long range with a high data rate and multiple-pulse (*ie* high probability) detections, as in Type 984. There are nine stacked beams in each group, which is generated by a stream of nine 3-microsecond pulses, spread in frequency to give a total spread of 6° for the group. Successive groups elevate from 0° to 45° An alternative burn-through (antijam) mode allows transmission for the whole 27 microseconds at one frequency. Maximum instrumented (low-angle) range is 230nm. SPS-48 is rated at a 220nm range with a minimum ceiling of 100,000ft. Accuracy is 230yds and $\frac{1}{6}$° in both bearing and elevation, and resolution 500yds and 2°. Scan rates are 7.5 or 15rpm. The 17ft × 17.5ft antenna weighs some 4500lb out of a total system

SPS-48. Note ground plane to separate the antenna from SPS-10 above.

weight of about 22,000. In 1965 it was officially estimated that one SPS-48 was equivalent to ten SPS-8s or to two SPS-2s.

The first production set went to sea aboard the cruiser *Worden* in March 1965. SPS-48C differs from -48A primarily in that it incorporates ADT, a modification which is to be extended to all SPS-48s in service. SPS-48E is to be fitted to missile cruisers as part of the CG/NTU (New Threat Upgrade) program. It doubles effective power radiated, reduces side lobes, and increases receiver sensitivity. ITT-Gilfillan.

SPS-49

A long-range two-dimensional air search radar, unusual in that it is stabilized. It was evaluated aboard the missile cruiser *Dale* in 1976, and its first application is aboard the *Perry* class missile frigates; ultimately it will replace the P-band long range search sets of the SPS-37/43 series. It was originally developed as part of a BuShips frequency-diversity program: the allowed bands at 1.5m and at 76cm had already been filled, and the band at 600mc discarded, leaving an allowed band at 890–942mc.

SPS-49 (XN), the prototype, was tested aboard the destroyer *Gyatt* between July 1964 and January 1965, and this system figured in many of the paper studies of the late 1960s. Improved versions of SPS-40, which was far less expensive, were adopted instead, and the SPS-49 was revived only in 1970, in much modified form.

The data which follow apply only to the original radar, and indicate the state of US radar development in the mid-1960s; they do not apply to the service system. At that time SPS-49 was a 280kW pulse-compression radar, transmitting 125-microsecond pulses corresponding to a 2.5-microsecond resolution with PRF fixed at 285 or variable between 142 and 285. The original design requirement was detection of a 1m² target at up to 250nm and 150,000ft up to a maximum elevation angle of 20°; lower PRFs permitted unambiguous detection of targets at greater ranges. Frequency was 850–942mc, and the 24ft × 14ft antenna produced a 3° × 10° beam. It weighed 1700lb, with an additional 5860lb in the radar room and 1419lb of power regulator equipment. Scanning rate was 6rpm.

SPS-49 air search radar at the Sperry Land Test Site for the Perry *class combat system. The radome is a Mk 92 system.* Sperry

In tests, MTBF was about 60 hours and SPS-49 was criticized for deficiencies in detecting targets in clutter and at very short ranges. The current SPS-49 has a demonstrated MTBF of over 300 hours and a 24ft × 14ft 3in antenna weighing (with other above-decks equipment) 3210lb; below-decks weights are now 13,791lb. Improvements include new clutter rejection features and ECCM provision, and the new radar is adapted to ADT. Raytheon.

SPS-50

This system was essentially the prototype SPS-49 modified to exploit the next step up in frequency, about 23 rather than 33cm, or the 1215–1400mc/s band. Had it not failed operational evaluation, it would have replaced the SPS-6 and -12 series.

SPS-50 was designed to detect a 1m² target at up to 250nm and 150,000ft, up to a maximum elevation angle of 20°. To overcome the minimum range limitation inherent in pulse-compression radars, it had three modes: long pulse (125 microseconds, compressed to 2.5), medium (2 microseconds) and short (0.2 microseconds), with PRFs matching those of SPS-49. The beam was 2–2.4° × 9.5°.

In the short-range mode (short pulse), minimum range was 500yds and maximum 25nm, timed to take advantage of the normal radar dead time, thus allowing normal long-range operation without degradation of maximum range. The short pulse was generated by a separate impulse generator, and bypassed the compression/expansion system. Another innovation was a medium-range (2-microsecond) mode, with a maximum range of 63nm, intended to improve sub-clutter visibility (SCV) at medium ranges. To achieve sufficient dead time to accommodate this mode, the original trigger logic had to be redesigned and maximum range in the long-range mode reduced to 180nm.

Signal-processing included coherent MTI. Processing options were MTR (Moving Target Resolver), AMTI (Area Moving Target Indicator), LIN (Linear), LOG (Logarithmic), SHT RNG (Short Range), and MTI (Medium Range). In the MTR mode, successive pulses were added together before compression, with doppler information extracted during pulse-compression. This type of processing also eliminated jamming signals. In AMTR, signals with the same bearing and range from scan to scan were eliminated; those with sufficient radial velocity (about 150kts) moved far enough from scan to scan to escape their previously recorded positions. Finally there was a Jamming Direction-Finder (JDF) circuit.

Compared to SPS-49, SPS-50 incorporated a fast-switching klystron modulator in which all tubes were replaced by transistors, as well as modified internal logic. As evaluated, SPS-50 was considered unreliable, with fault location extremely difficult. The short-range mode was considered unneccessary, particularly in view of the complications it imposed.

SPS-51

Coast Guard X-band (75kW) surface search radar, with high resolution. Range scales: ½, 1, 2, 4, 8, 30 and 40nm. Lavoie Laboratories.

SPS-52

A development of SPS-39, incorporating a new planar antenna, a parametric amplifier and a wide-pulse feature for longer range. Stabilization is digital rather than analog as in SPS-39. SPS-52 can detect a small jet at 60nm using a 4-second data rate; range on a similar target in the wide-pulse mode is about 245nm. SPS-52 is considerably lighter than SPS-48, and it is externally indistinguishable from SPS-39 with the Series III field change. Both are widely deployed aboard the smaller missile ships.

The current SPS-52C has four modes: high angle (moderate range and data rate); long range (high energy pulsewidth for target acquisition at long range); high data rate (quick detection and rapid update of close-in and 'pop-up' targets); and MTI (clutter rejection). Medium-range coverage employs a

SPS-52C on test, showing the current planar array antenna, which differs considerably from earlier models.

6-second data rate, long range and MTI employ the 8-second rate, and short range the 4-second rate. The SPS-52C differs from the previous SPS-52B in reliability (MTBF of 216 hours *vs* 189, resulting from increased use of solid-state components and de-rating of components), maintainability (more extensive test features, better accessibility, and modular design), and in compatibility with the SYS-1 Integrated Automatic Detection and Tracking (IADT) system. Normally IADT takes inputs from various sensors, such as SPS-52C and SPS-40, integrates them to form target-track files, and passes them to NTDS, which passes this data on to a Weapons Designation System (WDS) to provide target acquisition data for gun and missile fire control systems. There is an alternative Quick Reaction Mode in which the IADT passes data directly to the WDS in response to 'pop-up' threats. It is claimed that SYS-1 improves overall reaction time by 116 per cent.

SPS-52C is the primary sensor of the IADT system and can function as its only sensor, although the use of multiple radars at different frequencies is considered an important ECCM feature of the system as a whole. In effect the SPS-52C consists of an improved SPS-52B with a Video Extractor and Control Group (VECG) added to permit it to interface with SYS-1, via a UYK-20 computer. Land-based testing of the SPS-52C and IADT began in October 1977 at Johns Hopkins University's Applied Physics Laboratory, and these systems were installed aboard the missile destroyer *Towers* in December 1977, Technical Evaluation being completed successfully on 28 April 1978 with a formal Operational Evaluation from 6 June through 22 September 1978. Hughes.

SPS-72, next-generation SPS-52 antenna, on test. HUGHES

SPS-53
A Coast Guard high resolution X-band radar; also aboard naval vessels, including the battleship *New Jersey* as refitted, ocean and coastal minesweepers, fleet tugs, and gunboats. A 5ft slotted array provides a 1.6° × 20° beam, with 40kW pulses of 0.5 and 0.1 microseconds and PRFs of, respectively, 750 and 1500, for long and short ranges. Scan rate is 15rpm. In some ships it replaced SPS-5. Sperry.

SPS-54
A surface search radar evaluated aboard the destroyer *Gyatt* in February 1968. It was intended to supplement or replace SPS-5, -21, -35, -36, -41 and 46 and the commercial navigational radars then in use in the fleet. Peak power was 10kW, using 0.1- or 0.5-microsecond pulses (PRF 3000, or 1500 without jitter). The 2° × 20° beam scanned at 20rpm. Accuracy was 25yds plus or minus 2 per cent for the first range marker, 1 per cent for the others, and 1 per cent at 8000–50,000yds. Although evaluation was successful, production was not ordered. Sperry.

SPS-55
An X-band replacement for the C-band SPS-10, the new frequency having been chosen to avoid interference with standard C-band missile target-trackers. SPS-55 is a slotted-array search radar, with 0.12-microsecond pulses, PRF 2250 for short range/high precision or 1-microsecond pulses, PRF 750 for long range. Peak power is 130kW. The end-fed array actually consists of two arrays back to back, so that the operator can choose either circular or horizontal polarization. The 6ft, 80lb antenna forms a 1.5° × 20° beam and can detect objects about 50yds away from it; it has a gain of 1260. Developed by Raytheon, but production is by Cardion Electronics, the first contract being in June 1971. Its first large-scale application is aboard the *Spruance* class destroyers.

SPS-56
An Army surface search radar, later redesignated SPS-64(V)5. It operates in the X-band and is effective between 50ft and 40nm. Radiomarine commercial model CRM-NID 75. One version, operating in S-band (peak power 60kW) was manufactured by Raytheon as commercial model 1650B.

SPS-57
Small-craft surface search radar for the Coast Guard: X-band, peak power 3kW, 0.1-microsecond pulses (PRF 2000) for 0.5–2nm scales, or 0.2-microsecond pulses (PRF 1000) for 4–16nm. It produces a 1.9° × 25° beam scanning at 25rpm. Range scales are 0.5, 1, 2, 4, 8 and 16nm and minimum range is 25yds on the 2nm scale, 100yds on others. Ridge Electronics.

SPS-58/65
L-band pulse-doppler air search target acquisition radar to work with the Point Defense Missile System.It was developed at high priority after the sinking of the Israeli destroyer *Eilat* by Egyptian 'Styx' missiles in 1967, and was reportedly patterned on a West German Siemens radar with excellent anitclutter capabilities. That was probably the MPDR-45, one of a series of Mobile Pulse Doppler Radars, with a 4.4° × 11° cosecant-squared beam to 40° in elevation and scanning at 19rpm. Throughout the desig the emphasis was on simplicity and low cost, in contrast to the Hughes Target Acquisition System. Construction is modular for maximum standardization, with two alternative antennas: a large stabilized, 16ft, horn-fed ellipse and a lightweight antenna, essentially SPS-10 with a dual feed permitting duplex operation. SPS-58 and -58B use the large antenna, the latter having no display but rather feeding directly into NTDS; -58A and -58C are analogous systems duplexing with SPS-10. There is also a new 9ft 6in planar array antenna. MTI is used to exclude clutter, and improvements in that direction continue to be made. Current production versions are designated SPS-65. Westinghouse.

SPS-59
An Army radar designed to provide anti-collision, piloting and position data for inland waterway navigation. It is the X-band Raytheon 1900/ND. SPS-59(V)1 (X-band, 10kW, 0.05- or 0.5-microsecond pulses) is manufactured by Canadian Marconi.

SPS-60
A solid-state version of the vacuum-tube SPS-53E, using a standardized NAVSEC 8ft antenna. Sperry.

SPS-61
Teledyne-Ryan anti-low-flyer search radar, an unsuccessful competitor to SPS-58.

SPS-62
A developmental L-band air search radar, designed to detect low flyers and surface threats in a highly cluttered environment for designation to a point-defense missile system. It was originally described as an 'alarm radar', and shared many SPS-58 components – an identical transmitter and multiplexer, a modified receiver-processor, and a new antenna and radar set control. Its 114in planar array generated a 5.5° × 21° beam with 12kW pulses, 5 microseconds long, and with a PRF of either 2280 or 3040. Procurement of 9 sets was proposed in 1973, but in fact development was dropped, and some components were incorporated instead in advanced versions of SPS-58. The long pulse (which suggests a pulse compression mode) and the high PRF give a relatively high average power equivalent, for example, to 240kW for 1-microsecond pulses at a PRF of 1750. Westinghouse.

SPS-63
The US version of the Italian 3RM-20H navigational radar, used on the patrol hydrofoils of the *Pegasus* class. It is an X-band slotted array scanning at 25rpm with 20kW peak power, and has a minimum range of 20yds and a maximum of 40nm. Accuracy is equal to 2 per cent of the range scale; resolution is 10yds at 0.25nm on the range scale, 100yds at 10nm, and 400yds at 40nm. Bearing accuracy is 1° and resolution 1.2°. Total system weight is 217lb. The display incorporates true motion capability which stabilizes a display of fixed targets against own-ship forward motion. Dynell Electronics.

SPS-64
Marine navigational radar for Army, Coast Guard and Navy use. It is a modular system, with three optional transmitters, 20kW fixed or tunable X-band, 50kW X-band and 60kW S-band and has four alternative antennas (all end-fed slotted arrays), 4ft for X-band only, and 6ft, 9ft and 12ft for X- or S-band. SPS-64(V)1 has a single 20kW transmitter; SPS-64(V)2 and -3 have two each; SPS-64(V)4 has one X- and one S-band transmitter; SPS-64(V)5 has a tunable 20kW transmitter; and SPS-64(V)6, for vessels over 270ft in length, has the 50kW X-band transmitter. X-band beam dimensions, respectively, are 1.9° × 22°, 1.2° × 23°, 0.9° × 23° and 0.7° × 23°. Comparable S-band characteristics for the 6ft antenna are 2° × 25°, and all scan at 33rpm. There are three pulsewidths (PRFs in parentheses): 0.06 (3600), 0.5 (1800) and 1.0 (900) microseconds. Range scales are 0.25–3, 6, 12, 48 and 64nm. Raytheon.

SPS-65
See SPS-58.

SPS-66
Small-craft search and navigational radar, a 7kW X-band type with range scales of 0.5, 2, 4, 8, 16 and 32nm. Raytheon model CRP-3100.

SPS-67
A solid-state replacement for SPS-10, using the same antenna but incorporating Standard Electronic Module (SEM) technology for simpler repair and maintenance. Two prototypes were ordered under the FY77 program, and the first installation will be aboard the refitted cruiser *Long Beach*. Norden Division of United Technologies Corporation.

Besides the two operating modes of SPS-10, it incorporates an ultra-short pulse mode for navigation (0.10 microseconds, in addition to the usual 1.0 and 0.25) with a correspondingly increased PRF (2400 rather than, respectively, 750 and 1200).

C. Postwar Submarine Search Radars (AN/BPS series)

BPS-1
Short-exposure X-band submarine attack radar, designed to allow interconnection of its indicator to SV-3, and a replacement for SJ. It used a linear array of horns in a streamlined watertight shell about 40in × 10in, to minimize drag when stowed. The corrugated high-pressure plexiglass cover for the horns maintained a near-constant impedance match when under water, partially submerged or in air, so that the radar could operate without waveguide breakdown when its antenna was elevated to just break the surface. It operated in X-band at 100kW with 0.5-microsecond pulses and a PRF of 600, and had a 2.6° × 16° beam. Reliable range was 12nm on a destroyer, and accuracy was 25yds plus or minus 0.1 per cent of range, 0.25 per cent in bearing. Total weight was 3580lb. First deliveries were made in December 1950. It was the first submarine attack radar usable at periscope depth. Modified SS-1. Western Electric.

BPS-2
An L-band air search radar for radar-picket submarines, replacing SR-2. Intended to detect both aircraft (up to 70,000ft) and missiles, the latter with a bearing accuracy of 0.5°. The latter probably refers to the Regulus guidance mission. It was comparable to the SPS-6 series: 15ft × 5ft antenna (4° × 10° beam). An output of 500kW (1 microsecond with PRF 600 or 4 microseconds with PRF 150) gave a range of 70nm on a bomber. Total weight was 4950lb. First delivered January 1953. Raytheon.

BPS-3
An S-band 5cm, 125kW height-finder using 0.37- or 1.3-microsecond pulses with a PRF of 625-650 for radar-picket submarines. It was unique among US search radars in using a microwave lens: this was 6ft 6in wide, with multiple feedhorns in vertical array to form a 2° × 1.8° beam. It could track a 20m² target at 40nm, up to 35° in elevation. The sector from 9° to 35° could be scanned at 3–7 scans per second (rotation at 0–7rpm). PRF-limited range was 131nm. BPS-3 replaced SV-6, and became extinct when the SSRs were all discarded or modified as conventional attack submarines (1959–61). Western Electric.

BPS-4
Air search radar for the *Tang* (SS-563) class, similar to SV-3. It operated in S-band using a 4ft × 2ft antenna to generate a 5.3° × 50° beam. Output was 500kW, with 1-microsecond pulses and a PRF of 400. Reliable range on a bomber was 15nm, and accuracy was

200yds plus or minus 3 per cent in range and 1 per cent in bearing. Total weight was 2875lb. Firet delivered November 1952 but now extinct. Westinghouse.

BPS-5
An X-band surface search radar for 250-ton SST class submarines with a 3ft × 9in non-retractable horn antenna generating a 2.6° × 16° beam and 100kW pulses of 0.5-microsecond duration with a PRF of 600. Reliable range on a destroyer was 12nm, and accuracy up to 60,000yds was 25yds plus or minus 0.1 per cent and up to 120,000yds 2 per cent of range. Bearing accuracy was 0.25°. Range limits were 200–160,000yds, and total weight 2050lb. First delivered in November 1953. BPS-5 and -5A were later installed aboard some *Thresher* (SSN-593) class submarines. Lockheed Electronics.

BPS-6
Modified BPS-2; canceled. Raytheon.

BPS-7
Canceled. Was to have been a Raytheon submarine height-finder.

BPS-8
A short-range 5mm radar, using separate transmitting and receiving. Two antennas, one of parabolic section with two feeds, on the periscope (10in × 4in, 1° × 3° beam), and one for search (26in × 2.6in, 0.5° × 5° beam). Abortive.

BPS-9, -9A, -9B
Replacement for SS-2; referred to in 1963 as a reproduction of the SS-2 with more modern components. An X-band (75–110kW, 0.5-microsecond pulses, PRF 600) surface search radar, also useful for detection of low flyers, and first installed in December 1958 (SSBNs). Total weight is 4000lb and there are multiple antennas, horn, array, and parabolic, forming a 16ft, 2.6° × 16° antenna aperture. BPS-9A has a retractable antenna. Western Electric.

BPS-10
Three-dimensional FRESCAN radar planned for the radar-picket submarine *Triton*. Canceled February 1960 as 'not needed' with the demise of the pickets. Hughes.

BPS-11, -11A
An X-band surface search radar for the SSBN-616 class. Search only when surfaced. This radar is a redesign of SS-1/BPS-4/BPS-1, similar to BPS-5 except for its antenna. It uses a horn antenna (3ft parabola in BPS-11A). Peak power is 75–100kW (0.5-microsecond pulses, PRF 600) and it has both a parabolic antenna with a 2.6° × 16° beam and an ST (periscope) antenna of 30° × 10°. Range limits are 200 and 160,000yds, and accuracy is as in BPS-5. Western Electric.

BPS-12
An X-band surface search and navigational radar, very similar to BPS-13 and -14. It uses both an ST (periscope) and a conventional (2.6° × 12°) antenna, and peak power is 75kW (0.5-microsecond pulses, PRF 600) for range limits of 400 and 80,000yds. Bearing accuracy is 1° and range 1 per cent out to 40,000yds and 2 per cent beyond that. It is a redesigned BPS-5. Fairchild Camera and Instrument Corporation.

BPS-13
An X-band surface search/navigational radar, using a horn antenna. It differs from BPS-12 primarily in having no inboard antenna drive. Only one was in use in 1973. Fairchild.

BPS-14
An X-band surface search/navigational radar, using a horn antenna, differing from BPS-12 in that its antenna drive is designed for use from within the submarine. Twelve were in use in 1973. Fairchild.

BPS-15
An X-band low-power (35kW *vs* 110 of other sets) surface search/navigational radar, using a horn-array antenna with a 40in aperture. It uses both 0.1- and 0.5-microsecond pulses with PRFs of, respectively, 1500 and 750, and a 3° × 13° beam to achieve a minimum range of 25yds and a range resolution of 30yds in the short-pulse and 100yds in the long-pulse modes. Rotates at up to 9.5rpm. Sperry.

D. Aegis Radar

AN/SPY-1
This four-facet S-band phased array is the successor to SPG-59/Typhon. It comprises fore and aft pairs of flat arrays angled to look to port and starboard for full 360° coverage. Development of the

SPY-1 radar for the Aegis system, RCA Land Test Site, 1980. RCA

new long-range air defense system (ASMS) began in 1965, and a contract for the entire Aegis system was let to RCA in 1969. Each pair of 4080-element 12ft × 12ft panels is driven by its own radio-frequency generator. Full phased-array operation permits very precise tracking, of both targets and defensive missiles, and the radar functions in a TWS mode. It detects targets automatically and commands Standard (SM-2) missiles to fly into engagement range. Terminal homing is by semiactive illumination provided by slaved illuminators (SPG-62). Ultimately the Aegis system is to include command links to other ships, so that an Aegis cruiser will be able to direct not only its own but other ships' missiles.

The system was first tested at sea aboard the *Norton Sound*, 1974. In a May 1974 experiment SPY-1 acquired and then automatically tracked twenty targets simultaneously; later a land-based version maintained full track on all aircraft using the busy Washington–New York corridor. The projected platform is the *Ticonderoga* class missile cruiser (CG-47), the Carter Administration rejecting the larger nuclear-powered Aegis strike cruiser (CSGN) as well as an Aegis variant of the *California* class (CGN-42). Modified and simplified versions of SPY-1 are contenders for the 1981 DDGX Multi-Function Array Radar (MFAR) award.

POSTWAR NAVIGATIONAL RADARS (AN/SPN SERIES)*

SPN-1
Radar beacon (transponder), also referred to as an IFF and buoy beacon.

SPN-2
Air speed indicator in experimental CCA installations aboard the carriers *Valley Forge* and *Philippine Sea* in 1947. This role was later filled by SPN-12. A pair of modified aircraft radars (APG-17) was used to send and receive a CW signal for doppler speed measurement. The 48in parabolic dishes generated a 12° beam at 1500mc/s, and could pick up an airplane at about 1nm.

SPN-3
An X-band approach-control radar in the experimental CCA installation aboard the carriers *Valley Forge* and *Philippine Sea* in 1947, replaced by SPN-8 in production installations. It was a modified APQ-7 (Eagle) airborne radar in a wing-shaped radome. The 0.4° × 20° beam scanned over 30° in the horizontal at 90 scans per minute at an output of 500kW, with 0.357-microsecond pulses and a PRF of 1200. Range was 5nm, and presentation by B-scope and Elevation-Position (EPI) scope. A peculiarity was that with the ship rolling and yawing, twin targets would sometimes appear on the displays. In operation it was a very early form of FRESCAN, a moving plate in a waveguide, varying effective wavelength.

SPN-4
An S-band navigational radar for MDAP and MSTS. 15kW, 0.4-microsecond pulses (PRF 1000) gave a minimum range of 100yds, with range scales of 1.5, 5, 15 and 50nm. Accuracy was 100yds at 100–5000yds, 2 per cent or better at 2.5–50nm, and 2°; resolution was 100yds and 5°. Beam size was 3.5° × 11.5°, scanning at 7rpm. These details indicate the level of commercial S-band radar technology about 1950. Raytheon (Pathfinder CX-1086 with modification for true-bearing reading).

* In many cases these are LORAN receivers, *ie* passive devices only. Many radars were for MDAP (foreign transfers) and for MSTS (Military Sea Transportation Service).

SPN-5
An X-band navigational radar for MDAP and MSTS. This set was roughly contemporary with SPN-4, and appeared in several versions. Peak power was 30kW, with variable pulse rate and length: 0.25 microseconds and PRF 3000 for short range, 1.0 and 750 for 20 and 40nm scales. The set was considered effective between 80yds and 40nm. Range scales were 1.5, 4, 8, 20 and 40nm. The 1.8° × 19° beam allowed an accuracy of 1 per cent in range and 2° in bearing. RCA (Radiomarine), with SPN-5A the CR-101-A commercial radar.

SPN-6
The S-band air search radar of the CCA system. In effect it replaced SG-7, and in fact it employed many SG-3 components plus a new stabilized antenna (12ft parabola with a gain of 5000 and a 2° × 2.5° beam) and a standarized display (VK). It employed 500kW pulses of 0.33 or 1.25 microseconds, with a PRF of 760. SPN-6 was introduced in the early 1950s and is now being replaced by SPN-43, attack carrier sets being transferred to the helicopter carriers. In 1978 the only surviving sets are aboard the training carrier *Lexington* and seven helicopter carriers. Raytheon.

SPN-7
LORAN receiver.

SPN-8
An X-band CCA approach control radar, the successor to SPN-3. Operational evaluation was completed in 1951 aboard *Coral Sea*. For a time in the 1950s many carriers had a limited CCA installation consisting of SPN-8 (in a small pointed-top radome) and SPN-12, without SPN-6. Beginning in 1960, SPN-10 replaced it. SPN-8 was designed to guide aircraft from a distance of about 6nm to within a minimum of 200ft from touchdown, by displaying an aircraft's position relative to an ideal approach path on offset sector PPIs; these displays were corrected to cancel out ship roll, pitch and yaw. The small parabola produced a 1.5° × 6° beam with 40–60kW pulses, 0.25 microseconds in length and with a PRF of 4000, and it scanned a 100° sector (looking aft) 3 times per second. The antenna was stabilized, with a maximum tilt of 5° up and 2° down from horizontal. Bendix.

SPN-9
Officially listed as a radio navigation set for use in conjunction with the Transit Navigating System, for US Navy surface ships, receiving signals from the satellites and the computing position; it was also known as TRIDON (Transit Integrating Doppler Navigator). The designation was applied in 1962, which makes one suspect that it replaced some earlier equipment, or that it was a misprint somewhere for SRN-9. Applied Physics Laboratory, Johns Hopkins University.

SPN-10
Landing-control central portion of the CCA system. The Carrier Automatic Landing System (CALS) included a ship-motion sensor and predictor, and it could control an approaching airplane through a link to its autopilot. Alternative modes were semiautomatic transmission to the pilot coming in, and talk-down. CALS could also issue wave-offs at any time from acquisition down to a few seconds before touchdown.

SPN-10 and -12 antennas, USS America, *December 1964.* BY COURTESY OF A D BAKER III

The system was capable of simultaneous automatic control of four aircraft; and it could control take-offs as well as landings from the moving deck of a carrier under all weather conditions. The controller would generally guide a pilot to the capture point, about 4nm from the carrier. Aircraft entered SPN-35 or -35A control at about 10nm (at up to 25nm in optimum weather); in the absence of SPN-10 they could be controlled into about ¾nm from the carrier, after which the pilots switched to a visual (Fresnel lens) approach. The radar portion of the system consisted of a pair of 4ft K-band conical scanning dishes (gain 7200), forming 0.57° × 0.57° beams with 50kW, 0.2-microsecond pulses with a PRF of 2000 for very high precision. These systems are being replaced by the very similar SPN-42. Bell Aerosystems.

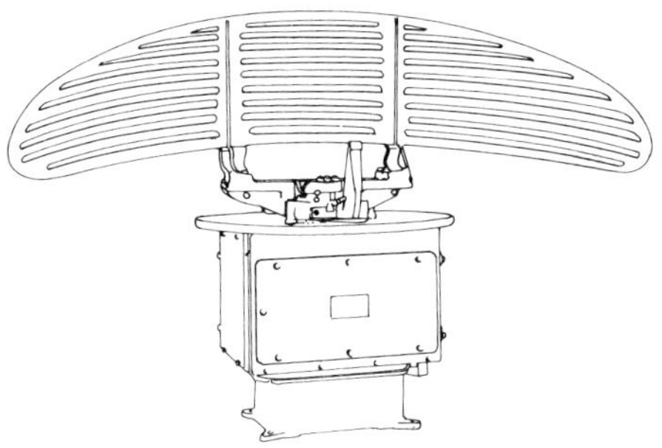

SPN-11 antenna.

SPN-11
Radiomarine X-band navigational radar for MDAP and MSTS use about 1951 and also an important Coast Guard set. Designated CR-103 in commercial service, it used 30kW pulses at 0.4 microseconds and a PRF of 1000, and a parabolic reflector with a 1.9° × 20° beam to achieve a range of from 75yds to 20nm. The 4ft × 2.7ft antenna had a gain of 316 and rotated at 17rpm. Accuracy was 2° in bearing, and resolution 75yds.

SPN-12
Airspeed indicator for the CCA system, a 2ft X-band continuous-wave (non-pulsed) dish measuring speed by doppler shift. It was considered effective within 2nm of the ship, and could measure both true air speed and air speed relative to the ship. If it were trained into the wind, it could also measure wind speed relative the carrier. Now being replaced by SPN-44. Raytheon.

SPN-13
S-band (20kW) and X-band (50kW) radar for MDAP and MSTS, using a common PRF of 1200 and, respectively, 1.25- and 0.25-microsecond pulses, to produce a beam 3° or 1° × 15°. Range scales were 1.5, 5, 10, 25 and 50nm, and this set used a horn and slotted line antenna. Nomenclature was assigned in 1951, and this set had disappeared from most official radar lists a decade later. Band choice was at the operator's discretion. General Electric.

SPN-14
LORAN receiver with continuous indication.

SPN-15
Designation canceled; no information available.

SPN-16
Automated LORAN receiver, *circa* 1954.

SPN-17
Navigational radar, redesignated SPS-21.

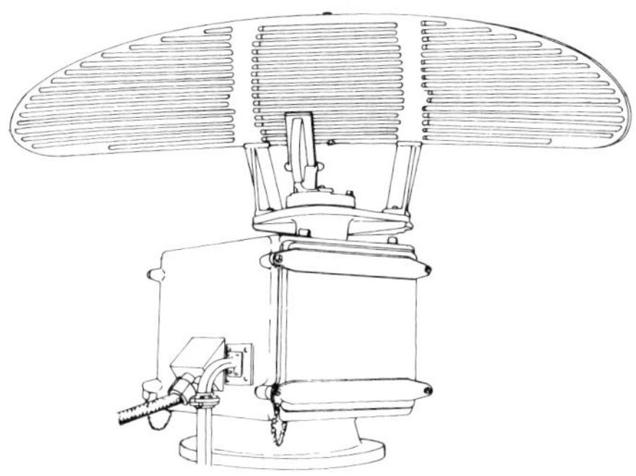

SPN-18 antenna.

SPN-18
Radiomarine X-band navigational radar, designated CR-104 in commercial service. Very similar to SPN-11 but rotated at 8.6 rather than 17rpm. The 40kW pulses had a PRF of 2000 for the 1, 2 and 4nm range scales, 800 for the 8, 20 and 40nm scales, with, respectively, pulsewidths of 0.25 and 0.65 microseconds, and provided an accuracy of 2 per cent over all ranges and 1° and a resolution of 55yds on the 1, 2 and 4nm scales and 1.5°. Minimum range was 55yds. The antenna duplicated that of SPN-11.

SPN-19
Radio beacon buoy.

SPN-20
Ship interrogator; omnidirectional beacon. SPN-24 is similar.

SPN-21
Raytheon X-band navigational radar (Pathfinder 1500), also known as SPS-35. Although suitable for such small craft as LCPLs, it was often mounted aboard larger ships for inshore navigation. It used 7kW pulses (0.2 microseconds, PRF 1500), rotating at 20rpm to achieve good definition; the 48in reflector produced a 3° × 15° beam.

SPN-22
Radiomarine Corporation CR-105 X-band navigational radar, comparable to SPN-21. It used 0.2-microsecond, 7kW pulses at a variable PRF: 2000 for 1, 2 and 4nm scales, or 1500 for 12nm and 1100 for 32nm. Minimum range was 50yds, and the 1.9° × 20° beam scanned at 20rpm.

SPN-23
Edo small-boat surface search and navigational radar (Edo model 320), redesignated SPS-36.

SPN-24
Radar beacon similar to SPN-20 (LORAN interrogator).

SPN-25, 27–34
LORAN receivers.

SPN-26
Replaced by SPS-35.

SPN-35
Aircraft approach control central, used in conjunction with an optical final-approach system. It displays both elevation and bearing, and consists of an AN/TPN-8 landing approach control radar mounted on a stable platform. There are two parabolic antennas, 2ft × 8ft (15° × 1.1°) for elevation, and 6ft 6in × 4ft 6in (1.3° × 15°) for bearing. Elevation scan coverage is 11° or 36° adjustable from −1° to +5° in 1° increments. Both antennas operate on X-band at 175kW, with a PRF of 1200 and 0.2- or 0.8-microsecond pulses. Overall search accuracy is 500ft or 2 per cent of range and 2°; in the precision mode the corresponding figures are 60ft and 0.5° (and 20ft of elevation, except that above 10° it is 20ft or 5 per cent of elevation angle). In the height-finding mode, SPN-35 is accurate to 0.4 per cent in range and 10 per cent in elevation.

The entire assembly vaguely resembles the old SX; it rotates beneath a dome looking aft from the island. Gross weight is 3186lb. This system was originally evaluated aboard the rarrier *Kearsarge* in 1962, a developmental model of the TPN-8 being mounted aboard an SPN-6 stable pedestal and secured to a modified SPN-8 platform. About a year later a prototype SPN-35 was evaluated aboard the carrier *Bon Homme Richard* as a replacement for SPN-8 aboard small attack carriers and all ASW carriers; it required the pilot to shift to a visual approach when he came out below the overcast. After the original SPN-35 operational evaluation, Commander Operational Test and Evaluation Force had found it superior to any existing carrier final approach radar. ITT-Gilfillan.

SPN-35A incorporates a more reliable stable platform, an electro-mechanical type in place of the mechanical-hydraulic type originally developed for SPN-6.

SPN-36
LORAN receiver.

SPN-37
Height monitor for hydrofoils, mounted under hull.

SPN-38–40
LORAN receivers.

SPN-41
A carrier-approach guidance Ku-band system transmitting angle errors to an approaching aircraft, using a pair (horizontal and vertical) of slotted waveguides 26.75in and 42in long, the former with a reflector. The horizontal azimuth element is housed under a radome, the vertical on a pedestal. Generally the elevation transmitter group and radome is on the port side of a carrier's angled deck near the touch-down point; the corresponding azimuth set is on the centerline below the fantail. The system has 2.2kW peak power and a range of 50nm. Cutler-Hammer.

SPN-42
Solid-state replacement for SPN-10, with similar performance. It interfaces with NTDS. MTBF is increased to about 225 hours, compared to 35 for SPN-10. Bell Aero-systems.

SPN-43
A new air traffic control search radar, replacing SPN-6, of which it is a much-modified and much-improved version. The new antenna produces a cosecant-squared fan beam for airspace-filling (high-altitude coverage), and there is a higher power transmitter of 850 rather than 400kW, with a PRF of 1125 rather than 760 and 0.6–0.25-microsecond pulses. The 12ft parabola produces a 1.5° × cosecant-squared to 45° beam, compared to 2° × 2.5° in SPN-6, for coverage to 30,000 rather than 8000ft; effective range in optimum weather is 50nm (compared to 30), and in adverse weather it is 35 (compared to 15). Tracking of low flyers is effective down to a minimum range of 250yds. There are two IFF antennas: a directional broadside 'hog trough' designated AT-1688 and also used in SPN-6 and SPS-48, and an onmidirectional 'back-fill' antenna for sidelobe suppression. SPN-43 is also effective as a back-up air search antenna. ITT-Gilfillan.

SPN-44
A solid-state replacement for SPN-12, with similar characteristics. Applied Devices Corporation.

SPN-45
LORAN-C receiver.

WEAPON CONTROL SETS

A. Fire Control Radars (Mk and AN/SPG series)

Mark 1 (CXAS)
Originally designated FA and first delivered to *Wichita* in June 1941. Used more for surface search than gunnery, it suffered from a short (75-hour) oscillator tube life. Bell Laboratories developed for it the first lobing systems and also worked out a system for locating controls and display in the gun director. Ten were built for use with Mk 34 directors, and were superseded by Mk 3. It operated at 40cm (40kW, 1–5 microseconds, PRF 1640) and with an 8ft 5in × 6ft 1in antenna weighing 800lb and producing a 20° × 30° beam could pick up surface or air targets at 5nm. Prototype performance was even better: in 1941 *Wichita* picked up two transports at 55,000yds and resolved them at 44,000. Accuracy was 200yds and 9 mils, resolution 400yds and 20°. Western Electric.

Mark 2
FB: an FA with magnetron power tube. Never produced, as a much better transmitter soon appeared, and was incorporated into the new Mk 3.

Mark 3
FC, the first important US main battery fire control set, operating at 40cm with a power of 15–20kW, 1.5-microsecond pulses and a PRF of 1640. There were two configurations: a square (cylindrical section) Mod 1 or 3, 6ft × 6ft and producing a 12° × 12° beam, and an oblong Mod 0 or 2, 12ft × 3ft, and with a 6° × 30° beam. Mod 2 was mounted in most battleships and cruisers, Mod 3 on the single-purpose destroyers (1850-ton 'leaders'). This set introduced lobing to US practice.

At an antenna height of 80ft, reliable ranges were: 16in splash at 20,000yds; surfaced submarine at 12,000; and bomber (10,000ft) at 45,000yds. Accuracy was (Mod 0 and 2) 40yds and 2 mils, and (Mod 1 and 3) 4 mils; resolution was 400yds and 10° respectively. Production of 139 sets was completed in 1944; the replacement was Mk 8. The first installation was aboard the light cruiser *Philadelphia*, October 1941. Western Electric.

Mk 4 (FD) antenna atop Mk 37 director, USS McCalla, *January 1944.*

Mark 4
FD: first installed aboard the destroyer *Roe* in conjunction with the first GFCS* Mk 37, for 5in/38 dual-purpose guns, in September 1941; it was also used on Mk 33 directors. FD consisted, in effect, of two half Mk 3 Mod 2s stacked vertically. It was the first US fire control radar for dual-purpose guns, and proved quite successful, but it provided inadequate blind fire capability (hence Mk 12) and was ineffective against low-flying aircraft (hence Mk 22). Lobing pattern was right up, left down, using 12° × 12° beams, and overall antenna dimensions 6ft × 6ft. Reliable ranges were: 5ft splash at 12,000yds; battleship at 30,000; destroyer at 20,000; submarine at 12,000; and bomber at 40,000. Accuracy was 40yds and 4 mils and resolution 400yds and 10°. A total of 667 were built 1941–44, replaced by the very similar Mk 12 and (in some cases) the Mk 28 dish. Western Electric.

Mark 5 (CXAZ)
Experimental 5in fire control radar.

*Gun Fire Control System

Mods 0 and 1 were lobing, Mods 2 and 4 conical-scan; Mod 3 was abortive 15cm set. Four were built, 1942–44. Mk 5 introduced automatic range control and operated on 75cm, using 10kW pulses (1 microsecond, 656 pulses per second) and a 6ft × 5ft 2in antenna, to achieve ranges of 15nm on a ship and 25nm on an airplane; accuracy was 20yds and 12 minutes in bearing and elevation, and resolution 160yds and 10°. General Electric.

Mark 6 (CXBF)
Experimental 50cm, 100kW 0.5-microsecond pulse, PRF 4000 set for 5in fire control, using a 6ft × 6ft antenna weighing 310lb (compared to 250 for Mk 4) and having 30° × 30° lobing beam. One was installed briefly aboard *South Dakota*. Ranges were 15nm on a surface ship and 25 on an airplane; accuracy was 25yds and 15 minutes and resolution 80yds and 13°. RCA.

Mark 7 (CXBN)
Experimental S-band (Mod 0) and X-band (Mod 1) set, 1943. Tried conical, spiral and elliptical scanning. The S-band version used a 57in dish ((1.25° × 1.25° beam) and 175kW pulses (0.25 microseconds, PRF 1000); range was 11nm on a surface ship, minimum range was 500yds and accuracy 25yds and 1° bearing and elevation. Resolution was 50yds and 3°. Test of concepts, not intended for any particular GFCS. Western Electric.

Mark 8 (CXEM)
A 10cm main battery fire control set for battleships and cruisers, the successor to Mk 3. It was quite unusual in appearance, consisting of an array of 42 (14×3) elementary radiators (polyrods) scanned by phase-switching. A 2° × 3° beam was scanned over a 30° sector in this way, achieving accuracies of 15yds in range and 2 mils in bearing. Thus Mk 8 combined a wide field of view (for good target acquisition) with a narrow (high-resolution) beam, scanning fast enough to present the entire field of view almost continuously (10 scans per second). A recent Bell Laboratories history describes it as the first TWS radar; the polyrod design was adapted from the prewar Bell MUSA steerable array radio antenna.

Mk 8 used 15–20kW pulses (20–30 in Mods 1 and 2) 0.4 microseconds long and with a PRF of 1500. Both A- and B-scopes were provided, the latter showing a field 29° wide. The entire antenna was 10ft 2in × 3ft 4in. Reliable ranges were 40,000yds on a battleship, 31,000 on a destroyer, 10,000 on a submarine, and 30,000 on a bomber. By 1943 a total of 205 had been completed. Many were converted to Mod 3, which was equivalent to the later Mk 13. Mod 2 moved the transmitter below decks, hence achieved better shock-resistance. Mk 8 was succeeded by Mk 13. Western Electric.

Mk 8 radar antenna, USS Mobile, *18 July 1943.*

Mark 9
A 10cm radar for the abortive Mk 45 heavy AA machine-gun director. 30 were built, but they became obsolete when Mk 45 was canceled in 1943. Six were converted to Mk 18, others were adapted as Mk 19 for the Mk 48 AA director. Its 24in parabolic dish (gain 22db, 12° × 12° beam) executed a 4.5° conical scan at 30 cycles per second; peak power was 20kW, with 0.5-microsecond pulses and a PRF of 3000. Reliable range was 10,000yds on a bomber at 10,000ft or on a battleship or cruiser, and minimum range was 300. Accuracy was 15yds plus 0.1 per cent of range and 0.3°, and resolution 200yds and 10°. It was intended for use with optical tracking, and had aided range-tracking; on its Mk 45 director, the dish could elevate from −25° to +100°. Developed by Bell Laboratories, produced by Western Electric.

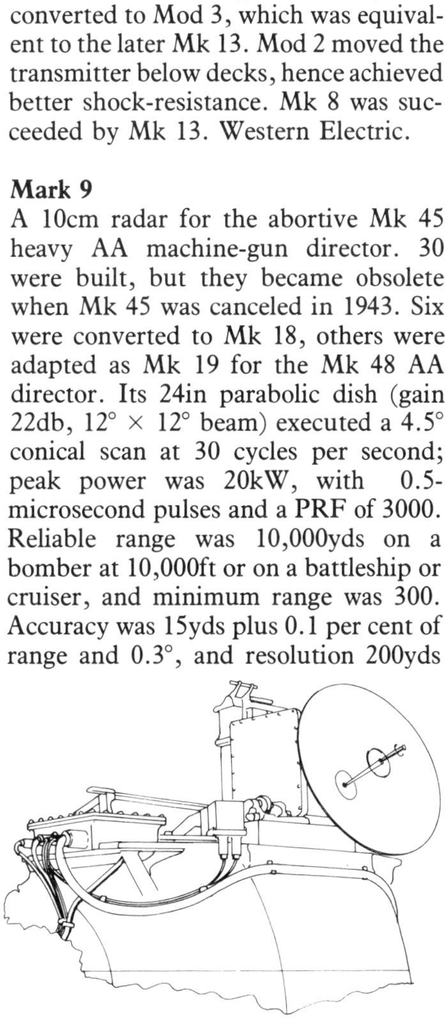
Mk 9 radar.

Mark 10
A 10cm radar for AA director Mk 50, used for 3in/50 fire control in the old battleships and *Omaha* class light cruisers. Only 20 reached the fleet due to cutbacks in the Mk 50 program; in addition 37 Mod 5s were built in 1944. Another conical-scanning type, the larger 45in dish (gain 830, 6° × 6° beam) giving better performance: 14,000yds on a bomber and 20,000 on a battleship. Output was 20–30kW, with 0.5-microsecond pulses and a PRF of 3600.

The arrangement was unusual in that the transmitter was mounted on the back of the dish, conical scan being 2° at 30 cycles per second. The antenna was stabilized in level and cross-level. Accuracy was 15yds plus 0.1 per cent of range and 2.3 mils, and resolution 20yds and 6°. Elevation was from −25° to +110°. The data here apply to Mods 3 and 5, the only ones produced, which used mesh dishes; Mods 0 and 1 used a 30in solid dish, and were intended for the abortive Mk 47 director. Western Electric.

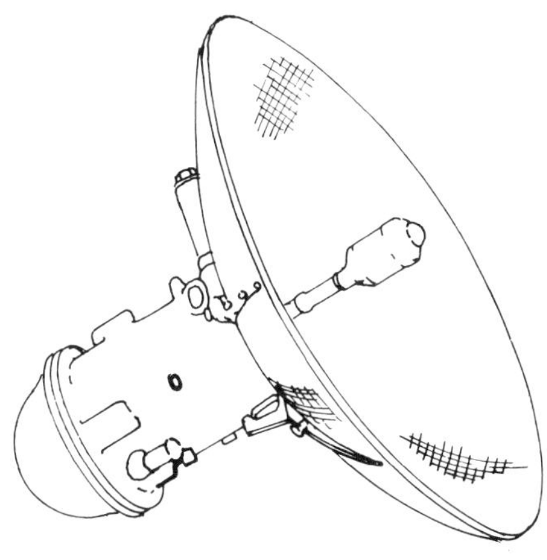
Mk 10 fire control radar antenna assembly.

Mark 11
A 10cm range-only radar for 1.1in and 40mm guns on warships from destroyer escorts up, for use with the Mark 49 AA director. Elevation limits were −30° and +110°. The antenna was a dipole in a 24in perforated parabolic dish (no conical scan, gain 240, 11° × 11° beam); it produced 40–60kW pulses 0.5 microseconds long and with a PRF of 600, to achieve a reliable range (antenna height

40ft) of 6000yds on a fighter, 12,000 on a large ship. Accuracy was 70yds (6000yd scale) or 200yds (30,000yd scale) and resolution 100yds. Mk 26, for use with Gun Director Mk 52, was similar. Developed by RCA, with production of Mod 0 by RCA and Mod 1 by GE. 1500 were ordered, but relatively few were produced, as the Mk 49 director did not prove successful.

Mark 12
The replacement for Mk 4, incorporating parasitic elements to reduce side lobes. Development began early 1942, and at first S-band operation was contemplated. However, there were doubts as to the range achievable, hence the decision in favor of a reduction only from the 40cm of Mk 4 to 33cm in this system. The other major improvement was automatic tracking in range, automatic measurement of range rate for transmission to the CFCS. Mod 2 was to have added automatic tracking in elevation and train; Mod 3 was an antijam modification.

Adoption of the shorter wavelength tightened the beam, for much the same size antenna, to 10° × 10°; it used 100–110kW pulses (0.5 microseconds, PRF 480) to achieve reliable ranges of 45,000yds on an air target and 40,000 on a ship. Accuracy was 20yds and 3 mils, and resolution 300yds and 7°. Production of 801 units began in 1943, and by the end of World War II a complete installation included the Mk 22 heightfinder. Western Electric.

Mark 13
A 3cm replacement for Mk 8 in battle-

Mk 12 radar antenna, with Mk 22 located alongside, Clarence K Bronson, *June 1945.*

Mk 13 surface fire control radar, USS Iowa *(July 1947), with SK-2 air search radar visible above and a TDY jammer below.*

ships and cruisers, the complex mechanically scanned polyrod array being replaced by an 8ft × 2ft parabola with a gain of 14,000 and a 0.9° × 3.5° beam, scanned by rocking it about its vertical axis ('rocking horse' motion) within a radome. Mk 13 Mod 2 (and Mk 8 Mod 4) were designed late in the war to substitute motion of the microwave feed ('Lewis feed') for the antenna motion. There was also to have been a higher-powered transmitter; neither of these advanced surface systems ever entered production.

Range and accuracy performance were comparable to that of Mk 8; Mk 13 used 50kW pulses (0.3 microseconds, PRF 1800) and scanned through a 5.75° angle. Mk 13 was the last main battery (heavy weapon) director radar to be produced for the US Navy, and by late 1945 most Mk 8 installations had either been replaced by Mk 13 or converted to the very similar Mk 8 Mod 3. Western Electric.

Mark 14
An abortive, higher powered (200kW, 0.25-microsecond, PRF 1800) version of Mk 10 Mod 5 for use with the Mk 50 fire control system. Western Electric, 1943.

Mark 15
A 1943 X-band radar bombsight, resdesignated Radar Bombsight Mk 15 in a new series separate from Radar Equipments. Mod 2 was used for Bat missile control, 1944–45.

Mark 16
A 1943 land-based 10cm search and fire control set for the Marine Corps. It was a

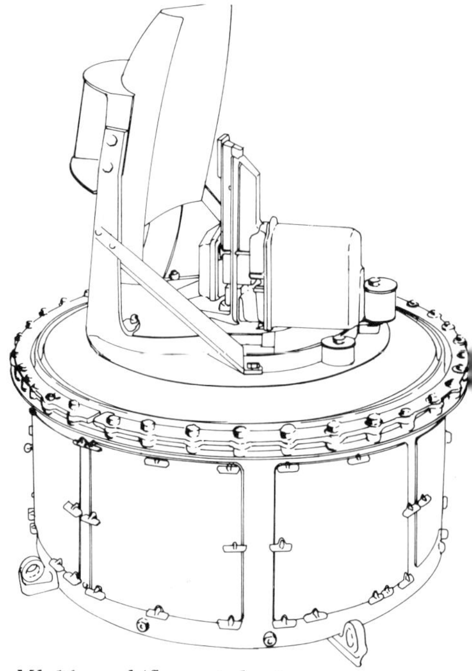

Mk 16 search/fire control radar.

trailer-mounted version of SH, developed by Bell Laboratories. Ten units (without trailers) had been delivered by January 1944.

Mark 17
Airborne computing sight for aircraft, 1943; redesignated AN/APG-5.

Mark 18
Modified Mk 9 for 5in GFCS Mk 28, six being installed aboard the three *New Mexico* class battleships in 1943. The antenna assembly was coupled in level and cross-level to the optical rangefinder of the Mk 28; the range operator was below deck. Aided range-tracking was provided. A 45in perforated parabola produced a 6.5° beam and a 2° conical scan (30 cycles per second); peak power was 20–30kW (0.5 microseconds, PRF 3600) for a range (92ft high antenna) of 18,000yds on a bomber at 10,000ft, 16,000 on a fighter and 35,000 on battleships and cruisers; minimum range was 400yds, and accuracy was 15yds plus 0.1 per cent and 3 mils and resolution 200yds and 6°. Western Electric.

Mark 19
A 10cm range, bearing and elevation radar for GFCS Mk 49 (light AA); modified Mk 9. When GFCS Mk 49 was canceled, Mk 19s were modified for use with GFCS Mk 57. 200 were built, some being operational in 1943–44. Elevation was from −30° to +90°. The antenna was a 24in perforated paraboloid with

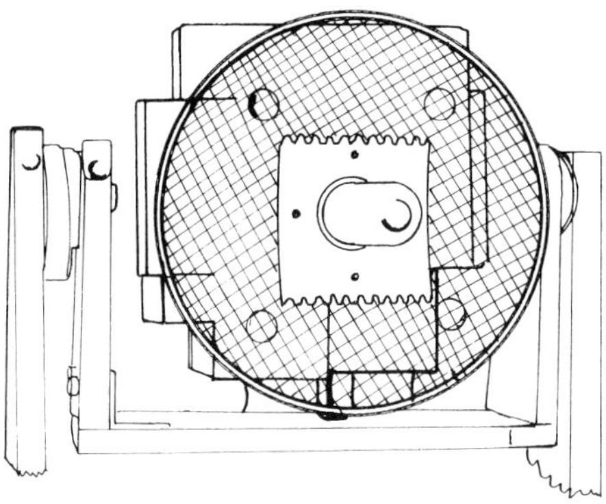

Mk 19 antenna.

rotating dipole (4.5° conical scan at 30 cycles per second). Peak power was 50kW, with 0.5-microsecond pulses and a PRF of 1800, for a range (40ft antenna height) of 8000yds on a bomber at 10,000ft or 16,000 on a battleship. Accuracy was 15yds plus 0.1 per cent of range and 4 mils and resolution 100–240yds and 10°. Western Electric.

Mark 20
Ground-based 28cm (24cm in Mod 1) searchlight-control radar. Mod 0 used a double parabolic cylinder like that of Mk 4, which lobe-switched both in azimuth and elevation (38in × 69in, for a 17.3° × 9° beam); peak power was 150kW with 1.6-microsecond pulses and a PRF of 480yds and 1°. Mod 1 used a dish antenna, with a spinning diode for conical scan: it had a 9° × 9° beam, a PRF of 225 and an accuracy of 300yds, but was otherwise the same. Western Electric.

Mark 21
Repackaged Mk 19 Mod 1 for GFCS Mk 49. Never produced.

Mark 22
Parabolic height-finder, for use in conjunction with Mks 4 and 12. Wavelength was reduced by 3cm to achieve a well-defined 5° × 1.3° beam from an 18in × 72in 'orange peel' by means of which aircraft as little as 0.8° above the horizon could be detected. The low-flyer problem was urgent because Japanese torpedo attack tactics emphasized low-level approach. Mk 22 was essentially an elevation-only radar (Mod 1 added range measurement) accurate to within 3 mils above 1° over the horizon. It could resolve ships 5° apart and aircraft 1.5° apart. Peak power was 25–35kW with 0.5-microsecond pulses; PRF was 480 or 1640, chosen to match the associated radar. Maximum ranges were comparable with those of Mk 4/12. Production of 995 began in 1943. Western Electric.

Mark 23
An abortive project to develop a 'standard packaged' radar for use with a small lead-computing sight director. Canceled in concept stage, due to greater urgency of other projects.

Mark 24
An abortive 'standard packaged' radar incorporating new servicing techniques.

Mark 25
A 60in conical-scanning X-band dish which replaced Mk 12/22. Peak power was 50kW, with 0.2-microsecond pulses and a PRF of 2000, and maximum range was 50,000yds, later improved to 100,000. Accuracy and resolution were better as well: 15yds and 40yds, compared to a resolution of 300 for Mk 12 and an accuracy of 25 for Mk 22. Accuracy in bearing was 0.1°. It began to replace Mk 12/22 postwar, being first tested in 1948; however, some of the earlier sets remained in service as late as the Korean War.

Mod 7 was the first US missile guidance radar, aboard the experimental ship *Mississippi* and the cruiser *Boston*; it used a microwave lens. Work on the Mod 0 version began in March 1944; it was to have combined a 1.25cm tracking radar with a 3cm beam scanning linearly in traverse and lobe-switching in elevation about the axis of the 1.25cm system, intensity-matching on the 3cm B-scope being used to estimate the elevation of a target with respect to the axis of the 1.25cm system. Although primarily intended for 5in fire control with GFCS Mk 37, this equipment could be used for fighter control. The 1.25cm radar would use a 4ft diameter circular waveguide lens, to produce a 0.65° × 0.65° beam with 0.2° conical scanning (30 cycles per second), its peak power 50kW. The X-band radar would use a rectangular waveguide lens, 6ft × 2ft, to produce a 1.2° × 4° beam, scanning 15 times per second over a 40° range, with lobe-switching 1.5° each way at 8 cycles per second; gain would be 37db, compared to 47 for the K-band system, and it would use 250kW pulses. Both were to use 0.2-microsecond pulses for high resolution (40yds; accuracy 10yds plus 1 per cent of range) at a common PRF of 2000.

Mk 25 radar on GFCS Mk 37.

Mk 25 Mod 0 was abandoned in view of the desire to have all controls below deck for an unmanned director. Mod 1 was begun but stopped as too complex, and the version actually produced, Mod 2, was first defined in March 1945. It was to incorporate a circular horizon scan to pick up island shorelines and direct fire at ground targets; a spiral scan for aircraft acquisitions; and a conical scan for automatic tracking of aircraft out of 50,000yds. The final test model was successful in May 1947, and over 400 were built. In Mod 3 power rose to 250kW, amplifiers were moved below deck, and only the indicators remained in the director proper. Range increased to 150,000yds, and projectiles could be tracked as they left the gun. All Mod 2s were converted to Mod 3 by 1952. Western Electric.

Mark 26
A 10cm range-only fire control radar for the Mk 52 AA director, a small 'handlebar' type for use with 3in and 5in guns aboard auxiliaries, DEs and PFs. Adapted from Mk 11, it had a 36in paraboloid with a gain of 400 and a 9° × 9° beam. Reliable range was 10,000yds on a bomber and 16,000 on a cruiser. Peak power was 40–60kW, using 0.5-microsecond pulses and a PRF of 600. Accuracy was 150yds. 450 Mod 3s and 328 Mod 4s were produced before cancellation in 1945. Deliveries began 1944, against orders for 555 from RCA and 670 from GE.

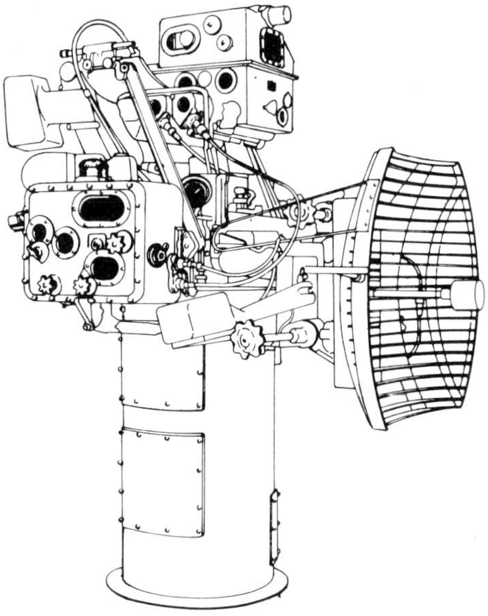

Mk 26 fire control radar.

Mk 27 fire control radar on conning tower of USS North Carolina *(memorial). This is the same antenna as SJ.* A D BAKER III

Mark 27
Standby fire-control radar for the main batteries of battleships and cruisers, mounted in conning towers. It was a modified SJ-1 and replaced the Mk 3 formerly mounted above conning towers in conjunction with Auxiliary Gun Director Mks 40 or 55, its small size showing the effect of passing from L- to S-band (10cm). The antenna was a 5ft × 3ft parabola (gain 1100, 8° × 8° beam); 50kW (0.3 microseconds, PRF 1500) sufficed to detect a merchant ship at 15nm. Displays included a PPI with 8000, 40,000 and 80,000yd scales, as well as a range indicator, and accuracy was improved by lobing (1720 times per minute). Accuracy was 30yds and 3 mils and resolution 160yds and 6.5°. Production of 41 began 1944. Western Electric.

Mark 28
10cm radar for 5in control with Mk 33 director, a lighter alternative to Mk 12/22 for directors with limited space. It was considered a great improvement over Mk 4, but was not as effective as Mk 12/22, although Mk 33 directors could not, in general, accommodate Mk 22. In addition some Mk 37 directors could not accommodate Mk 12 because their taper-back shields left inadequate space for its transmitter; in these Mk 4 was replaced by Mk 28/22. Mk 28 Mod 2 (antenna separated from transmitter) was used with GFCS Mk 63, mounted on the barrels of a twin or quad 40mm mount, and acquiring its target by nodding. Mod 3 involved a change of trunnion mounting holes.

The antenna was a 45in paraboloid with conical scanning, a gain of 800 and a 6.5° × 6.5° beam. The peak power of 30kW (0.5-microsecond pulses, PRF 1800) gave a range of 20,000yds on a bomber and 15,000 on a fighter; accuracy was 15yds plus 0.1 per cent of range and 4 mils. Production began in April 1944, and 103 Mod 0s and 3s and 247 Mod 2s were built. Mod 1 was an abortive X-band type. Production converted to Mk 34 Mod 2 in July 1945. Western Electric.

Mark 29
A 3cm radar for use with GFCS Mk 57 (AA machine-gun director) an interim for radar Mk 39. 102 were built, converted from Mk 19 Mod 0 by Section T, Office of Scientific Research and Development (OSRD). A 30in dish elevated from −15° to +85° (3° × 3° beam, 1.25° conical scan at 30 cycles per second). 30kW peak power (0.6 microseconds, PRF 1800) gave a range of 25,000yds on a bomber at 10,000ft and 20,000 on a fighter. Accuracy was 15yds plus 0.1 per cent of range and 2 mils and resolution 200yds and 2.75°. Conversion of Mk 19 by Section T, using SU transmitters, began in 1944. Mod 0 used a solid dish (24in), Mod 1 a larger one (45in). Neither was produced.

Mark 30
Army/Marine Corps shore-based system, similar to Mk 16. Also known as AN/MPG-1 and -2.

Mark 31
An experimental airborne radar for use with SWOD Mk 2 guided missiles, 1944. The project was begun in December 1943 and the first model was completed a year later. Western Electric.

Mark 32
IFF system for Mks 3, 4 and 12 radars.

Mark 33
A 3cm mobile coast-defense radar, using a 7ft × 3ft slotted 1° × 2.5° paraboloid and 50kW, 0.35-microsecond pulses using a PRF of 750–850 and scanning at 0–12rpm. Accuracy was 25yds and 0.1°. Production began in December 1944. General Electric.

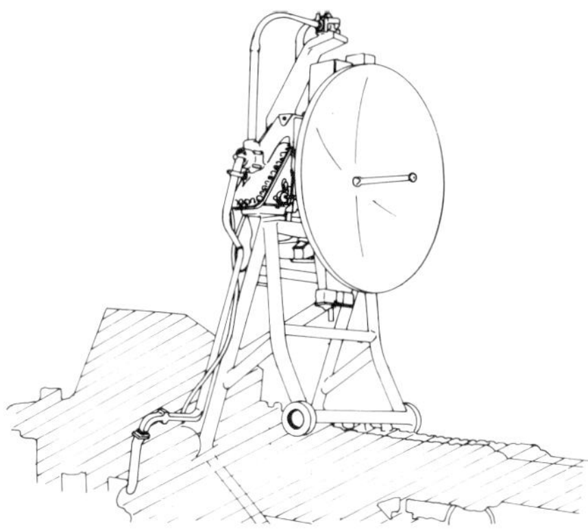

Mk 34 antenna assembly.

Mark 34
An X-band fire control radar for AA GFCS Mk 63 (Mods 1 and 2) and Mk 57 (Mods 3 and 4). It could elevate between −15° and +85°, and in Mods 1 and 2 could nod to acquire a target. There was a 30in dish with a 3° × 3° beam with 0.75° conical scan by nutating wave-guide – Cutler feed – at 30 cycles per second, stabilized against ship motion. Range was 25,000yds on bombers and accuracy was 15yds plus 0.1 per cent of range and 2 mils and resolution 200yds and 2.25° in bearing and elevation. Mod 0 was a proposed version using Mk 19 components and Mod 1 was a development model for Mod 2 which entered production, using an ST transmitter and Mk 28 Mod 2 below-decks components. Mod 3 was a production model using Mk 19 below-deck components, and Mod 4 was a production model using Mk 28 below-decks components. Widely used postwar.

SPG-34 differed in target designation and acquisition, and introduced a 40in fiberglass reflector similar to that of Mk 25 Mod 3: it achieved 12° coverage in spiral scan, and had a 2.2° conical scan; it did not have to nod to acquire. Mk 34 produced 25–35kW, 0.3-microsecond pulses at a PRF of 1800 and SPG-34 50kW, with the same pulse characteristics and with a 2.4° beam. SPG-34 was an improved Mod 17 with better Target

Acquisition and a new console. Western Electric.

Mark 35
A 3cm fire control radar for fully automatic tracking and blind firing with AA Gun Director Mk 56, using spiral scan for acquisition and conical scan for automatic tracking. There is true line-of-sight stabilization, and elevation is between −15° and +85°. For 3in/50 and 5in control, the 48in parabolic dish has nutating waveguide feed (6° spiral scan and 0.5° conical, at 24, 30 or 36 cycles per second; 2° beam), and a peak power of 50kW (0.1-microsecond pulses, PRF 3000), for a range of 30,000yds on a bomber at 10,000ft or a destroyer; accuracy is 10yds and 0.5 mil, and resolution 25yds and 1.5°. Operational in the 1950s. Preproduction models were manufactured by the Radiation Laboratory and production models by General Electric, and the first ones were delivered in August 1945.

Mk 35 radar on Mk 56 FCS, with a twin 40mm mount at left.

Mark 36
A 3cm fire control radar for use with GFCS Mk 60: automatic range tracker with nodding (plus or minus 20°) search mode, and conical scan (1.5°) for tracking. GFCS Mk 60 was a descendant of the 'handlebar' fire control systems: it tracked future position, the radar beam being displaced by lead angle. The entire system was to weigh 2500lb. It used a 24in parabolic dish emitting a 4.5° beam and had 25kW peak power with 0.5-microsecond pulses and a PRF of 1800 to achieve a range of 12,000yds on a bomber at 10,000ft; minimum range was 350yds. Accuracy was 50yds plus 2 per cent and 2 mils, and resolution 100yds and 4°. One development model was tested in the summer of 1944, and others were to be delivered early in 1945; the entire project was canceled that September. General Electric.

Mark 37
K-band (1.25cm) AA fire control set by Bell Laboratories, primarily for experimentation: one was built but not delivered. It could detect a single-engined plane at 6000yds and a twin-engined type at 12,000, using a rapid linear scan antenna. The latter was a 24in × 36in cylindrical parabola with a beam of 1.4° × 0.9°, with linear scan in elevation of 15° at 10 cycles per second, and lobe-switching by an oscillation plane mirror, 0.3° at 10 cycles per second. Peak power was 20kW (with 0.3-microsecond pulses and a PRF of 2400), and resolution was 50yds.

Mark 38
Radar for GFCS Mk 63, differing from Mk 34 primarily in having better below-decks packaging. The project was canceled in September 1945. Bell Laboratories.

Mark 39
A 3cm AA fire control radar for use with GFCS Mks 57 and 63; development ended at the end of World War II. The dish antenna (30in with nutating feed, 3° beam, 1.25° conical scan at 30 cycles per second) was suitable for both conical and nodding scan, the latter to bridge the gap between its narrow beam and the broad beam of a search radar. An E-scope was used for acquisition. An output of 35kW (0.5 microseconds, PRF 1800) was to give a range of 20,000yds on a bomber at 10,000ft, 20,000 on a fighter, 45,000 on a surface target. Accuracy was 30yds plus 0.5 per cent and resolution 80yds and 2.75°. Total installed weight would have been 1400lb (2300 for Mk 34) of which 70lb (100 or 130 for Mk 34 depending on model) was antenna. Submarine Signal Company.

Mark 40
An 8.5cm target acquisition and designation radar for giving target designations to fire control radars, providing rapid, positive, continuous target information. Presentation was shown on a PPI scope, all targets within range being visible at all stations; designation was by a coded signal on the PPI, one for each director station. There was a positive PPI repeat-back system. It used a 10ft × 10in linear array antenna with a 2° × 50° beam, and a 50kW, 0.5-microsecond pulse (PRF 1800) to achieve a range of 20,000yds on a bomber and 23,000 on a destroyer; accuracy was 200yds and $\frac{1}{4}$° and resolution 100yds and 2°. Scan was continuous at 50–60rpm, which was why very high power gave relatively short range (few pulses on target). Some descriptions, late in 1945, speak of two beams, an upper of 5° ×20–80°, and a lower of 2° × 0–20°, with 250kW. Developed by NRL and Raytheon, it was not produced.

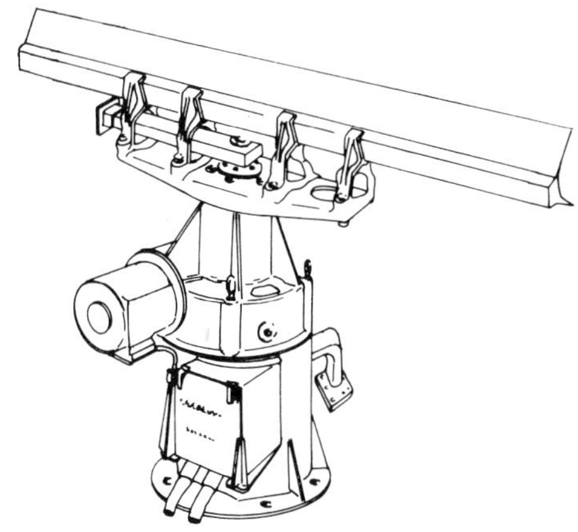

Mk 40 antenna assembly.

Mark 41
Experimental 1945 target designator to interconnect with fire control radar.

Mark 42
IFF for Mk V/UNB system, analogous to Mk 32 radar.

Mark 43
Abortive 3cm radar for GFCS Mk 61, intended to control 6in/47, 5in/54, 5in/38 and 40mm guns. Redesignated Mk 47.

Mark 44
A submarine fire control radar, associated with FCS Mk 101 and featuring A- and B-scopes and an 8in × 40in antenna and operating on Ku-band (1.87cm) with rapid antenna rotation. In 1948 it was expected that a pilot model would be available in July 1948 and preproduction models by June 1951, but Mk 44 was never produced. Capabilities were to have included scanning of a 15° sector at 30 scans per second and automatic tracking of low-flying aircraft. The 1.5° × 9° beam and 50kW pulses (0.2 microseconds with PRF 3200–9950 or 0.8 microseconds with PRF 900–1100) gave an instrumented range of 30nm (30,000yds on a destroyer), and minimum range was 200yds. The sectional parabolic antenna had lobe-switching for accuracy and sector scan for acquisition. Total weight was 1520lb, of which

550 was in the antenna. Only one appears to have been built.

Mark 45
K-band radar to supply target position data for aircraft Bomb Director Mk 2 Mod 0. 1945.

Mark 46
K-band radar, part of Bombsight Mk 22, 1945.

Mark 47
X-band fire control radar for GFCS Mk 61, superseding Mk 43. Development began late in 1945 and was completed in 1947. However, only eleven sets were built owing to a cut-back in the GFCS Mk 61 program. It had automatic tracking in bearing and elevation and aided tracking in range. Instrumented range was 15nm with a minimum of 300yds; resolution was 100yds and 2°. The 48in dish (2° × 2° beam) used conical-scanning for tracking and elliptical scan (3.5° × 12°, 27.5–30.5 cycles per second) for acquisition. 25–35kW (0.5 microseconds, PRF 1800) sufficed for detection of air targets at 20,000yds. Mod 1 was designed for 5in/38 control (cams cut to provide suitable outputs), Mod 2 for 5in/54, Mod 3 for 6in/47 and Mod 4 for 40mm.

SPG-48
Bell Laboratories fire control radar, an X-band gun-mounted dish type for GUNAR 1, 2 and 3 and GFCS Mk 69. It was a repackaged Mk 35 designed for reduced space and weight, and employed the 250kW Mk 25 Mod 3 transmitter and a reflector similar to that of Mk 25 Mod 2. Maximum target range was 40,000yds with acquisition at 25,000, using a PRF of 3000. First R&D deliveries were made in October and November 1950, production having been authorized in August; the last deliveries were made in June 1958.

SPG-49
Talos guidance radar, a C-band monopulse target-tracker incorporating a big microwave lens. It operated in conjunction with the SPW-2 fire control set which generated the beam along which the missile rode. Both became obsolete with Talos itself at the end of 1979. SPG-49A incorporated CW Injection to guide RIM-8E (operational in January 1962) and later versions of Talos. The original design called for mounting on a

SPG-49A (**1**), *SPW-2A* (**2**) *and SPS-30* (**3**), *USS* Oklahoma City, *November 1976.*
GIORGIO ARRA

Mk 37 (GFCS) pedestal, SPG-49 being considered a universal long-range fire control system, for guns as well as missiles.

It was designed in 1947, before the advent of high-power rotary joints. The requirement for tracking through the zenith led to a three-axis system in which all the radio-frequency generating equipment was on the mount proper, where servicing was difficult and exposure to weather damage considerable. The original specification (1948) called for an instrumented range of 400,000yds, with the capability to track a 1m² target at 200,000. Later these figures were relaxed to 300,000yds and track on a 4.5m² target. Other design requirements were a tracking rate of 50° per second, an acceleration of 200°/sec² in elevation, with an antenna beamwidth of 1.6° and an acquisition field of view of 11° × 1.6° or 11° × 5°. Power was to be 3MW. In fact, with this peak power (3-microsecond pulses, PRF of 500 for acquisition and 450 for tracking), the radar had a theoretical range of 294,000yds on a 5m² target, using a 10° × 5° acquisition window. Range accuracy was 15yds plus or minus 0.025 per cent of range, and angular accuracy was 0.25 mils. The CW Injection power was 5kW, using a separate transmitter. Range resolution was 150ft, and processing was analog.

This was an excellent performance, but the lack of sophistication in design showed in severe degradation in rain, acquisition range falling to 29nm. Much worse, MTBF was less than 30 hours, and by 1969 MTTR (Mean Time To Repair) had doubled in only three years, to about 1.5 hours. At that time the proposed replacement was SPG-51E, which was described as an austere solution; MTBF would rise to 89 hours, and MTTR would be less than an hour. Improvements in performance would include better clutter rejection, ECCM and automatic target detection. Considerable weight would be saved, and sensitive components would be moved below decks. In fact this program was canceled, and SPG-49 shortcomings led to the demise of the Talos missile. SPG-51E was the third proposed replacement, following SPG-56 and SPG-61. SPG-49 was the basis of the original Terrier guidance radar, SPQ-5. Sperry.

SPG-50
An updated SPG-34, repackaged for GFCS Mk 63 with a 40in X-band dish (gain 4000, 2.5° beam), using 40–50kW, 0.25-microsecond pulses (PRF 2000) and having a 12.6° spiral scan for acquisition and 0.75° conical scan for tracking (29 cycles per second, 2 for spiral scan). Accuracy was 50yds plus 0.025 per cent of range, and the system could follow a closing target at 800kts or an opening one at 350. A total of 196 were ordered, and the first was delivered in February 1956. Western Electric.

SPG-51
Tartar tracking and illuminating radar, part of the Mk 74 Gun/Missile Fire Control System. The original Mk 51 was a navalized version of the very successful Army SCR-584, converted to automatic tracking in range, azimuth and elevation to 168,000yds, on targets with speeds as great as 1000mph, and with azimuth and elevation rates of 10rpm and 5rpm respectively (June 1947). It was a 6ft S-band dish developed by the Radiation Laboratory, General Electric and Westinghouse. In the naval version an 8ft parabola with a sector scan unit was substituted for the original dish. Mk 51 Mod 1 added a new auto-tracker, and was

SPG-51 Tartar tracker-illuminator, showing feed details. BY COURTESY OF NORMAN POLMAR emplaced at the Point Mugu missile range in October 1948. SPG-51 was a fresh development begun in 1952 by Raytheon, tracking by C-band pulse-doppler radar, with X-band CW Injection for target illumination. It differed from other high-PRF systems in the use of multiple medium PRFs to permit conventional range tracking combined with doppler clutter rejection.

C/X-band operation required a combined feed system to exploit a single high-gain antenna. In recent versions two separate but overlaid antennas are used to permit the use of two separate feeds, so that the tracking feed assembly can be made more nearly optimum for monopulse operation. The X-band reflector consists of parallel wires supported over a solid C-band skin. In early versions the transmitter was mounted behind the antenna, but later ones incorporated below-decks electronics. SPG-51C had automatic acquisition and tracking, improved ECCM features, clutter rejection and multiple target resolution as well as a better MTBF.

SPG-51D, the current type, is described as dual channel. An Ordalt provides (electronic) emission control on command. SPG-51E was designed as a 'universal fire control radar', capable of controlling Talos as well as Tartar. The original SPG-51 was a minaturized, repackaged SPG-48, but later models are entirely different. Raytheon.

SPG-52
A range-only Ka-band fire control radar with optical tracking, for GFCS Mk 70.

SPG-53
A 5in fire control radar for use with GFCS Mk 68, in effect the successor to Mk 25/GFCS Mk 37, SPG-53 is SPG-48(XN-4) adapted to the Mk 48 director. It incorporates both spiral (12° for acquisition) and conical (3° for tracking) scan, out to 120,000yds. There is also a 4000yd precision sweep. The transition from spiral to conical scanning takes about 5 seconds. SPG-53A has a 250kW X-band transmitter; accuracy is 10yds and resolution 80yds. The current -53F incorporates Angle Error Tracking of projectiles to display relative shell and target positions.

From SPG-53E (or earlier) there has been provision for monopulse tracking. This model also has low-elevation angle tracking capability via clutter cancellation and employs a double-pull mono-

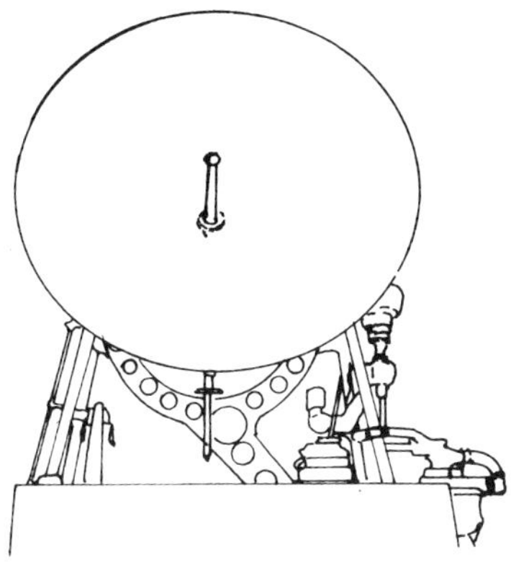

SPG-53A antenna.

pulse technique. It has surface splash-spotting and a missile launch alarm for quick reaction. In addition, at least -53E and 53F incorporate simulated ECM for training. Some SPG-53 series radars also incorporate CW Injection for secondary missile control. Western Electric.

SPG-54
An abortive missile control system, 1958.

SPG-55
Terrier C-band monopulse guidance radar, the successor to SPQ-5. It is distinguishable from the latter by its far shorter length, the pointed radome of SPG-55 revealing its Cassegrain feed system, in which power fed from the rear of the antenna is reflected off a secondary antenna in front of the main reflector. Such a system blocks less of the antenna than does a more conventional feed. SPG-55 must control not only beam-riding Terriers, in particular at present the nuclear Terrier BT(N), but also the newer semiactive types.

A typical SPG-55 consists of a main antenna with a capture antenna for CCM operation and a Custer horn antenna above the doppler antenna. In beam-riding operation, the capture antenna conically scans by means of a nutating feed. Tracking is either pulse, via the main antenna, or CW doppler; there is also a special low-elevation track mode. In a heavy jamming environment the system can operate passively, using the jammer as a beacon (Passive Angle Track mode); it can also accept external range data, for example from NTDS. A Local Range Track mode is used when the operator can distinguish the target through the jamming but cannot achieve range location; he manually tracks the target by operating a range rate control,

SPG-55B Terrier fire control radar.

and the computer of the fire control system maintains the range rate at the value last entered before Local Range Track was initiated. Finally, there is a Coast mode in which tracking is given up, but values of range and range rate are projected from the last pre-jamming figures.

Doppler tracking employs the same CW illuminator signals which are used to control semiactive Terriers in flight; the reflected signal is received by the doppler antenna, and a shield between it and the main antenna minimizes spillover from the transmitter. Maximum range in this mode is reportedly 200,000yds. In beam-rider operation, the broad capture beam nutates about an axis close to that of the main radar, to gain control of the missile within about 4 seconds; it steers the missile into the 1.6° main beam, which nutates about 30 cycles per second in sychronism with the capture beam, and the nutation axis of the guidance beam coincides with the track beam locked on the target. In semiactive operation, two X-band beams are radiated: a narrow beam, emitted at the point of the radome and focussed by the main reflector, illuminates the target for the missile seeker. The other side of the X-band feed emits into space, where it illuminates the rear of the missile to establish a rear reference signal for the guidance computer in the missile. The missile compares CW energy reflected by the target to this reference signal. Both of these modes are effective mainly at considerable ranges. However, the recent development of antiship missiles requires an auxiliary short-range system. In the case of current SPG-55s, this is provided by the Custer horn, a pillbox antenna (12° × 3° beam) intended to illuminate short-range, low-elevation targets while at the same time providing the necessary rear reference signal. The Custer horn is used in Sector Scan Engagement and Continuous Boat Track modes of operation.

Peak power in C-band in reportedly 1MW, with alternative pulsewidths of 0.1 microsecond compressed from 12 microseconds; of 1 microsecond compressed from 13; and of 0.1 microsecond uncompressed, with a PRF of 427 and an instrumented range of 300,000yds. The CW illuminator reportedly has an average power of 5kW and a 0.8° × 0.8° beam. Sperry and RCA.

SPG-56

A Sperry C-band Talos track and illumination radar, intended as the successor to SPG-49 but not procured. It was first proposed in November 1956, a prototype being ordered in January 1957; installations were planned for heavy cruiser Talos conversions (CG-13–15) which in fact were not carried out. Design requirements included reliability, effectiveness against countermeasures, ease of maintenance, and a greater range than that given by the existing SPG-49. Sperry's solution was a high-powered pulse-compression transmitter employing frequency comparison for accurate ranging, and monopulse processing for precise tracking. The dish antenna was unusual among contemporary missile guidance radars in being dome-enclosed. There were three separate transmitters: one C-band for tracking, one C-band for pulse-illumination, and one C-band for CW Injection. The sole prototype was employed by the White Sands Missile Range.

SPG-57

A Terrier missile tracking radar similar to SPG-55 (8ft C-band Cassegrain monopulse tracker with X-band CW Injection) but without the beam-riding feature. None was built.

SPG-58

Missile guidance and tracking radar for the Mk 86 FCS. Rejected in favor of SPG-60.

SPG-59

The main radar of the abortive Typhon system, employing a Luneberg lens for electronic beam-steering. Primary requirements included large bandwidth and frequency independence in steering the beam, to avoid several types of countermeasure. The cylindrical Luneberg lens focussed a beam projected onto it from any point on its surface. In the complete system, a lens deep in the ship was used, in effect, as a computer to generate signals with the appropriate phase relationships, for transmission from a sphere atop the cylindrical housing of the radar. It proved necessary to amplify these signals to overcome losses, and 356 travelling-wave tube amplifiers (TWTs), originally not part of the design, had to be interposed between the lens and the radiating sphere. Three more spheres were used for reception.

One problem in system design was that the spheres had a very small effective radiating aperture, less than 4ft, which had to be balanced by high radiated power. Altogether, 1800 medium-power amplifiers and associated waveguides were required for the 'small ship' system, with proportionately more for the 10,000-element 'large ship' system. It was tested aboard *Norton Sound* after the cancellation of the system proper. As installed aboard the test ship, it had a maximum peak power of 8.7MW (pulselengths of 2 microseconds for search, 0.1 microsecond for tracking, with jittered PRFs of, respectively, 200,000 and 400,000 on average burst) in C-band. Average power was 200MW, and pulse doppler operation provided high velocity resolution (100fs in search, 9fs in track). Range resolution was 20ft in search and 2ft in track, and acquisition range was stated as 165nm on a 1m² target. In principle the system could handle up to ten targets at a 0.1-second data rate for high-precision (3.5-mil) tracking, and could control up to 30 defensive missiles at the same time. At a lower data rate (4 seconds) it could maintain track on up to 120 targets (TWS).

SPG-59 was designated to carry out a high-power hemispherical search, with acquisition on 1m² targets (with a 6–14 second data rate, and a detection probability of 0.5) at 165nm (horizon to 5.1°), 155nm (to 8.5°) or 105nm (to 80°) and a low-power horizon search (37nm on a 0.5m² target with a detection probability of 0.9 at a 1-second data rate). Like the current Aegis, Typhon envisaged command control of defensive missiles.

SPG-59, installed aboard the test ship Norton Sound.

SPG-60

An X-band pulse-doppler monopulse fire control radar, part of the Mk 86 Fire Control System (*Spruance*, *Virginia* and *Tarawa* classes). It generally acquires targets designated by SPQ-9 or by a search radar, searching the designated area and then tracking at ranges of up to 50nm. FCS Mk 86 can use SPG-60 to track AA missile targets, and CW Injection can be used to illuminate targets for shipboard missiles such as Standard and Sea Sparrow. Pulsewidth and PRF can

SPG-60, tracking radar of the Mk 86 Fire Control System.

be varied under computer control to resolve the range ambiguities incident to pulse-doppler operation. SPG-60 is also part of the DDG Upgrade program, as part of the Mk 86 FCS which will replace the current Mk 68 with its SPG-53 radar. Lockheed Electronics.

SPG-61
Proposed replacement for SPG-49 under a Talos project of 1966. It was to have had a 100-hour MTBF (compared with less than 20), and to be repairable in under one hour. Instrumented range would be increased to over 225nm (compared to 147), and range resolution reduced to 50ft (from 150). Digital signal processing would improve resistance to ECM and to rain interference, acquisition range in rain being about 90 rather than 27nm. Alternative C- and X-band CW Injection systems were proposed. SPG-61 was to appear in service in 1971, with all SPG-49s replaced by 1974, but in fact only one prototype was built. RCA.

SPG-62
Aegis X-band CW (10kW average power) illuminator, slaved to the SPY-1 phased array. It forms part of FCS Mk 99 (Director Group Mk 79). RCA.

SPG-63
Dual-channel radar, using C-band (42.5kW, PRF 10,400–18,100) for target acquisition and tracking, and X-band (10kW CW) for illumination. Part of FCS Mk 99; can accept designation from other equipment. Same antenna as SPG-62. RCA.

B. Postwar Multi-Function Weapon Control Radars (Surface Ships)

SPQ-1
Noise-jammer developed during World War II ('Pimpernel').

SPQ-2
Regulus I 'Trounce' system tracking radar, a modified SP with stabilization and provision for tracking a beacon on the missile. Guidance commands were transmitted back to the missile by the radar, hence the multipurpose ('Q') designator.

SPQ-3
Designation canceled March 1951 and reassigned as Countermeasures Set SLQ-1.

SPQ-4
Cloud-base and -top measuring radar, transmitting a pulse and receiving two echoes which indicate the distances to base and top, using Ka-band (40kW pulses of 1- and 0.1-microsecond length, with PRFs of 492 and 4920). SPQ-4 is effective from zero to 60,000ft.

SPQ-5, showing capture antenna.

SPQ-5
Terrier monopulse C-band beam-rider guidance radar, derived from SPG-49. The prototype, tested aboard *Norton Sound*, was an SPG-49 modified for beam-riding as well as tracking, hence the multipurpose ('Q') suffix. It was first installed aboard the missile cruiser *Canberra*, replacing the Mk 25 Mod 7 of *Boston*. The complete system consists of a main beam-forming lens, a small dish to form the capture beam, and a telemetry antenna below the lens on the pedestal. There are four beams, capture, guidance, acquisition and track, and the entire system is line-of-sight stabilized. SPQ-5 is no longer active, having been replaced in Terrier ships by SPG-55.

SPQ-6 (P-6X)
An X-band mortar-locating radar, for the inshore fire support ship *Carronade* and possibly other amphibious fire support craft as well. An adaptation of the Marine Corps KPQ-1, it does not appear to have seen service. The (XN-1) prototype used 200kW pulses (1 microsecond, PRF 1200) to achieve a maximum range of 10,000yds on an 81mm mortar shell.

The principal feature of this radar was the use of a high-speed rotary waveguide switch operating at several thousand rpm, switching the radar to five antennas in sequence, to form an acquisition or search pattern of large angle for greater probability of early detection of the target; it incorporated a digital computer to determine firing point and detonation point even before the mortar shell reached zenith. Accuracy was within 50yds even at maximum range. Overall tracking rates were 5° per second during track and 20° during acquisition. The radar was to be stabilized against 25° of roll and 13° of pitch.

SPQ-7
Long-range precision tracker for satellites and ICBMs, mounted aboard missile range instrumentation ships. It corresponds to the land-based FPQ-6 and TPQ-18, but uses a 16ft Cassegrain antenna instead of their 29ft dish. Peak power is 3MW, and operation is in C-band. It is capable of obtaining signal level and position data on primary target (transponder) and on two secondary (skin reflector) targets simultaneously, for target cross-section measurements. Range (for the land-based versions) is reportedly 32,000nm, or 222,000 with modifications. RCA.

SPQ-8
Tracking radar for Pacific Missile Range ships. USAF, not Navy.

SPQ-9
A high-resolution pulse-compression TWS X-band radar, coordinated with SPG-60 as part of the FCS Mk 86. SPQ-9 is intended to operate out to 20nm, with a minimum range of 150yds. There are reportedly two versions: one for surface search and very limited (2000ft) air search, the other, with integral IFF, for air search up to an elevation of 25°. Only the surface version is dome-enclosed. Other reported features are a very high scan rate (60rpm) and a choice among five operating frequency ranges, using pulse-to-pulse frequency-agility as an ECCM measure. Digital MTI is also available. The complete Mk 86 system includes an electro-optical sensor, and it has been reported that the system can

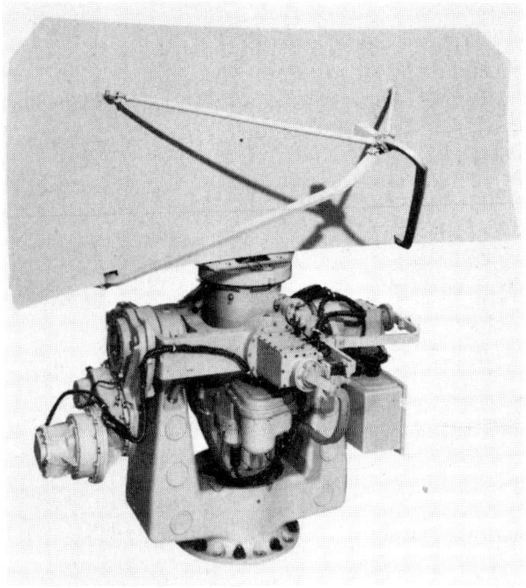

SPQ-9, with radome removed.

handle 120 tracks simultaneously. The prototype Mk 86 was delivered in March 1970.

SPQ-10
A Coast Guard weather balloon tracker, its computer providing constantly updated target direction and speed. For installation in large Coast Guard cutters, including the converted seaplane tenders (WAVP), it consists of Fairchild-Hiller components utilizing the Navy Mk 56 Mod 7 gun director antenna. This nomenclature was assigned in April 1966.

SPQ-11
An Air Force shipboard phased array radar ('Cobra Judy') supporting missile and space research. Input power is 1.1MW. Raytheon.

SPW-1
The prototype Talos guidance radar, projecting the beam along which the missile flew. It operated in C-band, producing a stream of triple pulses (two of 0.4 microseconds and one of 0.23 microseconds, or alternatively three of 0.23 microseconds) in a nutating beam, with a PRF of 900 (16kW peak power). Antenna position was controlled by a computer fed by the SPG-49 tracker. This system was replaced in production Talos ships by SPW-2. Sperry.

SPW-2
The projector of the beam along which Talos flew, now obsolete. It operated in conjunction with SPG-49 and was directed by a computer fed by that radar. In turn the receiver portion of the system

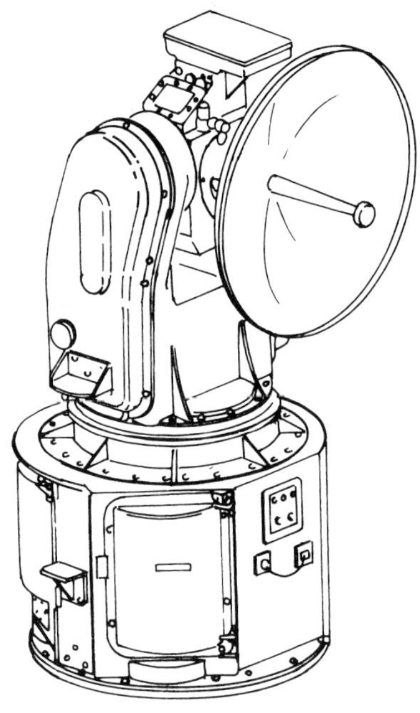

SPW-2B antenna assembly.

received transponder signals from the missile, so that missile range could be supplied to the computer directing missile flight. There were two pulse groups, one for missile guidance (triple-coded, 25kW, PRF 900) and one to arm a nuclear warhead or to switch on the semi-active homing feature of the missile, freeing it from the guidance beam. The small dish of SPW-2 produced a 4° beam, which nutated at 30 cycles per second. Range limits were 2000 and 212,000yds. SPW-2 remained part of proposals for improved Talos systems in which SPG-49 was replaced by improved (or more reliable) trackers. SPW-2B was redesigned to accommodate the nuclear-enable and CW-homing features, on Serials 7 through 21. Sperry.

C. Postwar Submarine Control Radars (AN/BPQ, AN/BPN Series)

BPQ-1
Regulus I Radar Course Directing Central, using the BPS-4 radar coupled to a specialized transmitter. It tracked a beacon on the missile and sent coded commands back. In the guidance mode BPS-4 transmitted 1-microsecond pulses (PRF 370), guidance was possible up to 400,000yds, search up to 160,000, and BPQ-1 could detect a bomber at 50nm. Total weight was 8870lb. BPQ-1 was the successor to the first Regulus guidance system, P-1X (a modified SV-1). Four were built, one each for NAMC Point Mugu and the missile submarines *Cusk, Carbonero* and *Tunny*. Installation began in 1953. P-1X was not an integrated system, and its radars were not standardized; in effect, BPQ-1 served as a preproduction version of the more fully engineered BPQ-2.

BPQ-2
Regulus I and II control and tracking radar, with an X-band surface search antenna similar to that of BPS-1 integral with an S-band missile operation and air search antenna and an L-band IFF. Characteristics were similar to those of BPQ-1, but weight rose to 13,550lb. The new air search antenna was 66in × 24in (3.5° × 17.5° beam, 500kW, 4 microseconds, PRF 200 for guidance, 200 or 400 for search); the surface search antenna produced 85kW pulses in a 3° × 17.5° beam. A 20m² target could be detected at 60nm and a missile guided at 350. There was a project for a BPQ-3 as an alternative guidance radar (second-source procurement).

BPN-1
'Trounce' radio beacon for Regulus guidance. Two were used to set up a hyperbolic grid, which established the missile's intended track through the target. Automatic ranging from the two BPN-1 units determined the 'dump point' at which the missile would dive. One problem was the need for guidance submarines close to an enemy coast, and the solution tried was 'Beagle', a beacon in a torpedo-tube launched buoy. Probably it is the radar beacon listed among the SPNs. The BPN-1 system incorporated no provision for control of missile altitude. Bendix.

OTHER US NAVAL SYSTEMS

The United States has recently deployed several naval radars which do not fall within the standard series of designations. They are generally known by their system names, rather than as individual units.

CAS

CAS is the Combined Antenna System of the Sperry Mk 92 Fire Control System, the US version of the Dutch M28. It is a combined monopulse tracking and illuminating X-band gun and missile control system, incorporating two antennas; an upper parabola with IFF for air and surface tracking, and a parabola with monopulse tracking and CWI fill-in horns below for air and surface tracking. Power is time-shared, so that full power can be devoted to search when the system is not tracking.

In the Mod 2 version (as in the *Perry* class frigates) it is combined with a STIR (Separate Tracking and Illuminating Radar) which provides a second missile guidance channel; STIR reportedly combines an SPG-60 pedestal and drive with a modified Mk 39 radar antenna, a dish with a monopulse cluster and CWI fill-in horns. The Mod 5 version of FCS Mk 92, which does not incorporate the STIR, is mounted on the Saudi Navy PCGs and PCCs of the US program.

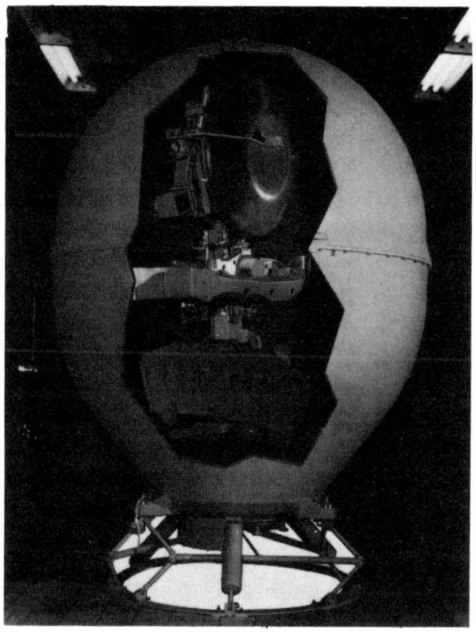

Mark 92 Fire Control System, showing search and track antennas inside radome. SPERRY

Mark 92 STIR (Separate Tracking and Illuminating Radar). SPERRY

Mark 115

Basic Point Defense Missile System (Sea Sparrow) director/illuminator, adapted from the Mk 51 director mount. It operates in CW only, and therefore requires two antennas, a parabolic dish for transmission and a flat plate planar array for reception. Tracking errors detected by comparison of transmitted and received signals are displayed in a gunsight atop the modified director pedestal; the operator tracks a target by moving 'handlebars' to keep the error signal centered in his optical display. Manual pointing was adopted for targets with high closing velocities but relatively little angular motion relative to the ship being attacked: Sea Sparrow functions as a point rather than an area defense weapon. However, the presence of the man in the loop makes for a relatively long reaction time. Fire control is carried out remotely from the CIC, where a fire control panel monitors the system. The operator there remotely energizes launcher and radar, and selects and fires the missiles. Interception range is described as over 4nm. Production began in 1967; the system had been under development since 1963.

Mark 91

Nato Sea Sparrow (US Improved Point Defense) Missile System fire control radar, operating in CW with separate receiving and transmitting antennas side by side. It exists in both Mod 0 (single) and Mod 1 (dual) configurations as used by the Danish, Norwegian, Italian, Japanese and US Navies, and there is also a special 'Netherlands' configurations used by the Belgian, West German, Dutch and Spanish navies, employing the Fire Control Transmitter Mk 73 linked to the Dutch M25 fire control system. In some NATO navies the radar director is designated EX-91; it is a tracker-illuminator typically designated automatically by SPS-58/65 or by the TAS. Raytheon.

Phalanx

The radar system of the Vulcan-Phalanx close-in (short-range) defense system, which employs a 20mm Gatling-type cannon firing depleted-uranium sabot rounds against incoming cruise missiles. The radar is a Ku-band pulse-doppler type employing a single transmitter with separate search and tracking radars; it operates autonomously, and is set on a stable platform which reportedly maintains system effectiveness even in Sea State 5. The track antenna is itself space-stabilized for rapid target handover and tracking acquisition. Basic functions are search, target acquisition, tracking and electronic spotting. In search, the radar uses unambiguous doppler/ambiguous range and switches PRF to achieve range resolution in three range-coverage zones. During target acquisition, the system again uses unambiguous doppler/ambiguous range

Phalanx (Close-In Weapon System). The two antennas are in the radome over the gun.

and a derived search range, and bearing and speed to search the range, angle and speed 'basket' provided by the search radar. Angle tracking is monopulse, with PRF switching to track across blind ranges. Electronic spotting uses variable PRF with selected spectral (frequency) line-tracking to measure the projectile stream angular error.

In operation, Phalanx is unusual in that it tracks both the incoming target *and* the outgoing rounds, detecting the angular error between them and continuously correcting to achieve a very high kill probability. Normally there is no operator intervention; the system searches continuously over a preselected sector, automatically engaging any high-speed incoming target unless a 'hold fire' button is pressed. Alternatively it can accept designation from other shipboard systems, or it can operate as a gun mount (*eg* against light surface craft) under the control of a shipboard fire control system.

The system was first proposed in 1968 as part of the early antiship missile defense program prompted by such incidents as the *Eilat* sinking; initial funding was under the Quick Reaction Contract (QRC) system then in use to obtain devices very rapidly, to keep pace with enemy devleopments. A feasibility contract was awarded to General Dynamics in 1969, and the first closed-loop system was demonstrated at White Sands the next year, after which a full-scale engineering development contract was issued. Although a prototype was installed aboard the missile frigate *King* in 1973, production was delayed as there were doubts that Phalanx could effectively destroy incoming cruise missiles. These were resolved, and full scale production was authorized in 1977, the first installation (aboard the carrier *America*) following early in 1980. The radar itself has been proposed for land applications as well. General Dynamics.

TAS

The Hughes Target Acquisition System (TAS) Mk 23 is designed to react automatically to incoming antiship cruise missiles, designating them to a Sea Sparrow illuminator (Mk 91 or Mk 115) and engaging them. Up to 54 targets can be tracked simultaneously; in effect the TAS is an antiaircraft combat system (excluding missile fire control directors, which it controls) for a small ship; it can also be integrated into the fire control system of a larger unit. Its L-band pulse-doppler radar has no separate designation, but the system as a whole goes considerably beyond that visible element.

Hughes TAS Mk 23 (land test). HUGHES

Originally TAS Mks 20 thorough 24 were designed, and all were tested ashore in 1974. Sophistication increased with the mark number, Mk 24 incorporating an infrared sensor as well as the radar. Mk 23 was tested at sea aboard the frigate *Downes* in 1975, and it was the one put into production. Mod 0 is the engineering development model aboard *Downes*; Mod 1, for the *Spruance* class, is integrated into NTDS; and Mod 2, for major auxiliaries (AOE, AOR) is a stand-alone type, with its own UYA-4 NTDS console.

The TAS radar is designed for short-range operation in severe clutter and jamming; it requires high precision for good designation accuracy and, in turn, rapid lock-on of weapon fire control systems, as well as a low false-target rate, since the number of defensive weapons is quite limited. There is also a requirement for very fast switching on and off (in fact in milliseconds) to permit fleet operation in EMCON (RF Emission Control). The linear array antenna scans at 30rpm; it is mounted back-to-back with an IFF antenna on a stabilized platform: it is 14ft wide, for a horizontal beamwidth of about 3°. Vertical beamwidth is 75°, for gapless coverage of diving and sea-skimming approaching missiles. Total weight is 10,000lb (2000lb topside).

There are four operating modes: normal, with an instrumented range of over 20nm, to employ point defense missiles; medium range, for radar surveillance, with an instrumented range of over 90nm (this also provides aircraft control capability); a mixed mode alternating normal and medium range scans so that early warning or air control can be combined with point defense; and EMCON, including effective operation over selected sectors, as well as use as an ESM detector. Once a target has been detected, the IFF system determines whether it is friendly, and the associated UYK-20 computer determines whether it is a threat and when it can be engaged. Future development will include the ability to correlate ESM signals with radar returns for more effective target classification. In the frigate *Downes* the TAS radar has completely replaced the SPS-40 air search radar formerly installed.

DDGX radar

In 1981 the major US Navy surface combatant program is the DDGX missile destroyer, with up to fifty units envisaged for construction from the FY85 program onwards. The goal is a combat system simpler than that of Aegis, but more capable than the one- or two-channel Tartar/Standard type of earlier ships. In principal it is to have a Multi-Function Array Radar for target acquisition and tracking, plus a Terminal Engagement Radar for terminal missile control; the weapon will presumably be some version of the Standard missile, fired from a vertical launcher. Details have not yet been settled, even to the extent of whether the TER will be slaved to the MFAR, as in Aegis. An alternative possibility is the use of interrupted CW Illumination (ICW) so that a single illuminator can be time-shared among several defensive missiles even during the 10–20 seconds of missile flight. Such a system would be a phased array illuminator which would allow multiple illumination; alternatively it might be a multifrequency radar.

In 1981, ICW is still in the laboratory stage, and the 1985 construction date is quite possibly too close for the first DDGXs to be fitted with it. Candidate engagement radars include the Sperry X-band system proposed for the FFG-7 upgrade and a Hughes FLEXAR radar loosely based on the AWG-9 of the Phoenix/F-14 airborne system.

As currently envisaged, FLEXAR (Flexible Adaptive Radar) employs a rotating single-face electronically scanning radar and searches both a zenith cone and the horizon. Over most of the hemisphere, which is covered by a conventional two-dimensional radar, it tracks and engages designated targets only. Its computer system is to provide TWS and multiple tracking, and the single antenna rotates rapidly. Pulse-doppler wave-forms reduce clutter.

For the primary MFAR radar there are several contenders. RCA has proposed a lightweight derivative of its SPY-1, which has the great advantage of already being in production, with software in place. Gilfillan proposes a dual-frequency S/L-band system. Sperry has proposed a C-band system, which it considers superior to S-band because the lower wavelength makes for narrower beams and hence for better resistance to ECM. In theory a C-band radar cannot reach S-band range, Sperry's view being that clear range is not nearly so important as is range in the presence of severe jamming. C-band also makes for greater accuracy in tracking, and may therefore eliminate the need for trackers and illuminators entirely. In that case possibilities include commanding a missile into an acquisition 'basket' so small that a K-band active seeker will suffice.

FFG-7 Upgrade

This is a Sperry private venture which received $20 million of Congressional Appropriation as a line item in the FY81 budget. Sperry's view was that the Perry class frigates would have to face the entire spectrum of Soviet weapons, including steeply diving air-launched missiles. In that case, in a severe jamming environment, conventional two-dimensional air search radars might well be unable to detect these weapons before their dive, and the lack of a zenith search or target acquisition capability might prove fatal. It was necessary to provide some system which could see at angles above 30°.

The Sperry proposal was to supplement the existing Mk 92 CAS system, which was designed specifically against the 'pop-up' threat with a 1-second data rate, with a phased array. X-band was chosen in view of space and weight constraints, and alternatives considered included back-to-back rotating radars, the new dome radar and a single face type. Ultimately a four-face system was chosen for the data rate required. Sperry proposed one break-in modification kit and six production kits for frigate refits, as well as an uplink permitting the system to fire the SM-2 (Aegis) missile. The Navy rejected the latter on the grounds that the uplink software would be too complex, but Congress appropriated the $20 million for research and engineering as part of a 3-year effort which Sperry estimated would cost $65–70 million.

The Soviet Union

DESIGNATIONS

As one of the two principal naval powers of today, the Soviet Union has developed a great variety of naval radars, many showing quite distinctive features. Due to Soviet secretiveness, their history and even, generally, their Soviet designations, cannot be described with any great authority. Moreover, although many of their signal characteristics are accessible to Western navies equipped with signal intercept equipment, they are not available to the public on an unclassified basis. This account must, therefore, be brief and incomplete. All of the names given are NATO codenames based, not upon radar function or characteristics, but rather upon antenna appearance.

RADARS

Soviet naval radar development reportedly began in 1944, when a copy of the British Type 285 gunnery radar was developed as Redan (NATO codename 'Sword Fish'). In addition, the Soviets received numerous Western ground and naval sets under Lend-Lease, including Type 271, SG, SF and SO for surface search, and SA, SK and Type 291 for air search. SO led directly to the standard Soviet torpedo-boat radar of the early 1950s, 'Skin Head'. Type 291 was produced as 'Cross Bird' (Soviet designation GUIS-2), evolving into the much more powerful, and more directional, 'Sea Gull' (GUIS-2M). The latter used a curved mesh reflector, fed by a pair of yagis emerging from it.

In retrospect it appears that Stalin considered the creation of a Soviet military radar industry a vital national priority. The Western powers were impressed by how rapidly the Soviets were able to put sophisticated sets into production, particularly for national air defense. The first native designs appeared about 1950, the replacement of foreign equipment being complete by 1957. At the time, such progress was considered a disquieting advance in Soviet technological competence. On the other hand, weaknesses in Soviet radar production continued (and, indeed, continue now) to be evident. Ships, such as the new cruiser *Sverdlov* which participated in the Spithead Naval Review of 1953, often deployed with major electronic equipment not yet installed, as evidenced by empty radar or countermeasures platforms.

Ship installation policy reflected Soviet tactical concepts, which emphasized traditional forms of surface warfare. Destroyers, for example, were provided with large X-band torpedo control (target designation) radars: 'Half Bow' (1952), succeeded by 'Long Bow' (1958) and by 'Top Bow', the latter being associated with the first Soviet antiship missile (in effect a long-range airborne torpedo), SSN-1. By way of contrast, the destroyer air search radar was 'Cross Bird', the equivalent of which the Royal Navy had discarded in favor of Type 293. However, the Soviets had no such high-definition air set.

The cruisers did have more elaborate air search systems. Generally they were completed with 'Sea Gull' on a main topmast, a large platform below it being reserved for a longer-range set. The latter materialized in 1952 as a naval version of the land-based P-8 (NATO designation 'Knife Rest-A'), operating at about 75mc/s. Peak power was about 100kW, with pulsewidths between 4 and 12 microseconds, and a 20° × 25° beam. The antenna consists of four long yagis, which is why it achieves much better directionality than British air search sets operating at similar frequencies; in additon it could probably achieve very limited height-finding by lobe-switching between its antennas. It appears to have been intended primarily for the *Sverdlovs*, although a few appeared aboard destroyers and aboard minesweepers (T-43 class) converted to radar-pickets. Since 'Knife Rest' was a standard Soviet land-based early warning radar of the time, the latter amounted to no more than moving part of the Soviet radar network out to sea. Land-based 'Knife Rests' are credited with a range of about 190nm, which might be increased by constructive interference from sea reflection. Most of these sets vanished from cruisers in the late 1950s and early 1960s, but in 1965 a successor to 'Knife Rest', 'Spoon Rest', appeared aboard some *Sverdlov* class cruisers. Its Soviet designation is P-12, and on land it is normally used in conjunction with the SA-2 missile. It consists of six vertical pairs of yagis, to form a 2.5° × 1° beam at 147–161mc/s, and with a peak power of 350kW it is credited with a range of about 150nm. It is far too large to fit on a mast, and instead must be accommodated on the superstructure abaft the mainmast.

At this time the Soviet Navy operated very large numbers of short-range land-based interceptors in defense of its bases (and, perhaps, of the Soviet coast, as the Soviets consider coast defense a naval function). It was often credited in the West with the ability to use these aircraft in direct defense of warships at sea. However, no height-finding radar, which Allied experience suggests is essential for naval fighter control, appeared at this time. The radar-pickets were a special case; there were even radar-picket submarines ('Whiskey Canvas Bag' conversions). It is not clear to what extent such ships were part of a separate program, entirely subordinate to Soviet national air defense forces (PVO-Strany). Certainly they were far less capable than were contemporary US DERs and YAGRs, given their minimal height-finding capability and their almost certain lack of fighter control facilities.

A new generation of Soviet naval air search radars began to appear from 1957 onwards in two bands, S and about 850mc/s. The major low-frequency set was 'Big Net', an open lattice paraboloid illuminated by a horn feed. It appeared aboard two *Sverdlov* class cruisers, some of the 'Kashin' class missile destroyers, and aboard 'Kresta-I' class missile cruis-

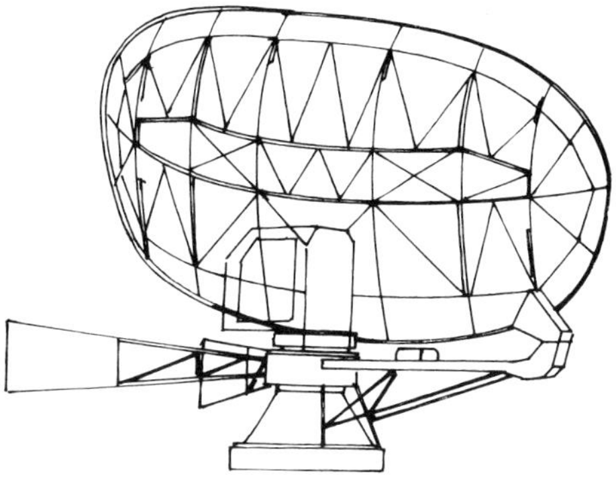

'Big Net' antenna.

The mainmast of a Sverdlov *class cruiser, fitted with a 'Top Trough' antenna, April 1971.*

ers. Another set in the same frequency range was 'Top Trough', which first appeared aboard a 'Skory' class destroyer in 1960, then aboard the cruisers *Kutuzov* and *Murmansk*, replacing 'Sea Gull' and 'Slim Net'. Like the British Type 983, it employed a slotted waveguide which fed a curved reflector. Later the same frequency range was extended with the appearance of the first Soviet three-dimensional radar, 'Top Sail'.

The new S-band series began with 'Hair Net', a general-purpose search radar originally fitted to *Kirov* class cruisers and to 'Tallinn' and 'Kotlin' class destroyers. Visually, it began the Soviet practice of providing prominent balancing (wind) vanes for their air search radars. 'Slim Net' succeeded it, beginning in 1957, appearing aboard many ships up through cruisers. It could be used for both air and surface search, varying its rotation rate (data rate) for the two alternative functions. Dimensions are about 5.5 × 1.8m, which for S-band should generate a very narrow beam, about 1° × 3°.

For smaller ships, the Soviets developed 'Head Net-A', which was

'Slim Net' (**1**) *and 'Hawk Screech'* (**2**) *radars aboard a 'Petya' class patrol craft.*

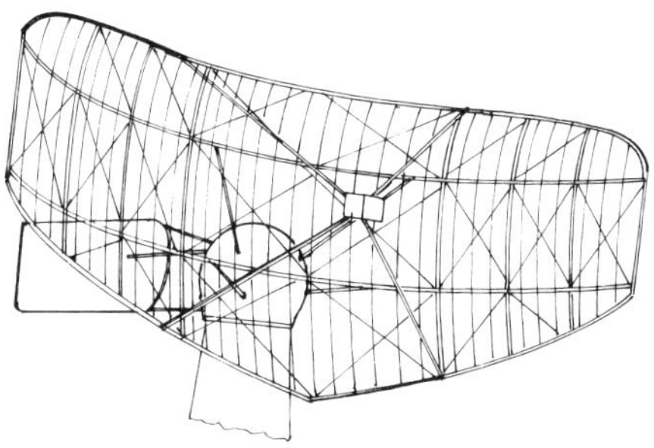

'Head Net-A' radar antenna.

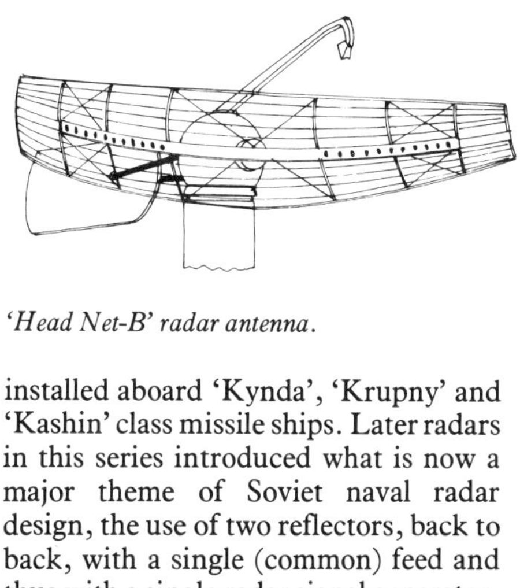

'Head Net-B' radar antenna.

installed aboard 'Kynda', 'Krupny' and 'Kashin' class missile ships. Later radars in this series introduced what is now a major theme of Soviet naval radar design, the use of two reflectors, back to back, with a single (common) feed and thus with a single radar signal generator. In 'Head Net-B' one reflector provides high, the other low, cover. In the widely used 'Head Net-C', one reflector is set at an angle of 30° to the horizontal, for vee-beam height-finding. The operator finds his target with the conventional antenna, marking it and then setting range gates on the second, nonhorizontal beam, to measure altitude. The system is quite simple, although not nearly as effective as a more conventional height-finder. It first entered service about 1963. Range on a large bomber is esti-

'Head Net-C' (**1**) *and 'Peel Group'* (**2**)*, as fitted to the 'SAM Kotlin' class guided missile destroyer* Nakhodchivy.

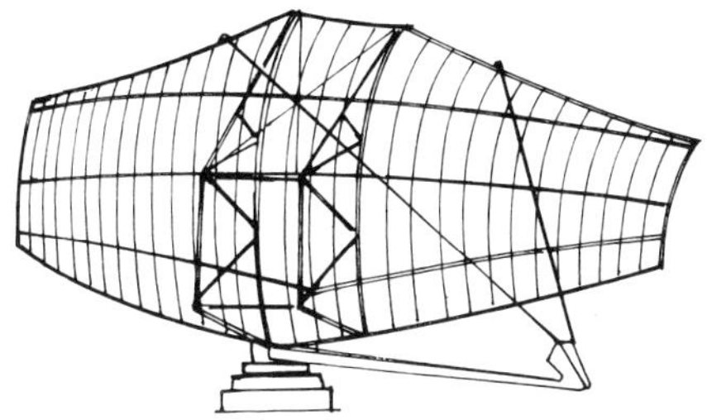

'Head Net-D' antenna.

mated as 60–70nm, compared to about 100 for 'Big Net'.

The other air search radar of this generation was 'Strut Curve', also operating at S-band, for smaller combatants such as 'Petya' and 'Mirka' class frigates and 'Poti' and 'Grisha' class corvettes. It reportedly operates at a somewhat higher frequency than the other S-band radars, and is unusual in not having a balancing vane behind its reflector. One might speculate that it is not stabilized, whereas most modern Soviet naval radars are. On this theory, the function of the vanes is to reduce the load on the stabilizer.

Probably the major Soviet naval radar development of the 1960s was the introduction of missiles, both surface-to-air and surface-to-surface. Two *Sverdlov* class cruisers were employed as test-beds, although only one of them, *Dzerzhinski*, operated outside home waters. She was fitted with the PVO SA-2 system, including a 'Fan Song' guidance radar and a 'High Lune' nodding height-finder, the only one of its kind ever in Soviet naval service. The system appears to have been unsuccessful, as the Soviets were willing to publish many of its details as early as the mid-1960s.

The SAN-1 of the 'Kashins' was much more compact, employing 'Head Net' as

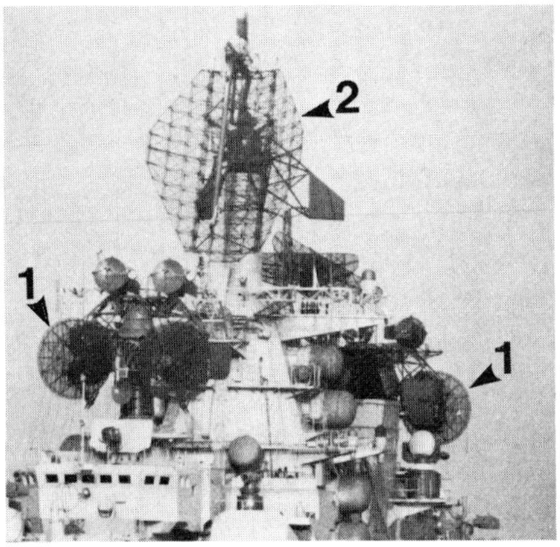

The 'Kresta-II' class destroyer Vasiliy Chapayev, *showing 'Head Lights'* (**1**) *and 'Top Sail'* (**2**).

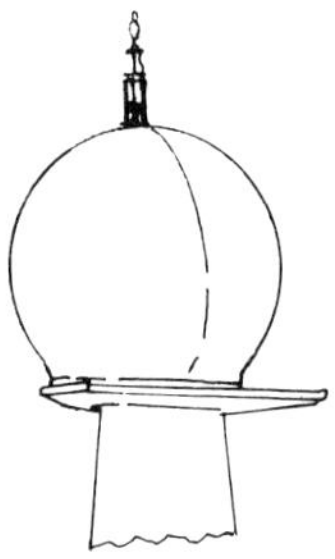
'Top Knot' masthead radome, wth 'High Pole-B' on top.

'Top Steer' aboard Sovremenny, *showing the spiral waveguide wrapped around the feed in front of the antenna.*

a search radar and 'Head Net-C' as a height-finder. As in later Soviet antiaircraft systems, SAN-1 is command-guided, separate dishes of its 'Peel Group' director tracking target and missile. Each director has two sets of dishes (to control two missiles flying towards the same target), plus a dish transmitting command signals. Similarly, the 'Head Lights' director of the larger SAN-3 system of 'Kara' and 'Kresta-II' class cruisers and the *Moskva* and *Kiev* classes shows two sets of radars mounted on a single director, suggesting a standard firing policy of two missiles per target. The longer-range SAN-3 apparently requires a much better height-finder, the frequency-scanned 'Top Sail'. The other major command-guided missile, SAN-4, employs a 'Pop Group' director with only a single target tracker and a single missile tracker, although dual operation is still possible with some form of polarization diversity, two signals sharing one antenna. SAN-4 functions as a point-defense system, corresponding roughly to the US Sea Sparrow.

Surface-to-surface missiles were the other major weapon/radar development; the 'Kyndas' introduced SSN-3 ('Shaddock'), with the 'Scoop Pair' radar to guide it. SSN-3 is designed to attack targets well beyond the horizon, receiving mid-course guidance from its launching ship. It can transmit its own radar picture down to the ship, so that the latter can designate a particular target. Indeed, the entire system is, in effect, a means of overcoming the possibility that the target ship will move significantly while the missile is in flight. For example, a Mach 1 missile covers 250nm in about 25 minutes; during the same period a ship of 30kts moves 12nm, so that a missile cannot be preset to home properly, particularly if its target is one of a formation of ships. Contemporary SSN-3-armed submarines, of the 'Juliett' and 'Echo-II' classes, had their own substantial guidance systems, 'Front Piece'/Front Door'.

Such a system is not entirely attractive, particularly for a submarine, which would have to remain on the surface during a large fraction of the missile flight, communicating with the missile. Most recent Soviet long-range weapons appear to require much less in the way of guidance. Thus SSN-12 (*Kiev*) requires only a very small retractable 'Trap Door' antenna. One reason why may be that SSN-12 acquires its target by means of a sensor with a very wide inherent search width, such as an ESM device, employing active radar (with a narrow search width, to avoid confusion with non-targets) only in the terminal phase of its flight. In the new *Kirov* there is no guidance radar at all for the SSN-19, the successor to both SSN-3 and -12. One may speculate that the new weapon is so fast that mid-course guidance is unnecessary.

With the introduction of the SAN-3 missile system in 1967, the Soviets also introduced a large frequency-scanned three-dimensional radar, 'Top Sail', operating at about the same frequency as 'Big Net', *ie* well below that of the US systems which had appeared a decade or more earlier. 'Top Sail' also differs from US practice in the design of its delay-line waveguide feed. In American radars the delay which makes frequency-scanning possible is achieved by twisting the waveguide into a serpentine shape, which fits into a narrow fin. In 'Top Sail' an analogous effect is achieved by wrapping the waveguide around a long tube. Presumably this provides a greater length consistent with the greater wavelength of the Soviet radar. It also appears that the Soviets have not taken advantage of electronic scanning to obviate mechanical stabilization. If that is true, 'Top Sail' may have a fixed frequency-shifting program.

With the *Kiev* the Soviets revealed a new FRESCAN radar, this time operating in S-band: 'Top Steer'. Although it is similar in general form to 'Top Sail', 'Top Steer' incorporates a two-dimensional radar, 'Strut Pair', mounted back-to-back with it and sharing a common feed. 'Strut Pair' had already attracted attention as, apparently, the first Soviet pulse-compression radar, aboard a modified 'Kildin' class destroyer. Mounted with a FRESCAN, it suggests that the Soviets have chosen to use a single chirped pulse in the three-dimensional radar, rather than employing successive pulses at different frequencies. 'Top Steer' would then be analogous to the Plessey airfield control radar, in which a single pulse also searches through the elevation range of the radar. In effect, parts of the pulse returning from different elevation angles are coded by frequency. The *Kirov* takes this evolution one step further. Its 'Top Sail' is mounted back-to-back with a 'Big Net' antenna to form a new radar called 'Top Pair'; it is difficult to avoid the probability of pulse-compression in 'Big Net' combined with single-pulse operation in 'Top Sail'.

All of these back-to-back systems have the advantage of reducing structural interference to radar emissions. In addition, net data rate is improved, and common stabilization (as in the Dutch M20 series) should make designation from the two- to the three-dimensional radar faster and more effective.

The *Kirov* also introduced a new

surface-to-air missile, SAN-6, reportedly employing TVM guidance, based on a pair of cone-shaped phased-array radars roughly comparable to the US Patriot land-based system. At the same time, the Soviet fleet has finally introduced a semiactive homing missile, SAN-7, reportedly equivalent to the land-based SA-11, which in turn is an improved version of the land-based SA-6. Unlike its US equivalents, SAN-7 appears to require a large number of relatively small-diameter illuminators, located alongside the superstructure. Visually, they resemble the small radars associated with Soviet close-in defensive guns. No electronic details are available.

The Soviets have also developed the usual series of gun fire control and surface search radars, including special systems to control their Gatling guns for close-in defense. They do not appear to have developed any equivalent of the US Phalanx system. On the other hand, their continuing devotion to heavy-caliber shellfire showed, after World War II, in the development of range-

'Egg Cup' range-only fire control radar, Aleksandr Suvorov, *April 1970.*

only radars ('Egg Cup') to provide local control on gun mounts. Probably the best known of the major gun fire control radars are 'Hawk Screech' (1954) to control 45, 57 and 76.2mm guns, and its successor, 'Owl Screech' (about 1961); both are small dishes with a large housing for the radar transmitter and receiver, turning and stabilization gear mounted with the antenna. For larger guns, the Soviets developed a spherical 'Wasp Head' director carrying a 'Sun Visor' (1953) X-band radar; there is also a 'Kite Screech' to control some 100mm guns. The 6in cruiser guns are controlled by a director carrying the 'Top Bow' radar.

Visually, all of these older fire control

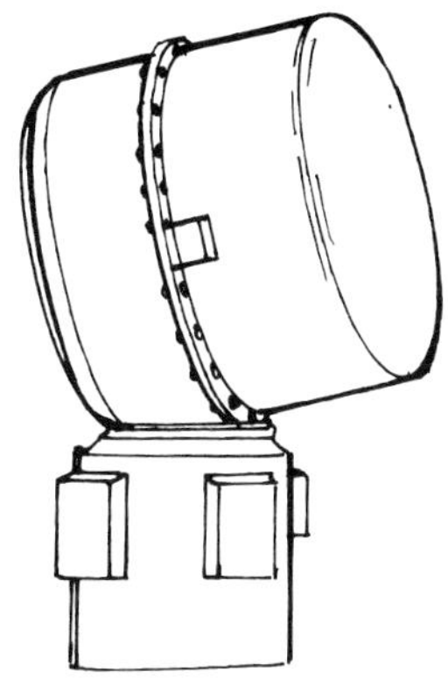

'Drum Tilt' fire control radome.

radars employ dishes or parabolic sections whose feeds are visible. In 1961, however, a new generation began to appear, consisting of radars in small drum-shaped radomes. The first was 'Drum Tilt', which controls the twin 30mm gun aboard 'Osas' and other combatants. It is superficially similar to 'Bass Tilt', which controls the Soviet Gatling gun, as well as 57mm weapons on 'Grisha-III' corvettes, and 76.2mm guns on 'Nanuchka III' corvettes and on 'Matka' class missile hydrofoils. 'Drum Tilt' and its associated guns were fitted to many Soviet ships from about 1973 onwards, presumably to counter low-flying aircraft and missiles; 'Bass Tilt' and its Gatling guns soon followed. The other major recent Soviet light AA control radar is 'Muff Cob' (57mm, liquid-cooled guns, 1962). All of these dates are approximate, reflecting the appearance of the earliest ship known to have carried the radar, and thus may well not reflect actual development dates.

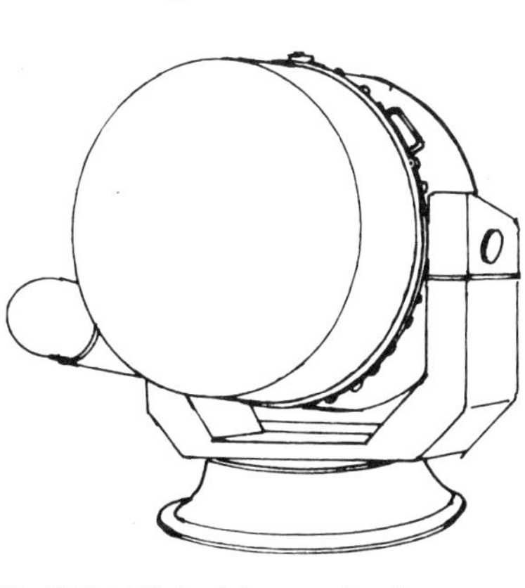

'Muff Cob' light AA control radar.

There are two other points worth making. First, the Soviets show a penchant for duplication in their radar systems on shipboard, surface search/navigational radars being a particularly noticeable case. It would be difficult to avoid guessing that unreliability is a serious problem. However, it should also be noted that the Soviet concept of shipboard maintenance differs materially from the Western one, in that the Soviets do not carry sufficient technicians (nor, probably, spare parts) to carry out the level of maintenance common in the West. In consequence, they may well have to accept the inconvenience of carrying duplicate equipment. Second, the Soviets appear to show far more appreciation of the virtues of ECM and ESM than do Western navies. Not only is ESM the primary targeting technique of the Soviet fleet (through land- and space-based sensors), but their ships have always shown a larger population of specialized electronic warfare radomes than have comparable Western units. Whether this reflects true concerns and investments is not, however, clear; all that is known is that the Soviets like to discuss electronic warfare in a far more positive sense than is common in the West.

Great Britain

DESIGNATIONS

The radar list which follows is by no means in chronological order, as it is arranged strictly by Type number. From Type 279 upwards the list was roughly chronological, except that with the advent of centimetric radar a new series began with Types 271–273. Centimetric equivalents of fire control radars (Types 284 and 285) became Types 274 and 275 in this series, but since 272 had already been taken, the centimetric equivalent of Type 282 became Type 262. No centimetric equivalent of the barrage director radar, Type 283, is known. Late in World War II, with the upper end of the 200 series already well occupied, new sets were given numbers in the 260 (*eg* 267 and 268) and 270 (*eg* 277) series; the former had begun as early as 1941 with the experimental Type 261 antiaircraft fire control radar.

About 1943 the Royal Navy began to designate its new radars in a 900, rather than a 200, series. It appears in retrospect that the 900s were coded by function in decades. Types 900 through (probably) 939 were reserved for gunfire control, so that the first postwar weapon control radar was Type 901, and the series of splash-spotting radars occupied Types 930 through 932. Types 940–959 were IFF interrogators and transponders, in analogy to the 240 and 250 series of the older system. Type 960 was a new air search radar, and the 960 series was primarily devoted to small-ship and carrier-controlled landing systems. The 970 series consists of navigational and surface search radars, the 980s were carrier air control systems, such as the massive Type 984 three-dimensional radar and the abortive Anglo-Dutch Type 988, and the 990s are target indicators similar in theory to the wartime Type 293, which enjoyed a long postwar career.

Suffixes to the Type numbers indicate modifications. The first was 'B', applied to the early air search sets (79, 279 and 281) when they were converted from two-mast to single-antenna operation by the addition of a TR box to become Types 79B, 279B, 281B. Major modifications to radars were indicated by the suffixes 'M', 'P' and 'Q' for the first, second and third; 'R' was used only as an additional suffix (to 271, 272 and 273, and in Type 271PR) to indicate the addition of a precise ranging panel. The letter 'U' denoted a set modified for fitting in coastal craft, and 'W' was a set adapted for submarine use. Finally 'X' (as in 79X and 261X) denoted an experimental set, with 'Y' and 'Z' indicating later modifications. In some cases suffixes were combined, as in Type 267QW, or Type 281BP.

Aerial (antenna) outfits were indicated by three-letter combinations beginning with 'A', as in AUK. This series included not only radars but also radio and countermeasures types. In many cases only the antenna outfit is indicated on a plan. Where available, these codes are indicated in the detailed radar lists below.

RADARS

Type 79

The first British naval radar, employing separate transmitting and receiving antennas. Work began in 1936, concentrated on a 75mc/s system, which was installed aboard the Signal School tender *Saltburn* late that year as Type 79X. With a wire antenna strung between her masts she detected an airplane at 500ft at 17nm, and work began at once on a turnable antenna. This was a single antenna with two parallel dipoles, one transmitting and one receiving, the entire assembly turned by hand. In July 1937 an airplane was detected at 8nm, which was disappointing. In March 1938 frequency was dropped to 43mc/s to take advantage of the greater transmitter power available. This became Type 79, which received the highest priority in the radar program at a 25 March 1938 review; the other requirements laid down at that time were for surface fire control and for antiaircraft fire control. They became the basis of Types 282 through 285.

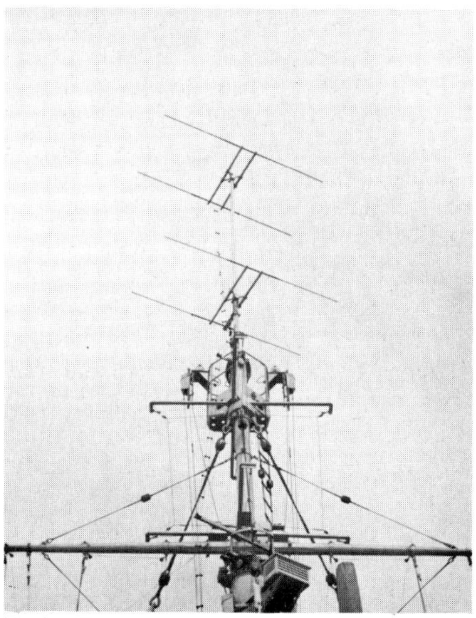

Radar Type 79Y, HMS Sheffield, *1940.*
CONWAY PICTURE LIBRARY

The new long range radar was to detect an aircraft at 50nm at 5000ft. Two of these Type 79Y were ordered in May 1938 for trials. The first, fitted in the cruiser *Sheffield*, managed 53nm at 10,000ft and 30nm at 5000ft in October 1938 on 15–20kW. The second set was installed in HMS *Rodney*. The beam was about 75° wide, but effective bearing accuracy was about 5°, using D/F methods (*ie* searching for a maximum return). Design work on an improved Type 79Z began in July 1939, the first being installed in HMS *Curlew* in September; thirty more were ordered. Peak power rose to 90kW (8–30-microsecond pulses, PRF 50); the antenna consisted of two active (transmitting or receiving) dipoles backed by two dipole reflectors, and measured approximately 14.5ft × 11ft (7m signals, with a half-wavelength of 11.5ft). Ranging accuracy was to the nearest quarter-mile up to 24nm, and to the nearest mile above that. There was no precision range scale, and the Royal Navy described Type 79 as useful only for warning, not for ranging, although it did have an A-scope display. Because it had simple open-wire feeders, the antenna could not turn freely; enough wire was provided for 400° of manual rotation.

Type 79B, introduced as a vertical coverage gap-filler for aircraft carriers, had a common aerial and operated at 70kW (1942). With an aerial at 110ft, range on an airplane was estimated as 90nm (20,000ft), 80nm (15,000ft), 65nm (10,000ft), 45nm (5000ft) and 20nm (1000ft). There were two range scales, 0–120nm (accuracy 1nm) and 0–24nm (accuracy ¼nm) and a bearing accuracy of 5° was claimed. In radio terms, this performance pertained to the 'sky wave'; there was also a 'ground wave' permitting limited detection of surface ships over a 2–6nm range scale.

Type 261X

Experimental fire control (AA) radar, 1941, operating at 50cm with 80kW peak power (1.7-microsecond pulses, PRF 500) with a range scales of 15,000 and 30,000yds, a range accuracy of better than 1 per cent, and bearing and elevation accuracies of 25°. Type 261Y was an AA direction radar using 10kW, 1.5m waves (2-microsecond pulses, PRF 200 to 800) intended to achieve a range of 30,000yds, with similar accuracy. Neither appears to have led to any operational radar, and neither appears to have been related to the postwar Type 262.

Type 262

Postwar light AA fire control radar, begun in 1942 and first delivered in 1946, in effect the centimetric successor to Type 282. It was designed for fitting on the automated twin 40mm STAAG (Stabilized Tachymetric Anti-Aircraft Gun) mounting, in a way a British equivalent of the US GUNAR, with radar, guns and computer all on the same mounting. Type 262 operated in X-band at 30kW peak power, using 20- and 200-microsecond pulses at a PRF of 1500, with a 5° beam, conically scanned, the dish spinning on an offset axis. It

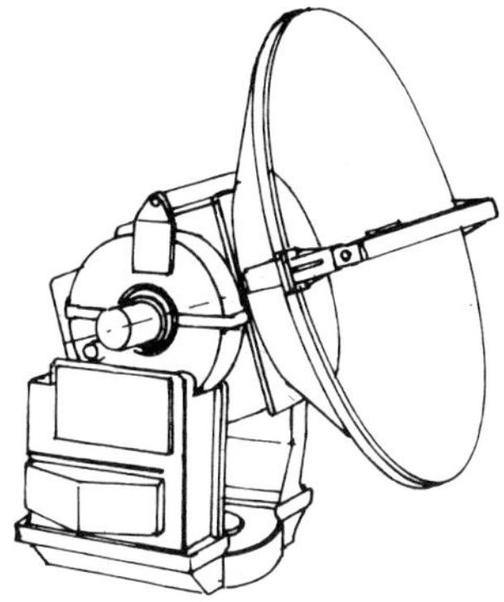

Type 262 antenna.

could auto-track at 5000yds, acquiring targets at 7000. Type 283 would initially acquire targets, and Type 262 would then scan 30° in azimuth in one second, elevating 3° at the end of each sweep, at the same time searching for 750yds on either side of the indicated range at 30 cycles per second.

In practice, Type 262 was badly affected by the vibration of firing, and also had a tendency to lock onto larger targets nearby, which discouraged target-towing pilots. The principal variant was Type 262Q, introduced in the *Daring* class in 1952, with the radar dish below the guns to reduce vibration. Type 262 was also fitted in a separate CRBF (Close Range Blind Fire) director, aboard Type 15 frigates and several aircraft carriers. In postwar terminology, this was Medium Range System (MRS)1.

Type 267

Dual-frequency (P- and X-band) radar for submarines, effectively a combination of Types 268 and 291, superseding 291W. The P-band dipoles could be used for air search when the submarine was at shallow depth (100kW pulses, 1 microsecond, PRF 500) and scanned at 5rpm. With the antenna at 25ft it could detect an airplane at 30nm (10,000ft), 25nm (5000ft), or 15nm (1000ft), and had a range accuracy of 100yds and a bearing accuracy of 3°. The X-band 'cheese' had a 2.5° × 30° beam and scanned at 5 or 10rpm (15–25kW, 1-microsecond pulses, PRF 500); with the antenna at 20ft it could detect a battleship at 15nm, a destroyer at 10, and a submarine at 4.5. Range accuracy was 100yds and bearing accuracy 0.5°. In 'S' class submarines the X-band antenna was outfit ANW instead of APS, Type 267QW rather than 267W.

Type 267MW dual-frequency radar (outfit ANW), HMS Scorcher, *May 1954.*
CONWAY PICTURE LIBRARY.

Type 267MW was an improved version with telescopic masts, so that operation submerged was possible. Surfaced performance was somewhat better than that of 267W, with detection of a destroyer at 10.5nm and of a submarine at 5.5. Submerged, aircraft could be detected at 17nm (10,000ft), 12nm (3000ft) or 4nm (1000ft), battleships at 7.5nm, destroyers at 5.5nm, and a submarine at 3nm. The usual three range scales (7500, 15,000 and 75,000yds) were supplemented by a 200yd scale on X-band and by a 250–500yd scale on P-band.

Type 267PW was an improved 267MW with a Master PPI and new range scales: 6000yds (accuracy 25yds), 3000yds (50) and 60,000yds (50) as well as a 600–800yd scale to exploit ground-wave effects.

Type 268

A 1944 X-band surface search and navigational radar developed in Canada for the Royal Navy. It was a small 'cheese', 30in wide, producing a 2.5° × 17° beam and scanning at 22rpm; a peak power of 40kW, using 0.75-microsecond pulses and a PRF of 500, permitted detection of a battleship at 9nm, a cruiser at 8, a destroyer at 7, a corvette at 6, an MTB at 5 and a submarine at 3nm. Range scales were 600yds (accuracy

50yds), 30,000yds (500), and 60,000yds (1000), and bearing accuracy was 2°. It was fitted in coastal forces craft and 'Hunt' class destroyers and minesweepers, and was an interim navigational set in cruisers and above.

Type 269

Modified AI Mk X night-fighter radar, operating in X-band.

Type 271

The first British centrimetric (S-band) naval radar, developed at very high priority for surface search use in the Battle of the Atlantic. The successful 10cm magnetron (the development of which Watson-Watt had not expected until after the war) was devised, by extraordinary good fortune, in February 1940, and by March 1941 the prototype 271 was aboard a corvette (HMS *Orchis*). It speaks much for British willingness to assist the United States even before the latter was at war that the roughly equivalent SG (albeit in 'breadboard' form) went to sea not so very much later.

The 10cm magnetron began with an output of 5kW, but this was later boosted to 90; it powered a 'cheese' surface search antenna, with separate receiving and transmitting antennas stacked one atop the other, with some of the transmitting and receiving elements directly in the back of the antenna, to overcome coaxial line losses. The receiver incorporated a crystal detector which was more sensitive than the previous tube types. The entire system was enclosed in a plastic 'lantern', the cheese antenna shape having been chosen because its broad fan beam would illuminate objects on or near the sea surface whatever the angle of roll of the ship carrying it.

In the original Type 271, power feed

Type 271 surface search radar 'lantern' (**1**) *and Type 72 aircraft homing beacon* (**2**) *aboard* Indomitable, *1943*. CONWAY PICTURE LIBRARY

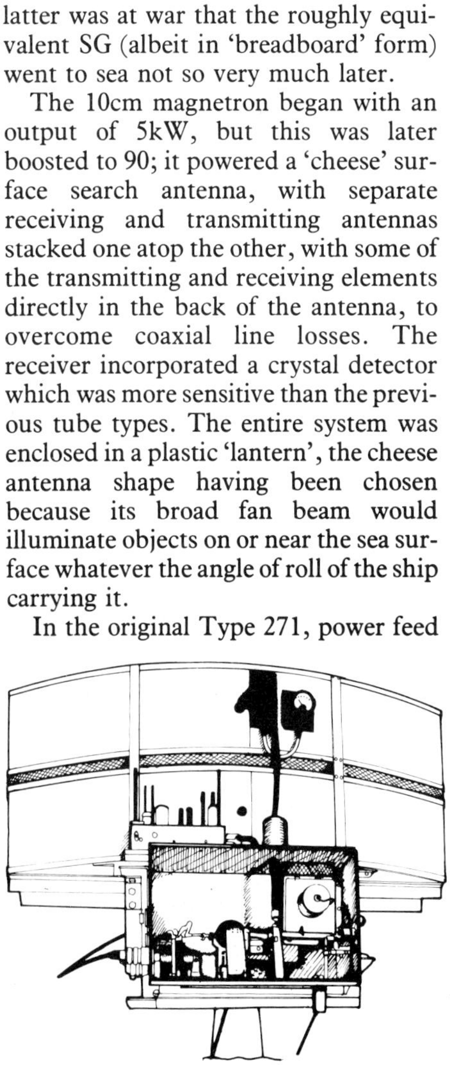

Type 271 'double-cheese' antenna.

to the antenna was by coaxial cable which limited rotation to 400°, and the display was an A-scope. The original Type 271 had a peak power of 7kW (1.5-microsecond pulses, PRF 500); on trials a surfaced submarine was detected at 5000yds and a periscope at 1300, and ships were detected at 6nm. Typical ranges were 11nm on a battleship, 10 on a cruiser, 8 on a destroyer, 6 on a corvette, 3.5 on an MTB and 3 on a submarine. Range scales were 7500yds (accuracy 200yds), 15,000yds (500) and 75,000yds (500), bearing accuracy was 2–3° and range resolution was 250yds.

As the original magnetron was boosted to higher power, waveguides were introduced to carry its power; the aerial system could be freed from the transmitter, making remote positioning possible. This was introduced in Type 271Q, which had both PPI and A-scope displays, and which had a 70kW magnetron, with similar pulse characteristics (alternative 0.7-microsecond pulse) for ranges, from a corvette, of 13nm on a battleship, 12 on a cruiser, 9 on a destroyer, 7 on a corvette, 5 on an MTB and 4–5nm on a submarine. In all Type 271 series sets, the addition of an Accurate Ranging Panel (L17 or L18) improved ranging accuracy to 25yds; the suffix 'R' was added, as in Type 271PR. An extra ranging panel and strobe unit were installed in the transmitting station to indicate the radar target. The principal successor set was Type 277.

Type 272

This was essentially Type 271 with a cable feed between the radar power source and the transmitter proper on the back of the aerial, so that the aerial could be as much as 40ft from the transmitter. In destroyers, for example, it could be placed atop a lattice mast. Radar characteristics matched those of Type 271 and 271P, although ultimately PPI displays (and power antenna rotation) were provided. Greater antenna height generally added about 15 per cent to radar range compared to Type 271.

Type 273

Large-ship surface search S-band radar of the Type 271 series, with higher gain aerials (two 4.5ft paraboloids side by side) which produced a much narrower beam. Installations had therefore to be restricted to large ships with relatively restricted or slow rolling. The hand-rotated antenna was enclosed with a plastic 'lantern' atop the radar office, just as 271 had to be located directly above its office. Typical ranges were 18nm on a battleship, 16 on a cruiser, 13 on a destroyer, 10 on a corvette, 7 on an MTB and 5 on a submarine. There were 7500yd (accuracy 200yds), 15,000yd (500) and 75,000yd (500) scales, and bearing accuracy was 3°. Type 273P was a modified version.

In Type 273Q (1942) a waveguide nozzle was substituted for a dipole feed of the earlier types, and power was

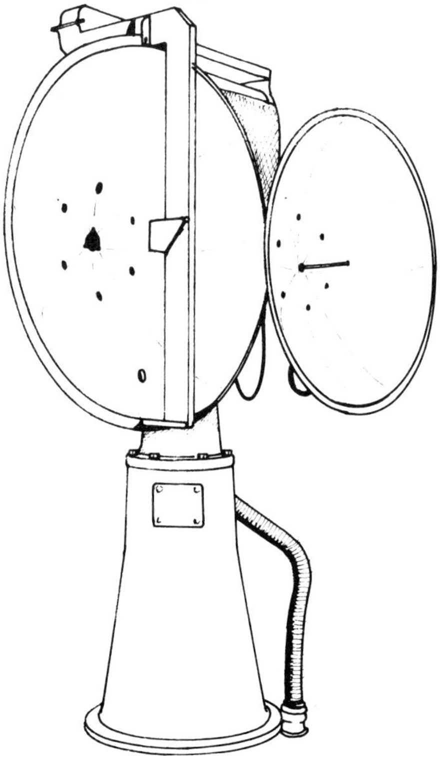

Type 273Q twin antenna.

increased as in 271Q. This version also introduced continuous rotation and a PPI, with a scan rate of 0–15rpm. The entire antenna was gyro-stabilized in the vertical plane, to eliminate the restriction to slow-rolling ships. Up to three remote PPIs could be fitted. Ranges improved to 23nm on a battleship, 20 on a cruiser, 17 on a destroyer, 12 on a corvette, 9 on an MTB and 7 on a submarine. Installation time for both 273 and 273Q was three weeks, as in 271 and 272. Also as in the earlier sets, an extra 'R' suffix indicated the precision ranging panel.

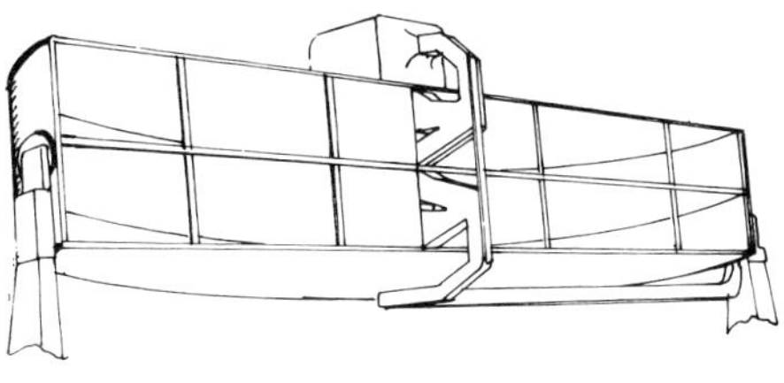

Type 274 antenna.

Type 274

S-band replacement for the Type 284 main-battery fire control radar. It used crystals for reception, but these were not amenable to single-antenna operation, and 274 employed two superimposed 'cheeses', 14ft × 7in, the upper one for reception. The lower, transmission 'cheese' was fed by a single horn, but two were used for reception, to permit beam-switching for improved bearing accuracy (3 minutes of arc), the beam itself being 2° × 14°. Peak power was 400kW (0.5-microsecond pulses, PRF 500) for an effective range of 40,000yds and an accuracy of 25yds.

Type 274 was designed in 1942, tested ashore in 1943, and first fitted that year. In contrast to the US Mk 8 and Mk 13, it had a very narrow fixed field of view, and so could not be used to spot splashes. Instead, small 930-series radars were appended to the Type 274, their 'cheeses' (and beams) at right angles to that of the main radar. They were discarded postwar as surface gunnery became less important.

Type 275

This was a dual-purpose fire control set, in effect a centimetric replacement for Type 285. It used the Type 274 radar transmitter, and was designed for the new Mk VI director; some were also fitted to Lend-Lease Mk 37s. As in the surface search set, separate transmitting

The frigate Leopard *(30 September 1958) showing Types 275* (**1**), *974* (**2**), *293Q short-range air search/target indicator* (**3**) *and 960 long-range air search* (**4**) *radars, and IFF Mk X for the Type 960* (**5**).
CONWAY PICTURE LIBRARY

and receiving antennas were fitted – in Type 275 they were 4ft dishes set within nacelles which could elevate – and conical scanning allowed the set to achieve an accuracy of 10 minutes of arc. Detection range was 36,000yds, with tracking at 30,000. Work began in 1942, and sea trials were held early in 1945.

Type 276

A surface search radar, the successor to 272 and briefly fitted in some destroyers in 1943–44. It was produced only in limited numbers, for ships which could not take Type 277, and postwar it gave way to Type 293. It introduced the 500kW S-band transmitter operating with 1.5- and 1.9-microsecond pulses at a PRF of 500, with an alternative short-pulse mode of 0.7 microseconds, which characterized Types 277, 293 and 980 through 983. The antenna was a single power-rotated 'cheese' stabilized in azimuth (outfit AUS) driving both A-scope and PPIs, scanning at 0–15rpm with a 3.5° beam. Typical ranges, with an antenna height of 150ft (70ft), were 22nm (16nm) on a battleship, 19 (14) on a cruiser, 16 (12) on a destroyer, 12 (10) on a corvette, 9 (7) on an MTB and 7 (5) on a submarine. Aircraft at 200–6000ft could be detected at 13–17nm. Range scales were 15,000yds (accuracy 50yds), 75,000yds (250) and 150,000yds (500).

Type 277

Sometimes described as the most successful wartime British naval radar, Type 277 served both as a surface search radar (its official function) and as an 'approximate' height-finder and low-

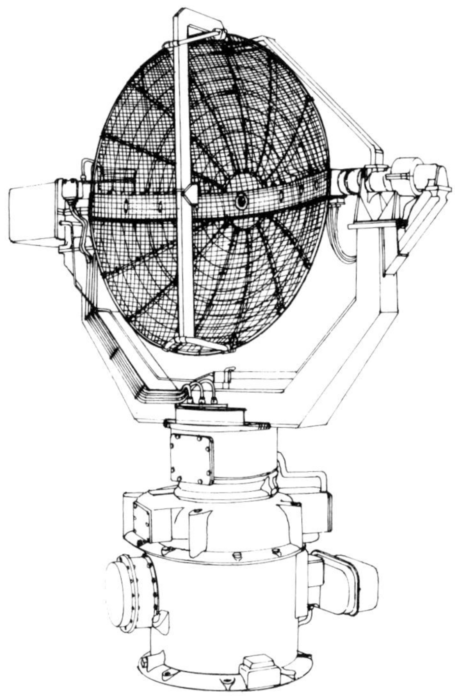

Type 277 antenna assembly.

flyer detector, and continued in service long after 1945. The antenna (outfit AUK) was a 4ft 6in parabolic dish stabilized in both elevation and azimuth, producing a 4.5° × 4.5° beam, and scanning at 15rpm in bearing. Transmitter and range scales (and range accuracy) matched those of Type 276, with a claimed height-finding accuracy of 1000–2000ft between 3.5° and 40° of elevation. With an antenna at 100ft (as in a battleship) ranges on surface ships were about 23nm on a battleship, 20 on a cruiser, 18 on a destroyer, 15 on a corvette, 11 on an MTB and 10 on a sub-

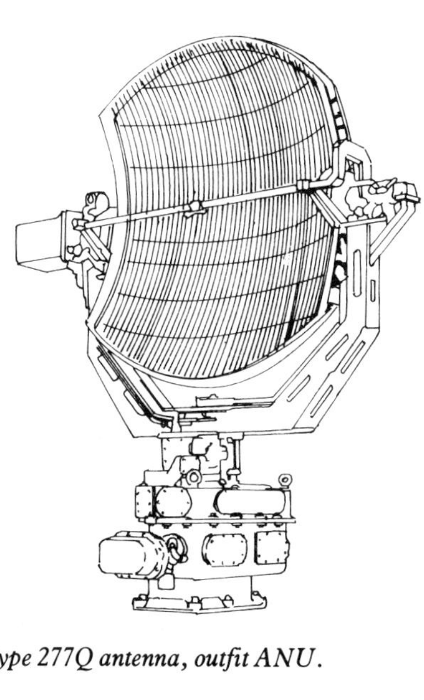

Type 277Q antenna, outfit ANU.

marine. Aircraft at up to 5000ft could be detected at 30nm, and height-finding could be done at up to 25nm.

The original design was intended for frigate/corvette surface search, but with the elevation feature Type 277 could serve as a height-finder and low-flyer detector aboard larger ships. It fed a PPI and, in larger ships, a sector display and a height–plan indicator (HPI). Elevation was by a handwheel. Type 277P was slightly modified for increased receiver sensitivity and slightly improved performance. The postwar Type 277Q incorporated a new antenna (outfit ANU), an 8ft spherical parabola with clipped sides (4.5° × 2.5° beam) for more precise height-finding. Performance against surface craft (0° elevation) matched that of 277P, but aircraft between 5000 and 20,000ft could be detected at 55nm, and height-finding range on an airplane at 20,000ft was 60nm. The scanning rate was halved, to 7.5rpm. Type 277 was closely associated with Type 293, and postwar Type 277Q was generally fitted in conjunction with Type 293Q.

Type 278

Postwar successor to Type 277, with ANU antenna, in 'County' class missile destroyers.

Type 279

Essentially Type 79 plus an Accurate Ranging Panel RBL10 (range 7nm), produced in 1940. It was credited with a range of 65–95nm on an airplane at 16,000ft, 27–40nm at 3000ft, 16–24nm at 1000ft and 5–7.5nm at 100ft. Range accuracy was 50yds on the detailed scale between 2000 and 14,000yds and bearing accuracy, as in Type 79, was 5°. As in the earlier radar, it used 7–30-microsecond pulses (peak power 70kW, PRF 50) and range discrimination was 500yds.

Type 280

Army GL1 gunlaying radar taken over for Navy use; it introduced the shorter wavelengths and the new antenna later incorporated in Type 281. The first, Type 280X, was installed in the cruiser *Carlisle* in 1939. It incorporated a precision ranging unit, with an expanded A-scan. Frequency was 54–84mc/s, peak power 50kW (3–4-microsecond pulses, PRF 100), and typical ranges were 65nm on an airplane at 16,000ft, 27 on one at 3000ft, 16 on one at 1000ft, 5 on one at 100ft, 6nm on a battleship, 5 on a cruiser and 3 on a destroyer. Accuracy was 50yds between 2000 and 14,000yds, and 3° in bearing. Type 280 was fitted to the armed merchant cruisers *Alynbank, Ariguani* and *Springbank*.

Type 281

The principal British wartime air search radar for heavy ships. It was designed to a 1939 specification calling for long-range surveillance as well as gunnery ranging (with an Accurate Ranging Panel RBL11) and employed two pulse lengths (short length for precision) and a double aerial system with four active dipoles, permitting beam-switching. All of this had to be done within the aerial dimensions of the earlier sets, so the frequency was doubled to about 90mc/s. Beamwidth was reduced to 35° and gain considerably increased. Typical range figures were 88–115nm on aircraft at 16,000ft, 38–50 on aircraft at 3000ft, 22–28 on aircraft at 1000ft, and 7–9 on aircraft at 100ft.

The first prototype was set up at Fastnet Fort East in June 1940, and the first ships fitted were the light cruiser *Dido* (October) followed by the *Prince of Wales*. Production sets were delivered from February 1941 onwards. Peak power in the long-pulse (15-microsecond) mode was 350kW and in the short-pulse (gunnery; 2-, later 1.7-microsecond) mode 1000kW; PRF was 50 in both cases. Range accuracy was 100yds between 14,000 and 28,000yds, and by means of beam-switching a bearing accuracy of 1° could be obtained. Beam-switching also permitted automatic tracking of a target in later versions. There were two pairs of scales: long range (20 and 100nm) and gunnery (14,000 and 28,000yds), with accuracies of, respectively, ¼nm and 1nm and 50 and 75yds. Generally the foremast antenna was used for transmission and the mainmast for reception. In the single-antenna Type 281B only the mainmast antenna was retained, and the beam-switching facility eliminated due to limitation on the TR box.

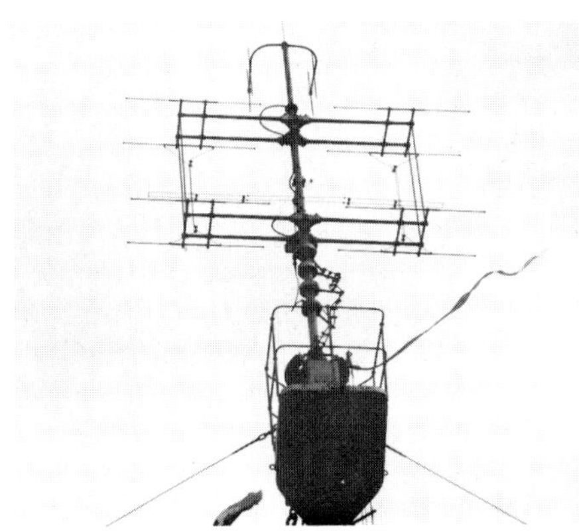

Type 281 air search radar (mainmast installation, looking forward; probably receivers). BY COURTESY OF JOHN ROBERTS

Type 281BP and Type 281BQ had improved receivers and eliminated the short-pulse feature (Accurate Ranging Panel). A new 150nm scale was added, and performance included ranges of 120nm on an airplane at 20,000ft (110 in Type 281), 90 at 10,000ft, and 65 (compared to 58 in Type 281) at 5000ft. In Type 281BQ power rotation (at 2 or 4rpm) was added, to drive a PPI display; the antenna was fed by a coaxial joint. Total aerial weight was 1344lb. In a typical installation, the main lobe of the radar pattern extended from 10,000 to 30,000ft; the second lobe was not considered adequate for the detection of fast jets. Type 281BQ was superseded by Type 960.

Type 282

The first of a series of 600mc/s (50cm) gunnery radars using a common transmitter; development began in March 1938. The effort was originally directed towards development of a surface gunnery set (which became Type 284), but late in 1939 the Admiralty stated an urgent requirement for a short-range antiaircraft gunnery set, and Type 282 was designed. Shore tests were carried out in February 1940 but ships were not fitted until 1941. In fact the first set to go to sea was the prototype 284, in HMS *Nelson*, late in 1939. The 50cm power tube produced 15kW (later 25kW) pulses at 1.7 microseconds and with a PRF of 500, and Type 282 was mounted on pompom directors Mk IV and also on Mk III in *Victorious* and *Roberts* and on close-range predictor Mk I in *Charybdis*. The antenna consisted of a pair of yagis, one to transmit, the other to receive: beamwidth was 40° × 43°, range 5000yds and accuracy 50yds.

The original set was not considered satisfactory, partly because with so wide a beam the operator could not be sure that he was ranging on the target at which his director was aimed. Wartime improvements were higher power (60 or 80kW, 1-microsecond pulses, PRF 500) with shorter pulses for greater accuracy, and a TR box which permitted both yagis to be used for sending and receiv-

Radar Type 282, on one of Prince of Wales' *pompom directors, 1941.* BY COURTESY OF JOHN ROBERTS

ing and so reduced the horizontal beamwidth to 29°; beam-switching could also be used for improved bearing accuracy. Thus the Type 282M1/2 and -3/4 used a range scale of 6000yds; accuracy was 50yds and 5°, improved to 1° in M3/4, and resolution was 330yds and 13°, improved to 160yds and 2°. Type 282 was incorporated in the Hazemeyer tachymetric predicting mounting for a twin Bofors gun which was actually fitted to some destroyers in 1944–45, but work on this radar ceased in 1943 in favor of the centimetric Type 262, which replaced it postwar.

Type 283

Gunnery radar for controlled long-range antiaircraft fire, using the Type 282 transmitter and antenna. It was introduced in June 1941, using a twin yagi with reflector. By 1945 peak power was 150kW (1-microsecond pulse, PRF 500), and beam was 60° wide. Display was an ABU (Auto-Barrage Unit), but this device was essentially an admission of the failure to produce an effective tracking AA system. Guns were preset to fire at a range between 1000 and 5000yds, and the ABU triggered them when the target came into range. It had no centimetric successor.

Type 284

Main battery surface gunnery radar, actually the first of the 282–285 series. The first production set was fitted to HMS *Nelson* in June 1940 after tests late in 1939 showed detection of a convoy at 30,000yds and of a cruiser at 18,000. The first production set was installed

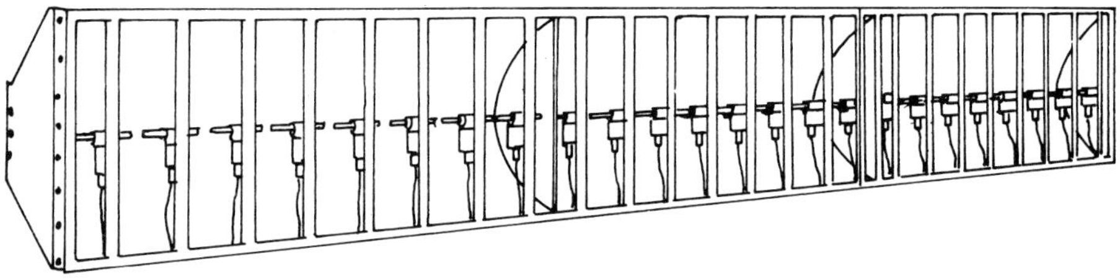

Type 284 antenna.

aboard *King George V*, with a total of 24 dipoles (12 in *Nelson*). The total installation consisted of a pair of 21ft × 2ft 6in trough reflectors, each with 24 dipoles (one to send, one to receive) fixed to the director; in some ships only 12 dipoles per antenna could be accommodated, in an 11ft installation. Peak power was 25kW (1.7-microsecond pulses, PRF 500) and claimed accuracy was 200yds, 1–2° on the 24,000yd scale and 500yds on the 48,000yd scale. The beam was 8° wide.

As in the other systems in the series, power was increased to 150kW and pulse-length reduced to 1 microsecond at the same time that a TR box was introduced; in Type 284, unlike the others in the series, this last improvement did not alter performance. Thus Type 284M of 1942 had an accuracy of 250yds and a resolution of 330yds and 5° in the M1/2 version on the 48,000yd scale; on the 24,000yd scale accuracy was 125yds. In the -M3/4 version accuracy was 25yds and 10–15 minutes of arc and resolution 160yds and 1°. Improvements included a new display panel and a range unit for better accuracy, as well as beam-switching within the 24-dipole array. The set was succeeded by Type 274.

Type 285

Secondary-battery gunnery radar, in effect an extension of Type 282 with a six-yagi antenna (three to transmit, three to receive) for a narrower 18° × 43° beam. Accuracy was 100yds on the 15,000yd scale and 250yds on the 30,000yd scale, with a bearing accuracy of 3–4°. Typical ranges on surface ships were 7nm on a large cruiser, 5nm on a destroyer, 4000yds on a submarine in surface trim (compared to, respectively, 15,000, 8000, and 5000yds for Type 284), whilst on aircraft the figures were 17,000yds at 5000ft, 18,000 at 2000ft (which indicates the lobe structure of the antenna pattern), 14,000 at 1000ft and 9500 at 600ft.

Types 285M and 285P used a TR Box

Type 285 gunnery control radar, HMS Aurora, *January 1948.* CONWAY PICTURE LIBRARY

(common aerial; 9.5° beam in 285P) and the higher power short-pulse transmitter; an ABU was also installed. Accuracy for 285M1/2 was 150yds and 2° and for -M3/4 50yds and 15 minutes. Resolution was, respectively, 330yds and 5–6°, and 160yds and 1.5°. Type 285P beam-switched for greater accuracy.

Type 285 was fundamentally flawed in that it could not follow aerial targets in elevation; it was replaced by the centimetric Type 275, which could.

Type 286

The first British small-ship radar, developed from an airborne (ASV Mk I) set on the basis of successful trials with a Walrus on a seaplane ramp in 1940. It was not particularly satisfactory. An Admiralty summary of May 1941 noted that 'the chief merits of this set are that it is available in quantity, can be fitted inside a week, and occupies little space. A crane is necessary for placing the aerials. The chief defects of the set are low range, high top weight for small craft, and poor sense of discrimination so that confusion is caused by "back echoes" from vessels to about 3 miles from right astern to the quarter. The set is fitted either in vessels whose size does not admit of a heavier aerial system (*ie* desttoyers, corvettes, MGBs, MTBs, MLs) or in any big ship where time cannot be spared to fit a more elaborate set.'

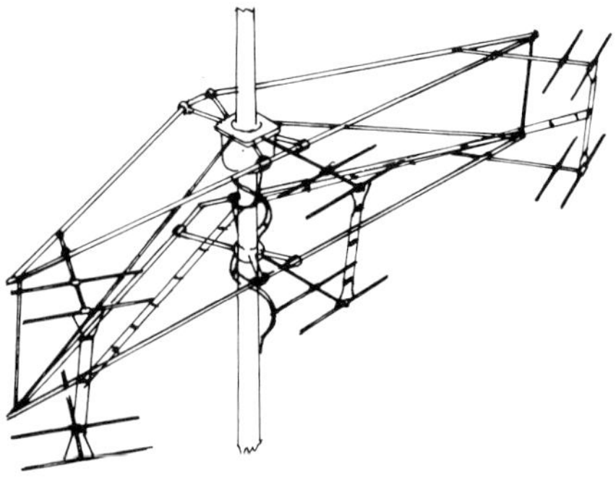

Type 286 antenna.

It operated at 214mc/s (1.4m) with 6kW peak power, 2-microsecond pulses and a PRF of 200–800, and employed a fixed antenna (two yagi aerials, each with six radiator-reflector dipole sets, paired vertically to produce three broad fixed beams) for surface detection and warning over 60° on either bow. Bearing accuracy was poor, except for dead ahead, but bearings could generally be determined within 10°. Typical ranges were 6–8nm on a cruiser, 4–7 on a destroyer, 1–1.5 on a trimmed-down submarine and 2nm on an 'E-boat'. It could detect an airplane at 1000ft at 15nm, and one at 16,000ft at 25. Range accuracy was 200yds between 1000 and 20,000yds, and half a mile between 10 and 25 miles. The improved Type 286M was adapted from ASV Mk II, and had higher power (7kW, 1.2-microsecond pulses, PRF 450–500 or 2000) and had similar range performance. Accuracy was 2–10 per cent and resolution 200–250yds, with either 9/36/90nm or 10/40/100nm range scales. Aerial outfit ATQ was a non-rotating fixed rectangular frame supporting a forward-facing yagi for transmission, with two arrays angled outwards for reception.

Type 286P had a manually rotated antenna for more precise bearings, achieving an accuracy of 3–5°. Typical ranges, when mounted in a destroyer at a height of 80ft, were 6nm on a battleship, 4 on a destroyer, 1.25 on a submarine, 19 on an airplane at 10,000ft, 15 on one at 5000ft, and 12 on one at 1000ft. Range scales were 9, 18 and 36nm, with an accuracy of 5 per cent of the range scale in use. Type 286PU was designed for coastal craft, with a fixed, forward-facing yagi for transmission and two yagis angled out for reception below it, for a fixed forward cover of 140° of bearing, with beam-switching for accuracy. Later it employed a lightweight rotating aerial. Typical ranges, for a set installed in a trawler, were 4.5nm on a battleship, 2.5 on a destroyer, 1.25 on a submarine, 11 on an airplane at 10,000ft, 14 on one at 5000, and 12 on one at 1000ft; the variation in figures presumably shows the effect of antenna height on the lobe structure of the radar signal. Both 286P and 286PU were replaced by Type 291.

Type 286PW was designed for submarine use, with a frame aerial on a telescopic mast between the periscopes; typical ranges, for an antenna height of 30ft (4ft) were 4nm (1.5nm) on a battleship, 2.5 (1) on a destroyer, 1 (0.5) on a submarine, 16 (10) on an airplane at 10,000ft, 15 (7.5) on an airplane at 5000, and 9.5 (3.5) on an airplane at 1000ft. It was replaced by 291W or 267W. Type 286PQ incorporated a more powerful 100kW transmitter of Type 291 type (1.2-microsecond pulses, PRF 500); it took three weeks rather than the 5 days of the earlier types to install. As fitted in a destroyer, 80ft above water, it could detect a battleship at 9nm, a destroyer at 6, a submarine at 2, an airplane at 10,000ft at 30, one at 5000ft at 22, and one at 1000ft at 16nm. Range scales and accuracies matched those of 286P.

All of these dual-purpose systems were replaced by a combination of Type 290 or 291 for air search and Type 271 or 272 for surface search. HMS *Vanoc* made the first U-boat kill attributable to radar after her 286 detected a U-boat at 1nm on 17 March 1941.

Type 287
A shore-based radar to give precision ranging against surface ships attempting to negotiate controlled minefields.

Type 288
High-angle gunnery set; Type 284 modified for use aboard armed merchant cruisers and ocean boarding vessels, 1940.

Type 289
Dutch-developed 70cm gunnery radar, installed in the destroyer *Isaac Sweers* for 40mm fire control.

Type 290
Replacement for Type 286, designed from the beginning as a naval set. It had a peak power of 100kW (1.7-microsecond pulses, PRF 500) with a range accuracy of 35yds on a 30,000yd scale and a bearing accuracy of 3°. It could detect an airplane at 30,000 yards and employed the rotating aerial later used in Type 291. It served only briefly (1941–42) before replacement by Type 291.

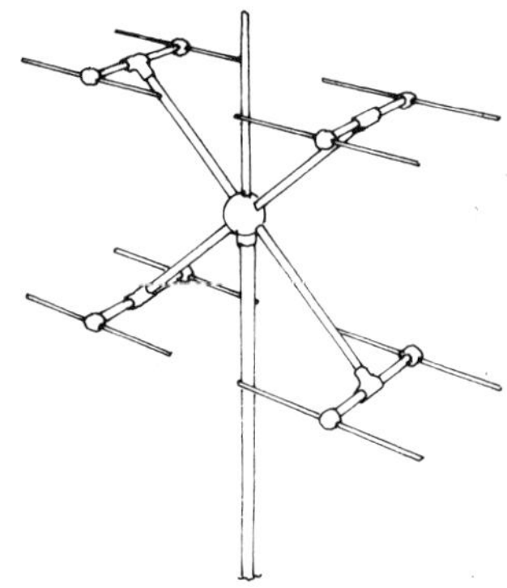

Type 291 antenna

Type 291
The final British 214mc/s (P-band) small-ship search radar, introduced in 1942. The earliest version required separate transmitting and receiving antennas, but a TR box was soon developed. The antenna was similar in concept to that of Type 281, but the dipoles were supported by an X-shaped structure; by 1944 Type 291 was fitted to nearly all British destroyers and lesser escorts, and one version was employed in submarines. Design was relatively simple, installation requiring 7 days.

Peak power was 100kW (1.1-microsecond pulses, PRF 500) and typical ranges (destroyer installation, 35ft above water) were 9nm on a battleship, 6 on a destroyer, 3 on a submarine, 35 on an airplane at 10,000ft, 30 on one at 5000ft, and 15nm on one at 1000ft. Beamwidth was 40°, accuracy 200yds and 3° (antenna outfit ATQ/R) or 5° (ATS), and resolution 160yds. Range scales were 7500, 15,000 and 75,000yds. 'M', 'P', and 'Q' versions had power rotation and PPI displays in addition to their A-scopes.

Type 291U, for coastal forces and trawlers, had a special lightweight aerial consisting of a pair of superimposed yagis. With the aerial at 20ft, its range was about 4.5nm on a battleship, 3 on a destroyer, 1.5 on a submarine, 15 on an airplane at 10,000ft, 22 on one at 5000ft and 12 on one at 1000ft. Again, the loss of range at 10,000ft seems attributable to the variation in antenna lobe pattern with antenna height. It was replaced by Type 268. Type 291W was designed for submarines, with a rotating aerial designed for watertightness under great pressure. Range, with the antenna at 30ft (4ft) was 5.5nm (2nm) on a battleship, 3.5 (1) on a destroyer, 2 (0.5) on a submarine, 30 (17) on an airplane at 10,000ft, 25 (12) on one at 5000ft and 15 (4) on one at 1000ft. Types 291U and 291W were limited to A-scopes. Type 291W was replaced in service by Type 267W. As for 291, it remained in service in destroyers until about 1952, after which destroyer air search was restricted to coverage provided by Type 293, the target indication radar.

Type 293
S-band target indicator, using the same transmitter as Type 277 but with a new stabilized 'cheese' antenna. Stabilization was necessary because otherwise the roll of the ship would tilt the fan beam, and air targets might be displayed at wildly wrong bearings. The antenna itself (AUR) elevated from the horizontal, with a 3.2° fan beam. Typically ships were fitted with 281B modified for PPI operation, as well as with 293 and with 277. Type 281 would be used to track a target until the latter appeared on the Type 293 PPI, which had a bearing cursor driven by Type 277. Thus the latter could be rotated by hand until it bore on the target, then elevated to find target altitude, displaying its data on a height-position scope. In AA frigates, Type 293 was used with an overhead optical system which could project the lines of bearing of gun directors onto its PPI for matching.

Typical detection ranges, when fitted in a destroyer at a height of 80ft, were 15nm on an airplane at 10,000ft, 12 on one at 5000ft, 10 one one at 1000ft, 17 on a battleship, 12 on a destroyer and 4 on a submarine. Range scales were 15,000yds and 150,000yds, with accuracies 50 and 500yds respectively, and bearing accuracy was 0.5° with a 6ft antenna. Provision was made for accurate ranging for fire control in fleet destroyers.

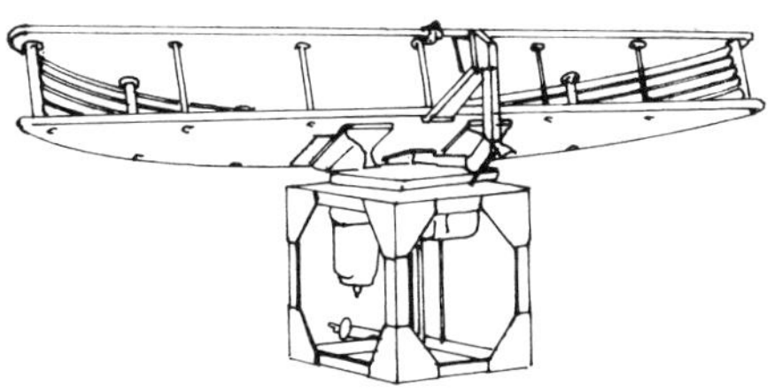

Radar Type 293P (outfit AQR).

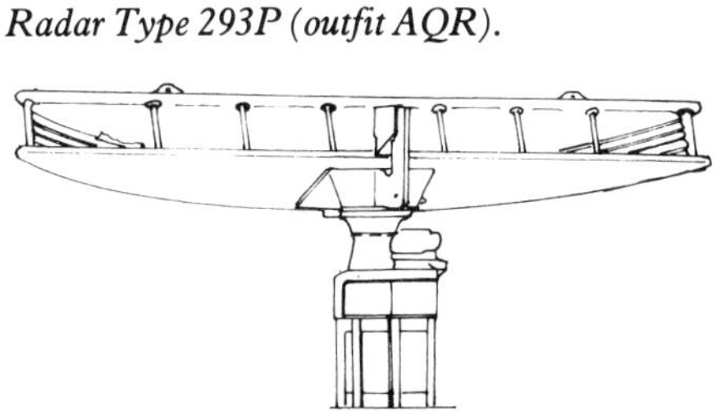

Radar Type 293Q (outfit ANS).

Type 293M of 1945 incorporated an 8ft antenna (outfit AQR, 2ft deep, scanning a 2.6° × 30° beam at 7.5 or 15rpm). Accuracy matched that of Type 293, but range improved to 18nm on an airplane at 10,000ft, 14 one one at 5000ft, 12 on one at 1000ft, 18nm on a battleship, 13 on a destroyer and 5 on a submarine. Bearing discrimination was improved. Type 293P was similar, but was modified for easier maintenance.

The postwar radar program included Type 293Q, with a new 12ft aerial (outfit ANS) for a beam reduced to 2° × 35°, scanning at 5, 10 or 15rpm. It was designed to operate with Type 277Q, and had a similar range, about 30–35nm; elsewhere the improvement over Type 293M was estimated as 25 per cent. These figures can be reconciled by reference to Type 293 performance against high altitude targets; the original Type 293 could detect an airplane at 25,000ft at 20nm.

Type 294
S-band combined air and surface warning and height-finding radar, intended to replace Type 277, with aerial outfit AUT. It had similar pulse characteristics, and scanned at 4–5rpm. Development was abandoned at the end of the war.

Type 295
This was intended to replace both 277 and 294, with a 12ft × 12ft stabilized antenna. It was abandoned at the end of the war.

Type 298
Abortive X-band surface warning radar of 1945.

Type 901
The beam-generating guidance radar of Seaslug, operating at X-band. Development began in 1945. It forms the basis for GMS1, Guided Missile (Fire Control) System Mk 1, and uses an elaborate combination of reflectors and a dielectric lens to generate a combination of acquisition and tracking (conical scan) beams: a pivoted reflector for the vertical motion of the acquisition scan, fed by an off-center horn which can itself rotate about an axis perpendicular to that of the radar proper. Together these motions generate a beam motion similar to the raster scan of a television tube: up and across. Originally developed as a universal long-range fire control radar for the LRS-1 system abandoned 1949, in analogy with SPG-49.

Seaslug guidance radar (Type 901), HMS Fife, *1973.* C & S TAYLOR

Type 903
Fire control radar for the MRS (Medium Range System) 3. MRS3 was based on the US Mk 56, and its radar was based on Mk 35. It was widely used.

Type 904
Antiaircraft fire control radar, which proved abortive.

Type 905
Surface gunnery radar for MRS4. Abortive.

Type 907
Range-only gunnery radar for SGS.

Type 908
Radar for the abortive Tachymetric One-Man (TOM) director.

Type 909
Tracking and illuminating radar for the semiactive Sea Dart missile system.

Type 910
Target tracking radar for the Seawolf point-defense missile.

Type 912
British designation for the Italian Selenia Orion RTN-10X. *Amazon* class (with Ferranti WSA-4 processor).

Type 909 radar, HMS Sheffield. C & S TAYLOR

Type 910 Seawolf tracker, aboard HMS Penelope, *October 1976. Note television tracker at left.* MoD(N)

Type 930
Canadian-built splash-spotting radar, operating in both X- and K-bands, with a 0.5° beamwidth and very short pulses for high precision – 150kW, 0.25 and 0.1 microseconds, PRF 200.

Type 931
Canadian-built K-band spalsh-spotter, scanning a 10° sector at 20 scans per second, as in US practice; its beamwidth was 0.67°. Peak power was 40kW, with 0.25-microsecond pulses and a PRF of 2000.

Type 932
Canadian-built splash-spotter.

Type 960
The principal British postwar long-range air search radar, first fitted to HMS *Vanguard*. Although it superficially resembled Type 281, in fact it was a new set with a very similar antenna, operating on the same frequency range (86, 88 or 90mc/s). Peak power rose to 450kW (5- and 15-microsecond pulses at a PRF reduced to 250 for long range) for a range of 170nm on an airplane at 40,000ft, 150nm on one at 30,000ft, 130nm on one at 20,000ft, 95nm on one at 10,000ft and 60nm on one at 5000ft. High-altitude detection was improved by a better lobe structure compared to that of Type 281; the first lobe reached 50,000ft and the second began at about 60,000. Horizontal beamwidth matched that of the older Type 281, 35°, and bearing accuracy was about 2°. Range scales were 0–100nm and 100–200nm and accuracy was 1nm. Scanning was clockwise at 0–8.5rpm, or counterclockwise at 0–30, and installation required 12–15 weeks, as in Type 281BQ (8–10 weeks for the simpler 79B, 279B and 281B).

Type 960M had a small mattress array; only one was built, and it was fitted ashore for tests. Type 960P had a large mattress antenna, 30ft × 15ft; it was expected to achieve ranges of 220nm at 50,000ft and 170nm at 30,000ft. It is probably the large mattress visible on plans of HMS *Malta* drawn in 1945. Only one example was planned, and it is not clear whether it was ever built. Both 960M and 960P operated over a wider frequency range, 80–90mc/s.

Type 961
Carrier-Controlled Approach radar, superseded by Type 963.

Type 962X
Experimental X-band CCA radar.

Type 963
CCA radar, superseded by the US SPN-35.

Type 964
Decca Type 45 surface search and navigational radar (see Type 978).

Type 965
In effect, this is the British equivalent of SPS-17, designed during the Korean War, its transmitter based on the American SR but its antenna entirely British in concept. The latter is designated Aerial Outfit AKE, and there are two types, AKE(2) consisting of two AKE(1), one atop the other. The 8 or 16 elements of the array are each shaped

Type 965 single-mattress (AKE-1) antenna, with IFF Mk 10 above. C & S TAYLOR

Type 965 double-mattress (AKE-2) antenna; Type 278 is below. 'County' class. C & S TAYLOR

as a reflector horn surrounding a loop dipole; the 26ft × 8ft 11in (or 16ft 9in) array is 6ft 1in deep and weighs 2425lb (or 5480lb). It operates in P-band, with a wavelength of about 1.4m, and a peak power of 450kW (3.8- or 10-microsecond pulses, PRF 200 or 400). The aerial produces a 12° × 40° beam and rotates at 10 or 8rpm, depending on whether the power supply is at 60 or 50 cycles per second. Generally it is surmounted by a Mk 10 IFF aerial fixed to the AKE outfit.

Type 965 was originally described officially as a small-ship air search radar, and indeed initially appeared aboard the radar-picket destroyers and the 'Tribal' class frigates. However, it ultimately replaced Type 960 in all British warships and it will probably be succeeded in turn by Type 1022 and similar systems such as the projected Type 1030 STIR. The unusual quarter-wave chokes ('ditches') between the radiating elements are designed to reduce coupling. The use of two superimposed antennas (rather than a wider antenna as in the US SPS-37A) has never been explained, as it does not improve beamwidth for greater range. Nonetheless, the introduction of a US-style, relatively high definition air search radar represented a major change in British thinking.

Type 966
Often described as a Surveillance Target Indicator Radar, *ie* as a replacement for both Type 965 and Type 992, this set was in fact a research project which was not yet ready for production, and it was discarded in favor of the Type 1022 STIR. One of its defects was that it was not precise enough to designate targets for Seawolf; the specialized Type 967 was needed instead.

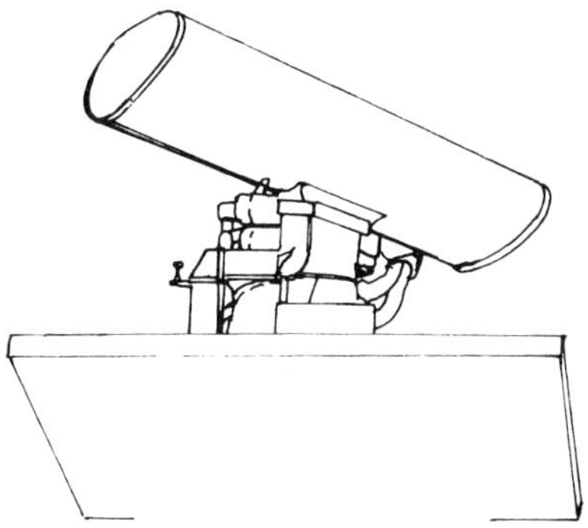

Type 967/968 antenna.

Type 967/968
Radars to support the Seawolf point-defense missile system. Both share a common mounting and designate to the Type 910 target-tracker. Type 967 is an L-band pulse-doppler air search radar; Type 968, mounted back-to-back with it, is an S-band radar for low air cover and surface search. Both rotate at 300rpm and are fully stabilized. The associated system initiates tracks, evaluates threats, and automatically engages targets. Marconi.

Type 970
Shipborne version of the RAF H2S (S-band) bomber-carried mapping radar, first mounted aboard *LCI(L)171* and five LCT(R)s in September 1943 and also used by some light forces in the Adriatic. It suffered from high side-lobes giving multiple returns. Peak power was 40kW or 50–70kW (1-microsecond pulses, PRF 500–1000) and even a small antenna gave a 6° beam, which scanned at 60rpm for a clear display. Range scales were 7000yds, 15,000yds and 50,000yds, with accuracies of 50, 100 and 500yds respectively; bearing accuracy was 3°. Typical ranges were 10nm on a battleship, 9 on a cruiser, 6 on a destroyer, 5 on a corvette, 3 on an MTB and 2 on a submarine. Installation required 4–5 weeks.

Type 971
Shipborne adaptation of the parallel X-band aircraft set, H2X, operating at 40–70kW (1-microsecond pulses, PRF 650 or 2000). The shorter wavelength shows in a narrower beam, 3°; scanning was at 60rpm. As in the case of Type 970, it was intended primarily for landing craft navigation (*eg* aboard control craft such as MLs), and was also available to light forces. Range performance showed a slight improvement on that of Type 970, and bearing accuracy improved to 2°.

Type 972
Navigational and surface search radar for survey ships, a modified Type 268 with a larger aerial (8ft 'cheese' on an AUK, or 277, pedestal giving vertical and azimuth stabilization: outfit AUZ) and azimuth-stabilized displays. It operated in X-band with 40kW peak power, and had range scales for 6000, 30,000 and 60,000yds and a bearing accuracy of 1°. Type 972M had in addition Panel L31 for accurate ranging (accuracy 10yds).

Type 973
Ranging radar for submarine periscopes (attack radar).

Type 974
High-definition surface search radar for antisubmarine craft and for navigation. It was based on Decca 12, which was X-band, with a peak power 7kW (0.1- and 0.2-microsecond pulses, PRF 1000), a rotation rate of 24rpm, and a horizontal beamwidth of 1.6°. Aerial outfit AKL or ATZ.

Type 975
X-band surface search and seaward defense radar, with 50kW peak power, using a 6ft or 10ft slotted waveguide antenna. Minimum detection range is less than 35yds, and beamwidth is about 1°. Type 975ZW is a minehunter radar with true motion display and with provision for marking sonar contacts on its PPI. Type 975, now being replaced by Type 1006, is manufactured by Kelvin Hughes.

Type 977
Shore-based radar to detect air-dropped mines in flight.

Type 978
X-band replacement for Type 974 in frigates and in larger ships. The antenna (outfit ATZ) is a 'double-cheese', the upper half transmitting and the lower half receiving; it is based on the commercial Decca 45, which has a peak power of 20kW (0.1- and 1.0-microsecond pulses, PRFs 1000 and 500), a rotation of rate 24rpm and a beamwidth of 1.2°.

Type 979
Navigational radar consisting of Type 978 with an additional display (B-scope). First installed in HMS *Echo* and HMS *Vidal* in 1969.

Type 980
An experimental wartime carrier interception radar, the forerunner of Type 982. It combined the Type 277 radar transmitter with a new 'triple-cheese' antenna (outfit AQS) to provide gapless, high-discrimination cover from the horizon to 50,000ft for interception control in combination with Types 960 and 981. The new antenna produced a 1.25° beam and scanned at 0–7rpm. Performance included reliable fadeless cover on aircraft up to 40,000ft and 40nm, and horizon range on surface craft. Range scales were 40nm and 80nm, with accuracies of 250yds and 500yds. Only one was built. Type 982 was a modified version.

Type 981
An experimental nodding-beam height-finder, the complement to Type 980, using the same radar transmitter. Its horn-fed vertical dish (outfit AQT) rotated continuously in the search mode, stopping on the relevant bearing to find height. Reliable height-finding was to be possible out to 50,000ft and 80nm, whilst reliable range on aircraft

Radar Types 982 (left) and 983 as fitted to the guided missile trials ship Girdle Ness, *1963.*
GIORGIO ARRA

up to 8000ft (antenna at 0° elevation) was to be possible out to 80nm. Range scales and accuracies matched those of Type 980. Only one was built, as it was superseded by Type 983.

Type 982
The standard postwar high-resolution air search radar, providing accurate bearing data in conjunction with a long-range air search set such as Type 960. It and Type 983 (below) were two elements of a 1944 program to replace Types 277/293, to achieve effective ranges as great as 60nm. Both reportedly used the Type 277 radar transmitter, in modified form.

The first Type 982 antenna was a large stabilized 'treble-cheese' installed aboard HMS *Eagle* in 1951. It was also fitted to the carriers *Centaur* and *Ark Royal*, and achieved a beamwidth of about 2° as well as the specified range. However, it was not successful and in 1955 was superseded by the 26.5ft × 3ft antenna. It consisted of a slotted waveguide producing a sharply defined beam which was fed into a cosecant-squared cylindrical reflector which in turn produced the narrowly defined low-altitude search beam. This S-band radar was limited in its altitude coverage because of 'squint': the waveguide produced not a narrow fan beam, but rather a cone, so that aircraft at different altitudes registered different bearings. Squint is a conseqeunce of internal reflections within the waveguide, and was a natural consequence of the use of the only long-base S-band radar feed available at the time, the slotted waveguide. A horizontal beamwidth of 1° was achieved. Type 982 will be succeeded by the new Type 1022.

Type 983
An S-band nodding height-finder, succeeding Type 981 and combining the Type 277P panels with the AQT antenna. It complemented Type 982 as part of the standard Royal Navy carrier radar outfit of 960/982/983. Its beamwidth was about 1.5° and its antenna was stabilized. Both Type 982 and 983 scanned together at 0–7rpm.

Type 984
A very complex and ambitious three-dimensional radar, contemporary with the US SPS-2. It was designed from the first to support fighter control, and therefore emphasized data rate and accurate continuous height-finding. Design requirements included effective tracking of a fighter out to 75nm, and warning at twice that range. Type 960 had better range (*eg* 175nm on a Mosquito at 35,000–40,000ft) but was considered easily jammable; moreover, it gave no precision tracking or height-finding. The original limits imposed on the design were a turning circle of less than 25ft, and a weight not too much more than existing sets (*ie* about 15 tons, which was considerably exceeded in practice).

Many projects of the early postwar period, such as carrier reconstruction and the Fleet Air Defense Escort (FADE), showed pairs of 984, although in fact only single sets were fitted to the three carriers (*Victorious, Hermes,* and *Eagle*) which installed it. S-band was chosen as a compromise between X-band, (compact, but power would be insufficient) and L-band (which would make range easier to obtain, would reduce cloud clutter, but which would provide insufficient precision given practicable lens sizes). Multiple, simultaneously scanning beams were chosen in favor of a stacked-beam system to give higher precision and slightly better range, with greater simplicity. As the beams swept up simultaneously they could cover a full 25° of elevation (five 1.7° beams, each scanning a 5° sector at 16 cycles per second) quite rapidly; the use of five separate scanners was justified on the ground that the antenna could then rotate five times as fast for the same number of pulses on target (6rpm). The top feed was a fixed horn (1.7° beam) which scanned the horizon for long-range search. The entire system was roll-stabilized, using trunnions at either side. In order to avoid high-power rotary joints, the transmitter was mounted on the carriage proper, connected directly to the scanners. Net peak power was comparable to that of SPS-2, but three separate water-cooled magnetrons, each generating about 3MW, were used.

The massive 3-D Type 984, here shown aboard the carrier Victorious, *1958.*
CONWAY PICTURE LIBRARY

Given the complexity of the scanning system, it was natural to use a lens antenna; sources differ as to whether it was 14ft or 14ft 6in in diameter. Lens design was a major problem, as conventional dielectric lenses could not be used given their immense weight and thickness. Instead a waveguide lens (*ie* with a refractive index dependent on frequency) was used, and the restriction in bandwidth (2–3 per cent) accepted.

Development began in the late 1940s at the Admiralty Surface Weapons Establishment, and was transferred to Marconi about 1950. The first installation was made in HMS *Victorious* in 1958, and Type 984 made a great impression as a fighter control radar in HMS *Eagle* the following year during a visit to the United States. It was specially designed for integration with the Comprehensive Display System (CDS) developed by the Royal Navy and, in effect, replaced the combination of Types 960, 982 and 983 characteristic of postwar British carrier practice.

Type 985
A fixed-array radar system for carriers, in the design stage about 1959, and thus a rough equivalent of the contemporary US SPS-32/33 system.

Types 986, 987
Planned successors to Types 982 and 983. It is not clear whether either was produced or deployed.

Type 988
'Broomstick', the planned Anglo-Dutch three-dimensional radar for *Bristol* and the abortive CVA-01 carrier. Its design probably survives in the Dutch MTTR radar.

Type 992 (**1**) *and Type 978* (**2**) *radars, HMS* Bristol. C & S TAYLOR

Type 992
S-band target-indicator, successor to Type 293Q, using a very similar antenna. Development began in 1948 and was completed in 1952; range on an air target was up to 30nm, depending upon altitude. Fittings included the cruisers of the *Tiger* class. The current version, Type 992Q, uses a slotted waveguide antenna, fully stabilized, and operates as the target-designation radar in missile ships such as the 'County' class and the Type 42 and 82 destroyers. It is the principal air search radar in the Type 21 and Type 22 frigates, in analogy to similar use of Type 293Q. In larger ships it can be synchronized with other radars in pulse and in rotation. Marconi.

Type 993
S-band target-indicator for frigates and similar craft, replacing Type 293Q, using a 'quarter-cheese' antenna. This has the advantage that the feed horn does not obscure the main radar beam, and the beam is better defined, even over a considerable turning band. It is being displaced in service by Type 994.

Type 994
Modernized Type 993, with the same antenna but with a new transceiver based on that of the Plessey AWS-4; performance is said to exceed that of the earlier system. Installation began in 1978.

Type 994 (**1**) *and 978* (**2**) *radars, HMS* Hermione, *1976.* C & S TAYLOR

Type 1000, 1001
Submarine radars. No details available.

Type 1002
Submarine radar in *Porpoise* class.

Type 1006
Kelvin-Hughes navigational radar, successor to Type 975. There are two alternative antenna systems: AZJ, a 2.4m slotted waveguide (1° × 18° beam) and AZK, a 3.1m slotted waveguide (0.75° beam, for greater precision); both scan at 24rpm. Peak power is 25kW (X-band; 0.8- or 0.25-microsecond pulses at a PRF of 1600, or 0.75-microsecond pulses at a PRF of 800). There is a submarine antenna variant as well, but details are not available. The PPI can also display helicopter transponder data for use in ASW. Development began in 1969, with a Kelvin-Hughes response to an Admiralty request for a solid-state successor to tube-type systems, with sea trials in 1970 aboard HMS *Grenville* and HMS *Otus*. Production began in 1971.

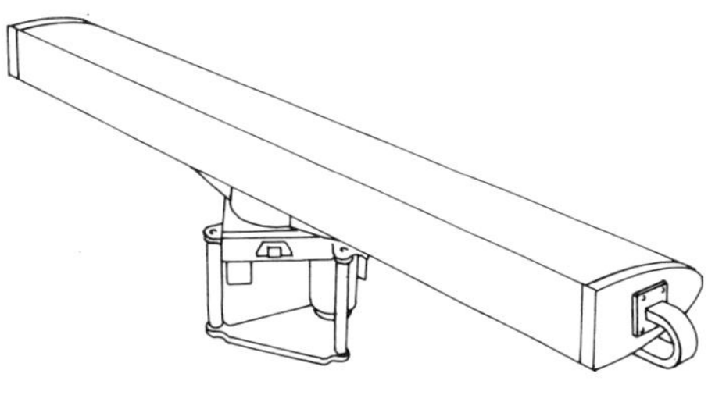
Type 1006 antenna.

Type 1010
Cossor Mk 10 IFF.

Type 1022
Surveillance and Target Indication Radar (STIR), to replace both Type 965 and Type 992Q. It incorporates a new Marconi antenna and the Dutch LW-08 radar transceiver. The Marconi antenna is similar in principle to that of Type 982, with a broader and less steeply angled reflector, which produces a beam better adapted to air search. The feed is the new 'squintless' type developed by Marconi to replace the slotted waveguide of Type 982 – that is, a conventional slotted waveguide produces a conical rather than the planar fan beam of a 'cheese' so that targets at different altitudes register at different bearings. On the other hand, really long 'cheeses' (for very narrow beams and high precision) are difficult to build. The Marconi system provides equal path length feed (for squintless operation) in a form which is relatively easy to machine, and which makes the Target Indicator much more effective as a surveillance radar. First installed aboard the escort carrier *Invincible*, 1979.

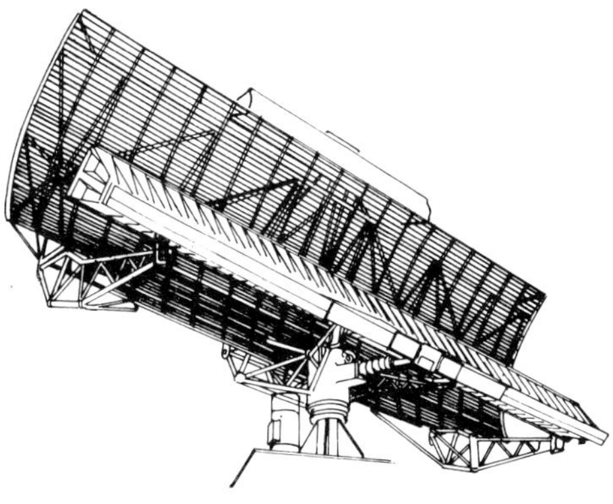
Type 1022 antenna.

Type 1030
This is the full STIR, incorporating a Marconi antenna and transceiver. It is scheduled for operation in 1985, and presumably outwardly resembles Type 1022.

SW-1C
This Canadian radar was installed in several US and British warships early in World War II the largest being HMS *Malaya*. Like Type 286, it was an adapted ASV system; however, unlike 286, it used a very simple yagi aerial, and so sometimes accidentally picked up targets on reciprocal bearings. In Febru-

ary 1941 the Canadian National Research Council consulted with the RCN concerning a small-ship radar to detect surfaced submarines; arrangements to develop what was then called 'Canadian Submarine Control' (SC) were fixed on 19 March, and as an intermediate step a conversion of the Canadian ASV, corresponding to Type 286, was proposed. Ultimately 214mc (1.5m) was chosen, and the yagi two-way antenna selected, with a Canadian TR box (which was later replaced by a British type).

The first unit passed its sea trials in May 1941 and in June one was fitted to HMS *Malaya* as a stop-gap; two more were later supplied to the Royal Navy for similar purposes. In the fall of 1941 the designation was changed to SW-1E; SW-2C and -3C were slightly modified versions. As of December 1941, fifteen corvettes had either CSC or SW-1C, as well as three auxiliary cruisers. Two hundred were on order, with production running at six per week.

The antenna was mounted at the masthead on a small gearbox, rotated by means of a flexible drive shaft leading to the operator's position below, and fed by a coaxial; open wires were rejected in view of the greater strength of the coaxial. Performance was similar to that of Type 286; a corvette could expect to detect a surfaced submarine at 3000–4000yds, another corvette at 5000, or a cruiser at up to 14,000. Aircraft could be seen out to 25,000yds (*Malaya* detected Swordfish flying at 500ft at out to 20,000). Range scales were 5 and 20nm, and the ground wave was effective out to about 4.8nm. Accuracy was 4° and 200yds, but targets could not be accurately ranged inside 1000yds.

SW-1C was later modified to operate on a wavelength of 1.4m so that it could be used in conjunction with existing IFF; it then became SW-2C. In December 1942 the RCN asked for a modified SW-2C suitable for fitting to MTBs. The yagi was considered too bulky, so a split-beam system was used, with a fixed aerial ahead and two receiving aerials, one to each side, with a switch. This was not satisfactory, and a lightweight rotating aerial had to be designed; although not as effective as X- and S-band sets, this SW-3C remained in service to the end of the war.

EXPORT RADARS (CURRENT)

Marconi

SNW series

Private-venture air search radars developed in the mid-1950s and superficially reminiscent of the American SC/SR series. SNW-10 operated at 1.5m (450kW, 3.8 microsecond pulses, scanning at 10rpm with a 14ft × 7ft antenna producing a 19° beam). A maximum range of 125nm was quoted, with an antenna weight of 336lb. SNW-12 was a larger version, with twelve (4×3) rather than six (3×2) dipoles. Marconi offered a modular British-style radar system, in which these mattresses were complemented by SNW-20, a slotted-wavelength target indicaton radar, and by a medium caliber fire control radar in a system numbered SNG-20. The full combination was installed aboard the Chilean destroyers *Riveros* and *Williams*, completed in 1960; in the mid-1970s their P-band air search radars were replaced by Plessey AWS-1s. Other SNW-10 installations included the Egyptian 'Z' class destroyers, the Pakistani light cruiser *Babur* and destroyer *Badr*, and the Spanish destroyer *Oquendo*.

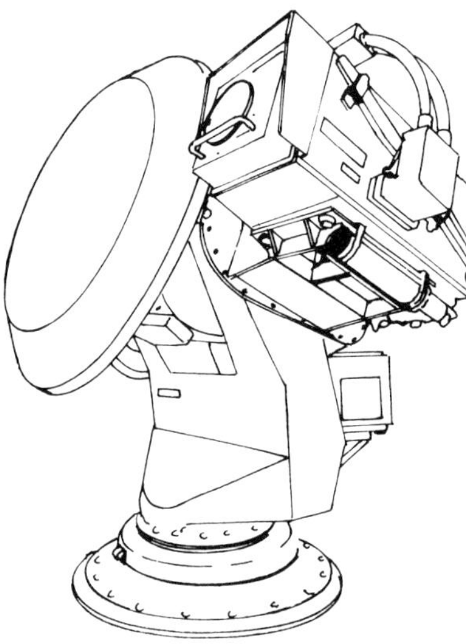

ST801/802 radar, showing television tracker mounted alongside.

ST801

A monopulse search tracking radar with MTI, intended as the auto-track element of a fire control system. It can search the horizon, scanning at 20rpm about a stabilized vertical. ST801 can control line-of-sight missiles such as Seacat, tracking both missile and target; it is also

Marconi ST802 radar scanner. MARCONI

fitted with a television camera for operation in a countermeasures environment. The antenna is a twist Cassegrain type, 1m in diameter, for a 2.4° beam in X-band. Peak power is 180kW (0.3-microsecond pulses, PRF either 3000 or 4400; in the MTI mode the PRF is staggered about 4400 pulses per second). The television camera can be used in an 'autogather' mode in which the television optically acquires a tracking flare in the tail of the ascending missile, and a data extraction unit measures the error between the missile position and the line of sight to the target; the command signal transmitter then automatically corrects. This is, for example, the principle of Seawolf operation. Claimed performance includes a detection range of 25km (13.5nm) on a 4m² target in a single scan and of 30km (16.2nm) on a fast patrol boat (limited by radar horizon), with tracking accuracies of 12m (13yds) and less than 3 minutes of arc. Total above-decks weight, including the television camera, is 1100lb, with an additional 1725lb below decks.

ST802

An autonomous version of 801, capable of functioning without a separate fire control system.

S810

An X-band stabilized search radar with a peak power of 200kW (0.67- or 0.33-microsecond pulses at PRFs of, respectively, 1500 and 4400), using a dome-enclosed 1.2m antenna (2.2° × 25° beam) rotating at 20 or 40rpm. Alternatively a 1.8m antenna (1.5° × 25° beam) can be used. Performance figures (those for the latter in parentheses) are, against a 5m² strike aircraft, range 28km at 2km altitude (32km at 3km) and greater than 20km (25km) at 8km altitude. There is also an S810P option giving a parabolic narrow beam for low coverage only, to detect a 0.1m target at 15km (8nm). S811 is similar, without the MTI processor.

S815

Similar to S810 but having a stabilized 3.5m × 0.9m (8ft 2½in × 3ft) antenna for improved air surveillance. S816 omits the MTI Processor.

S820

S-band (600kW, 1.2- or 0.6-microsecond pulses, PRF 750 or 1500 respectively) with a 2.5m antenna (3° × 30° beam) scanning at 24rpm to detect a 5m² target at 90km at 5km altitude, with height cover of up to 11km. MTI processing is at a pulsewidth of 0.6 microseconds (PRF 1470). Antenna weight, including the radome, is 600kg (1320lb), with another 760kg (about 1670lb) below decks.

S1820

At about the same time that it announced the Royal Navy's Type 1022 air search radar, Marconi announced another new series of commercial naval radars, the S1800 series; details of the small-ship S1820 have been announced. Like its predecessors, it is stabilized and radome-enclosed, and it is said to be suitable for ships of 300 tons and above, since above-decks weight is only 550kg (about 1210lb), with another 1080kg (about 2380lb) below decks. Marconi claims that radome construction saves weight and stabilization assures quick reaction by providing accurate target bearing data even from a tossing, rolling ship. For the first time in the series, the radar employs pulse compression for greater range. The antenna itself appears to be similar to that of S820, and one suspects that the 1800 series consists of 800 series radars modifed to employ pulse-compression.

Plessey

AWS-1

This set employed a 16ft × 6ft 6in antenna (1.5° × 40° cosecant-squared beam), a peak power of 750kW (0.35- and 1.5-microsecond pulses, PRFs 1000 and 400) and a scan rate of either 10–12 or 20rpm. Range on a 'small aircraft' is quoted as 60nm. The radar was designed as a private venture, beginning in 1956 as a navalized version of the AR-1 land-based air search radar – which explains the choice of S-band. The first prototype appeared in 1960. The very similar AWS-2 was begun as a private venture in 1965, the first prototype being completed in 1970. It is described as the first radar to go to sea with a digital MTI processor.

ASW-3

Similar to ASW-2 but has increased transmitter power and pulse compression. Range limits are about 140nm and 100,000ft, and the 4190lb (1900kg) stabilized antenna scans at 10 or 20rpm. Peak power is 1.1MW (PRF 380; pulses are 20 microseconds compressed to 0.1 microseconds). The 5.055 × 1.980m antenna produces a 1.45° × 40° (cosecant-squared) beam, with a gain of 1585 (S-band). Signal processing includes MTI, and the radar is frequency-agile.

AWS-4

This set features a new antenna, with a total weight of 450kg (about 990lb) in the unstabilized version, or 1200kg (about 2640lb) in the stabilized version, compared to 576lb for AWS-2. Range on a 4m² target is 104km (about 55nm). More precise details have not been released, but AWS-4 is described as a development of AWS-1 and -2; it incorporates pulse-compression. AWS-4 is combined with the Type 993 antenna to form Type 994 and with the 293 antenna to form the commercial Type 294. In addition, there is a lightweight linear array antenna (350kg unstabilized, 600kg stabilized) for installation in fast patrol craft as AWS-4B. Nominal elevation coverage is 30°.

Plessey AWS-5 surveillance radar, for Danish Niels Juel *class corvettes.* DEFENCE

AWS-5

This incorporates a new dual-beam (high and low) antenna weighing 1200kg (about 2640lb) including its stabilized mount. Beamwidth is 1.5°, the height beam is 30° wide, directed upwards at about a 45° angle; the low beam is the more usual cosecant-squared to 40°. With the low beam, a 4m² target can be detected at about 85nm and 50,000ft; a 0.1m² target (such as an antiship missile) at about 40nm and 25,000ft. The high beam peaks at about 220,000ft at 40nm, against a 4m² target. Cover in excess of 100,000ft extends up to 65° from the horizontal, for effective zenith search. No operational details have been released, but AWS-5 employs pulse-compression. Development began in 1972 with technical studies, and design work began in 1975. AWS-5 has been installed in the Danish KV-72 (*Nils Juel*) class corvettes and in the new Nigerian frigates (Blohm und Voss). As in AWS-4, there is a lightweight antenna variant, AWS-5B, employing a stabilized linear array antenna (580kg, about 1280lb) producing a 1.5° × 32° beam with a claimed azimuth squint of less than 0.1° over the operating frequency. It scans at 15 or 30rpm.

IFF SYSTEMS

Wartime British IFF systems were coded in two series, 240 for interrogators and 250 for responders and beacons. Type 241 itself operated at 214mc/s, with Type 252 operating as responder. Type 251, on the other hand, responded to the standard ASV radar. It was described as IFF Mk IIG, and swept 173–179mc/s in five seconds, responding for one second out of five, its signal chopped by a wheel to make it more distinguishable. A set in the destroyer *Fernie* gave ranges of up to 32nm on aircraft at 1700ft, and the carrier *Illustrious* reported up to 50nm. In addition, it operated a warning light and buzzer on the bridge, and could be used to send a Morse code message to an airplane. In all but corvettes it was ultimately replaced by Type 251M, with 2kW pulses at 176mc/s. Owing to its greater range and power, it was considered more useful to searching aircraft than was 251, and was fitted to independently routed merchant ships under the designation MAB. It was also fitted to three or more ships of each Convoy Escort Group.

Types 242 (179–182mc/s, 1kW, 6-microsecond pulses, omindirectional or 60° beam), 242M/P/Q (159–189mc/s, 2 or 10kW, 4-microsecond pulses, 88° beam) and 243 (10-microsecond pulses) were the interrogators of the British Mk III IFF system. Types 242 and 243 were fitted as outfit ASD with radars 291M and 268, and as ASW with radar 275. Type 242M was fitted as ASS with radars 276, 277 and 293. Type 242P was fitted in outfit ANR for 277P. Type 253 was the responder in each case, operating on 157–187mc/s, with aerial outfit ASH. There were three pulsewidths: 'N' (10 microseconds), 'W' (20), and 'E' (40). The remainder of the 250 series was devoted to beacons: 254, 255, 256 and 258; Type 257 was a Blind Approach Landing System (BABS).

Designations were moved to a 900 series late in the war, the 940 series being devoted to interrogators and the 950s to transponders. Reportedly the low numbers were all reserved for the late-war Mk V/UNB system, which was abandoned in 1945. Instead, the Royal Navy adopted Mk 10, fittings beginning about 1956. The interrogator generally transmitted on 1030mc/s at up to 3kW, with two pulses 0.8 microseconds in length (spaced 5 microseconds in Mode 1 and 8 in Mode 2) with a PRF of 200–450; the standard antenna, on radars such as 965, had a beamwidth of about 6° (about 13–14° in the smaller system employed in destroyers and frigates). Reported designations include 944 as an interrogator and 954 as a transponder; 957 was an aircraft beacon (TACAN). With the adoption of a new 1000 series for British naval radars, the Royal Navy adopted Type 1010 as its interrogator.

Germany

DESIGNATIONS

German naval radar designations went through three distinct stages. First there was a *Dete* (*Dezimeter-Teknik*) series in which the numbers indicated radar function: *Dete* 100 through 199 were the naval (*Seetakt*) sets operating at 80cm. Then there was a standard multiservice system in the FuMG or FMG (*Funk Mess-Gerate*, or radio measuring device) series in which the numbers indicated the year of sevice entry, with suffixes for manufacturer, frequency range, and type. For example FMG 39G(gB) was introduced in 1939, manufactured by GEMA, the principal naval radar firm ('G'), functioned at 355–430mc/s ('g') and was intended for mounting on land (B=*Boden*). The designation 'c' was 182–215mc/s and 'f' 120–150mc/s. After about 1934 there was yet another series, in which a prefix indicated the function and the number the series. For example, FuMO 12 was a fire control (O=*Ortung*) radar in the *Seetakt* (1 through 99) series. FuMO numbers were applied to all of the FMG sets, and the catalog which follows is therefore based on them.

The search receivers were numbered in an FuMB series, and many were given 'island' names, such as Bali, Timor, Sumatra and Palau. In the case of the *Prinz Eugen*, for example, they were far more variegated than were the radars proper – which suggests the type of war the German Navy had to fight.

RADARS

Seetakt

The first of the German naval radars, installed in 1936 aboard the 'pocket battleship' *Graf Spee*, operating at 80cm (368mc/s). Peak power was 7kW, for a range of 10nm against large ships (6 against cruisers). The antenna was a mattress fixed to the rangefinder, its upper part used for reception and its lower for transmission. The reception portion was divided in two for more accurate bearing measurement, and accuracy was 0.2°.

FuMO 21 with 4 × 2m mattress, DD-935 (ex-torpedoboot T35) *1945. Four fixed Sumatra search receiver dipoles are sited above.*
BY COURTESY OF ROBERT F SUMRALL

FuMO 21

Light cruiser and destroyer radar, originally designated FMG 39G(gL), on a bridge pedestal, with an effective range of about 10nm. First tested in the cruiser *Nürnberg*. It was similar in characteristics to the other 1939 radars, FuMO 22 and 23. Peak power was 8kW (PRF 500 5-microsecond pulses), and antenna dimensions 4 × 2m.

FuMO 22

The standard German capital ship radar of 1939, accurate to within 5° and capable of detecting a battleship at 13nm. Originally designated FMG 39G(gO).

FuMO 23

Radar for mounting on a fire-control director, originally designated FMG 39G(gP). Mounted aboard *Bismarck* and *Prinz Eugen*.

FuMO 24

First of the improved 1940 series of radars, designed to be mounted on a pedestal on a ship's bridge. Peak power was 8kW and antenna dimensions 6 × 2m. In 1944 many were upgraded to FuMO 32 by the replacement of their transmitters by 125kW units. The others in this series were FuMO 25 through 28, and they replaced the FuMO 21 series. However, naval radar production after 1941 was very limited, so that these 1940 sets remained through

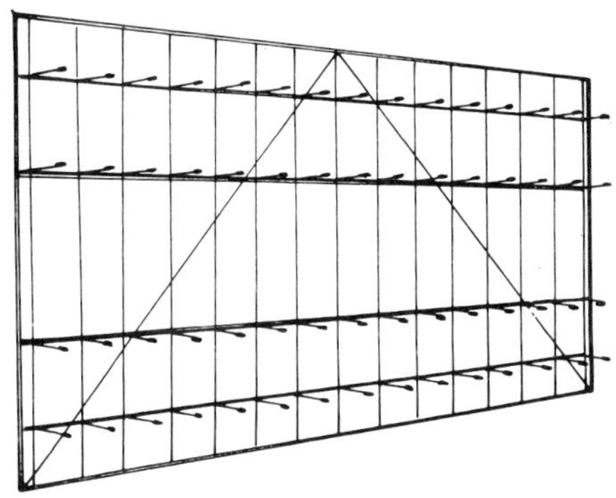

FuMO 22 antenna.

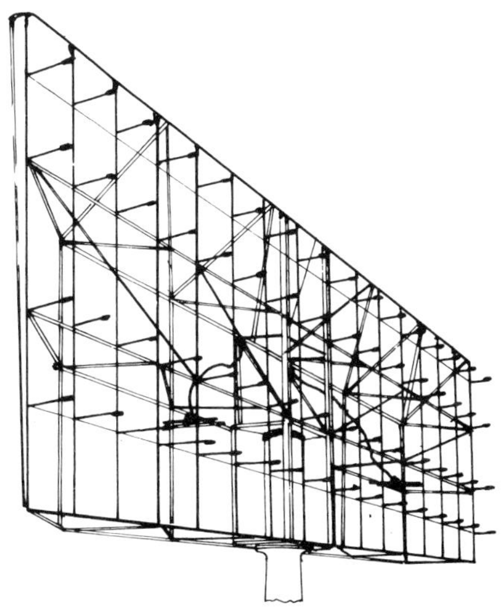

FuMO 24/25 antenna.

the war. By mid-April 1941 all German destroyers had either this set or its immediate predecessor, FuMO 21.

FuMO 25
Mast antenna, with a 6 × 2m or 4 × 2m antenna, otherwise equivalent to FuMO 24. Many were upgraded to FuMO 33 in 1944.

FuMO 26
Radar for fire control directors, using a new horizontally-polarized 6.6 × 3.2m antenna. It was credited with a range of 20–25km. By 1945 the set aboard *Prinz Eugen* had been upgraded to a peak power of 60kW (4-microsecond pulses) and had a range accuracy of 55yds and a bearing accuracy of 0.25° – the latter betraying its fire control origins. Some were upgraded to FuMO 34 (125kW) in 1944, range increasing to 40–50km.

FuMO 27
Combined optical rangefinder and radar of the 1940 series, with a 4 × 2m antenna.

FuMO 28
Radar for T-boats (seagoing torpedo-boats), with two fixed antennas, 2.6 × 2.4m.

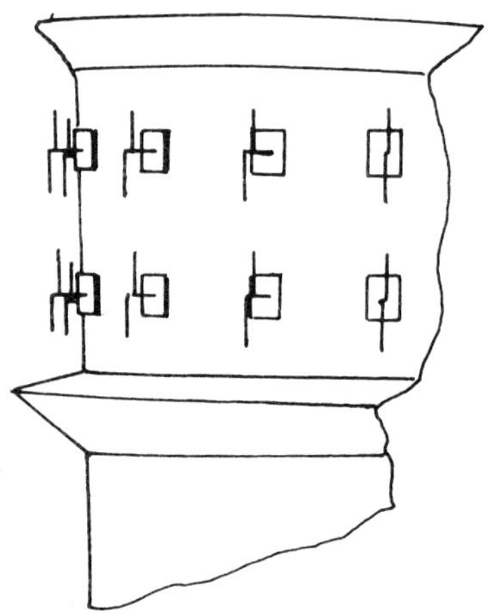

FuMo 29 radar (fitted to U-boat conning tower).

FuMO 29
U-boat radar, consisting of two 1m² antennas, one on either side of the brigde, with electrical phase-scanning. Originally designated FMG 41H(gU), it was first installed aboard the East Asia U-boats (*U156—158* of Type IXC40) and had a range of 2.5 – 3.5nm. In 1942 there was an alternative rotating mast with a 1.4 × 1m antenna and ESM gear.

FuMO 32–34
Upgraded FuMO 24–26, 1944, with peak power raised to 125kW.

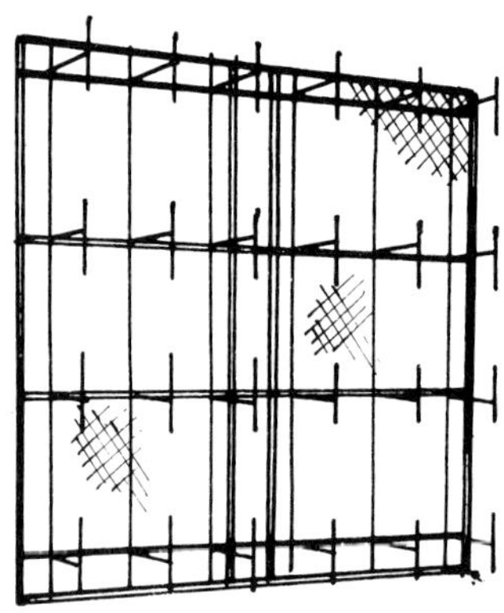

FuMo 61 antenna (Hohentwiel-U).

FuMO 61
U-boat radar (*Hohentwiel-U*), with a 1 × 1.4m rotating antenna, adapted from a *Luftwaffe* ASV-type radar (FuG 200, *Hohentwiel*) with a peak power of 30–40kW at 556mc/s (PRF 50, 1–5-microsecond pulses). Typical ranges were 8–10km. Low-flying aircraft could be detected at 15–20km.

FuMO 62
Hohentwiel-S, for MTBs, with a 1.5 × 1.6m rotating antenna, and typical ranges of 12–20km.

FuMO 63 (Hohentwiel-K) *radar antenna.*

FuMO 63
Hohentwiel-K, for large torpedo-boats and destroyers, each with two installations, as well as for cruisers. It used a rotating 2 × 2.4m antenna.

FuMO 64
Hohentweil-Land, for seacoast defense.

FuMO 65
Hohentwiel-U, for U-boats, with the 1 × 1.4m rotating antenna.

FuMO 71
Naval version of a Telefunken night-fighter radar (*Lichtenstein*, FuG 202, 490mc/s, 1.5kW) for MTBs (*Lichtenstein-B/C*), 1943. Its antenna consisted of four X-shaped arrays similar in appearance to the British Type 291, fixed to a framework. A destroyer could be detected at 2km, and a 6000-ton steamer at about 6km. A rotating version was designated FuMO 72, but a projected U-boat variant never materialized.

FuMO 81
Microwave search radar, derived from a series of radars produced after the Germans obtained a magnetron from a crashed Lancaster at Rotterdam on 2 February 1943. The first was *Berlin-A*, FuG 224. FuMO 81, *Berlin-S*, was a naval version, operating at 5300mc/s, and using an antenna consisting of four end-firing plastic rods. Peak power was 18kW (PRF 1500), for a reported accuracy of 110yds and 5°. A 500-ton ship could be detected at 20–30km. Other versions were *Berlin-K* (FuMO 82), *Berlin-UI* (FuMO 83), and *Berlin-UII* (FuMO 84), all designed for C-band operation; in 1945 work was proceeding on a *Berlin-E* at X-band. Although *Berlin-S* was originally designed for MTBs, in 1945 one was installed aboard the heavy cruiser *Prinz Eugen*. *Berlin-K*, for larger ships, had alternative 6- and 8-rod array antennas. A parabolic dish fed by a waveguide was designed for the U-boats. One of their commanders commented that it was as useful to him as an observation helicopter 200m above his boat, *ie* that it restored the kind of long-range vision which had been lost when the U-boats had abandoned their use of helicopter-kites

FuMO 391
A specialized onmidirectional air warning radar for U-boats, using a modified and miniaturized version of the land-based *Freya* radar transmitter (100kW, 125mc/s) and a single dipole antenna, for an effective range of 30km on an airplane. It would, therefore, provide about five minutes of warning for the submarine to dive. Accuracy was about 3 per cent of range. It appears that only one was built, in 1944.

Japan

DESIGNATIONS

Japanese designations were based on the year the set entered service, in terms of the Japanese imperial calender, in which 1940 was the year 2600 (*ie* the year zero in the single-digit form generally used). Within a year, mark numbers identified sets, so that Type 2 Mks 1 and 2 were two sets of 1942. Within each mark was a model ('Mod') series. Prototypes had their own mark series, as they received no type numbers until service introduction; airborne (as opposed to shore/ship) sets had a separate series. Changes within a Mod were identified by 'Kai' (modification) numbers, *ie* Type 2 Mk 2 Mod 1-Kai-3. As an added element of confusion, prototypes sometimes entered service *without* receiving type numbers (*cf* the US CXAM). The Japanese themselves often referred to sets by mark/model only, *eg* '21' for Mk 2 Mod 1 (without any reference to the type).

RADARS

Type 2 Mk 1

The first Japanese radar, a 3m, 5kW shore-based air search set (Mod 1), using two antennas to achieve a maximum effective range of about 80nm. Work began in April 1941, and was not based on captured Allied equipment.

Type 2 Mk 2

Work began in October 1941, on the basis of a 500W magnetron operating at 10cm. Development of the Mod 1 Kai-2 version was completed in August 1943, using a 1.5m wavelength, with a peak power of 5kW (10-microsecond pulses, PRF 1000). The antenna was a 4 × 3 dipole mattress (4×4 in the Kai-4 version) 6 ×2m in dimensions and with a 34° × 60° beam. Range was 70–100km (38–54nm); accuracy was 1–2km and resolution 2km and 20°; and weight was 840kg. The original Mk 2 Mod 1 was developed for converted mechant ships (January–April 1942); Kai-2 was ready in December 1942, and a Kai-3 for combined air and surface search was developed between August 1943 and February 1944 but was withdrawn from service. Typically Mk 2 was mounted on the foretop of a battleship, or in a rotating installation aboard carriers.

Type 3 Mk 1

Derived from a land-based set of 1943, development being completed in February 1944. It operated at 2m, with a peak power of 10kW (10-microsecond pulses, PRF 500), using a 2 × 4 dipole mattress (2 × 4m, with a 68° × 35° beam) to achieve a range of 50–100km (27–54nm). Accuracy was 2–3km and 10° and resolution 3km and 60°, with a total weight of 110kg. Typically it was mounted fixed on both sides of a battleship mainmast. Data refer to Mod 3.

Mk 2 Mod 2

A microwave surface search radar, using a horn antenna 0.4m in diameter, and operating at 10cm. It retained its original experimental designations; development was completed in December 1943. Peak power was 2kW, with a pulse-length of 10 microseconds and a PRF of 2500, and the beam was conical, 38°× 38°, with a range of 25km (13nm) on a battleship. Accuracy was 0.1–0.25km and 3°, and resolution 1.5km and 40°. Total weight was 1300kg. It was designed from the outset for shipboard use, but was not considered as reliable as the simpler metric Type 3 Mod 1. Mk 2 Mod 2 employed separate transmitting and receiving horns, and was originally designed for submarine use.

The Japanese Type 3 Mk 1 Mod 3 P-band air search radar, probably aboard a destroyer at the end of the war. All eight dipoles are visible, and the structure on the after side of the mast was presumably for intercept.

BY COURTESY OF A D BAKER III

Twin-horn Mk 2 Mod 2 Kai-4 S-band surface search radars (**1**) *and Type 2 Mk 2 Mod 1 mattress* (**2**) *aboard an* Agano *class cruiser.*

BY COURTESY OF A D BAKER III

POSTWAR RADARS

Japan resumed production of naval radars in the late 1950s, using the American system of designation, with 'O' replacing the initial 'S' in surface systems, so that the US SPS became OPS. For submarines, the initial letter became 'Z', so that the US BPS was rendered ZPS. Very few details of Japanese radars are available, because Japan does not export military equipment and so does not have to advertise the performance of her equipment.

OPS-1 and -2 appear to have been Japanese versions of SPS-6B or -6C; similarly, OPS-15 is described as a Japanese version of either SPS-5 or -10; OPS-16 is a modified version, and reportedly OPS-37 is SPS-5B. There is also reportedly a Japanese version of SPS-12. Domestic air search radar development produced OPS-11, an elaborate (8×4) array of yagis, the dimensions of which suggest operation at about 75cm. Given previous Japanese dependence on US naval radar technology, one might suspect that OPS-11 is related to SPS-40, differing primarily in its antenna design. Japan also uses the Hughes SPS-52 for missile control, but apparently does not manufacture it. Surface search sets, which externally resemble OPS-15, include OPS-16, -17 and -35.

Italy

DESIGNATIONS

Wartime Italian radars were prefixed by the letters 'EC', a special radar section of the Navy's Electronic and Communications Institute (RIEC) having been formed in 1936. However, from 1942 onwards the Italian Navy also received German naval radars, which continued to be supplied to ships remaining on the Axis side after the Armistice, principally destroyers and torpedo-boats. The cruiser *Abruzzi*, for example, had FuMO 31, and three destroyers and three torpedo-boats had FuMO 24. Ships operating with Allied forces were fitted with British radars, Type 286 being installed aboard the crusiers *Abruzzi*, *Garibaldi*, *Eugenio di Savoia* and *Duca d'Aosta* and Type 291 aboard *Montecuccoli*, the destroyers *Carabiniere* and *Grecale* and the torpedo-boat *Galliope*.

The designations of postwar Italian radars are somewhat confusing. Italy did take over the US system (*eg* SPS), prefacing the functional letters with 'MM', for 'Marina Militare' (Navy). The numbers are not, however, in any neat sequence but rather tend to be three-digit codes in a 700 series which presumably identifies Italy just as the 500 series identifies Canadian equipment. Examples include the MM/SPS-768 and MM/SPS-774 search sets, the MM/BPS-704 submarine radar, and the MM/SPN-703 navigational radar. As if to make matters more confusing, Italy adopted the designation MM/SPQ-2 for a major surface search radar. Selenia uses a different system for its own products, prefacing them with the letters 'RAN' for search, 'RTN' for weapon control and 'RAT' for land-based air search. There is also a numerical suffix which indicates the band in which the radar works: RTN-10X is an X-band fire control radar. Generally Selenia used the name 'Orion' for its fire control radar systems. SMA has been less systematic, although many of its products fall in the 'MM' series. It does, however, produce a 3cm navigational radar the designation for which is prefaced by '3RM'.

WARTIME RADARS

EC-1
Produced in 1936. Used FM transmission, with a parabolic antenna, and operated at 200mc/s, EC-1*bis* was a 1937 development.

EC-2
A 1937 radar operating at 1.7m and using dipole antennas.

EC-3
The principal wartime radar developed in 1940. EC-3 operated at 70cm, with a PRF of 7500 modulated at 600 cycles per second; detection was indicated by a 600-cycle note in an earphone. The antenna was a pair of mesh horns, one for transmission and one for reception. In the *bis* version (1941) a new Philips power tube boosted peak power to 1kW, PRF falling to 1000. This set was installed in the battleship *Littorio*, but proved unsatisfactory in November 1941 sea trials. It was particularly criticized for the absence of a visual display.

The production version was EC-3*ter*, in which peak power was boosted to 10kW (4-microsecond pulses, PRF 500) on a wavelength at 75cm (400mc/s). Beam dimensions were 20° × 10°, and the antenna scanned at 3rpm. Maximum range on a ship was 30km (antenna at 35m as in a battleship; 15km for antenna at 15m, as in a destroyer), and aircraft could be detected at 80km. The display was a J-scope, showing targets on a circular range scale (30 or 300km). The first installation was aboard a small armed boat, the ex-merchant *F14*. Of a series of fifty production sets on order, twelve had been delivered as of the Armistice, September 1943. EC-3*ter* was popularly known as 'Gufo'. Ships fitted included three battleships (*Littorio* class), the cruisers *Montecuccoli*, *Eugenio di Savoia* and *Scipione Africano*, and five destroyers.

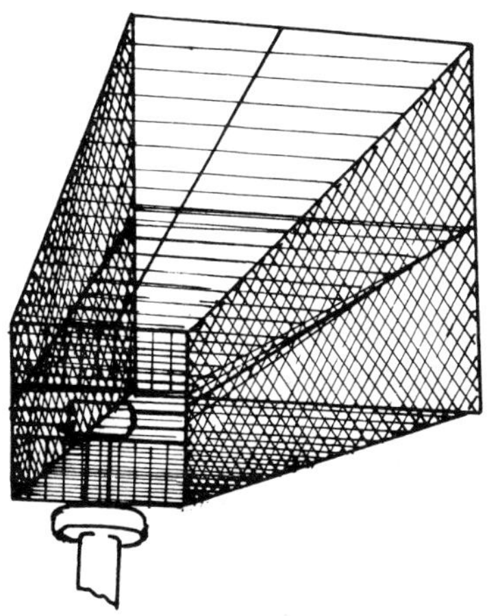

*EC-3*ter *('Gufo') antenna.*

POSTWAR RADARS

MLA-1
The first major postwar Italian naval radars were produced by a company no longer in business, Microlambda. There were three series: air search systems (MLA), navigational systems (MLN) and fire control radars (MLT). MLA-1 was not installed in Italian warships, but it did appear on the Italian *Almirante Clemente* class frigates exported to Indonesia, to Portugal and to Venezuela, as well as in Portuguese (*Dealey* class) and Spanish (*Canarias*, as well as many destroyers and corvettes) ships. It was a large L-band reflector roughly similar in size to SPS-6. Peak power was reportedly 250kW, with a maximum range of about 150nm.

MLN-1
The Microlambda navigational set, MLN-1, operated in X-band with a peak power of 20–50kW (PRF 500 or 1000) and a maximum range of about 80nm. It appeared in some small Italian warships as well as abroad.

MLT-1
Derivative of the US Mk 39 fire control radar.

MLT-2
US Mk 26.

MLT-2, -4
Improved version of the US Mk 39. Some of these MLT radars were employed by the Italian Navy.

Selenia RAN-3L search radar antenna.
BY COURTESY OF *DEFENCE*

RAN-3L (MM/SPS-768)
The new generation of Italian search radars was developed under the joint sponsorship of the Italian Defense Science Technical Committee (CSTD) and the Italian Navy, and consisted of a range of systems for ships of different sizes. RAN-3L is for large ships, and employs pulse compression; its 7.6m × 3.6m (24.9 × 11.8ft) antenna rotates at 6rpm and has a claimed accuracy of 70m and 0.4 ° and a range of about 150nm. Peak power data is not available.

RAN-10S (MM/SPS-774)
The smaller brother of RAN-3L, designed for frigates and corvettes. It employs a 43 × 0.7m (14.1ft × 2.3ft) antenna which rotates at 15 and 30 rpm. Accuracy of 20m an 0.35° is claimed, with a range of about 40nm. This system is currently being fitted aboard *Lupo* class frigates of the Italian, Peruvian and Venezuelan navies. Maximum unambiguous range is 90nm, and peak power is 125kW; the beam is 1.5° wide.

RAN-11L/X
A system for very small ships such as hydrofoils which radiates simultaneously in both bands. Its X-band transceiver is based on that of the MM/SPQ-2; the L-band system is a new pulse-doppler type. The antenna itself is unusual in that only the lower portion is filled in; the upper part, which is mesh, reflects in the L- but not in the X-band. Total width is 8ft, and the two beams provide resolutions of 500m and 6.6° (L) and 130 and 1.1° (X). The entire assembly rotates at 15 and at 30rpm. There is also an RAN-12L/X for small craft.

RAN-20S
A three-dimensional radar, probably a developmental system. No details available.

RAN-2C
A Selenia private-venture system, designed as the primary search radar for small craft, or as a secondary radar for larger ones. It has a single coherent transmitter and a single antenna (G2) which is roll- and pitch-stabilized, but there are two receivers, one for air and one for surface reception. The radar transmits simultaneously in two frequency bands, with different PRFs and pulsewidths.

RAN-7S
Developed in 1970 from the ATCR-3T series of land-based air search radars. Characteristics of the land-based radar include a peak power of up to 800kW (PRF up to 1000, pulsewidths about 0.6 microseconds). In the naval version, pulsewidth is either 1 microsecond for air search or 0.5 for surface search; horizontal beamwidth is 1.4°. There are two antennas; G6, a double curvature type with added high-angle cover; and G8, a modified cosecant-squared pattern with remotely controlled elevation adjustment (or stabilized). Features include a Selenia-developed digital MTI. Claimed performance includes a range of about 54nm (100 km) and 39,000ft (12,000m). The RAN-7S antenna system was supplied to the Danish Navy. Another Selenia private venture.

RAN-13X
A system which uses the X-band transmitter of the RAN-11L/X with a G15 double-curvature reflector, for fast patrol and missile boats. In 1976 it was reportedly obsolescent, having been superseded by later designs. Selenia.

RAN-14X
A private-venture small ship radar, development of which was begun in 1968; it appears aboard the Danish frigate *Peder Skram*. A range of about 27nm (50km) on an airplane was expected. Selenia.

Orion RTN-10X (RN Type 912), as fitted aboard a British Type 21 frigate. C & S TAYLOR

RTN-10X
The Selenia fire control radars all fall in the Orion RTN series: RTN-10X, -20X, and -30X. RTN-10X is a conical-scan pulse radar employing the RAN-11L/X X-band radar transmitter; it is designed particularly for one-man operation, and for integration into the Albatros fire control system. Development began in the early 1960s for the Italian Navy, and the current Orion-10X was developed in 1970 (series production began in 1972).

RTN-20X
A private venture begun in 1973, and intended particularly for the Dardo close-in weapon system, for automatic target acquisition and for spotting rounds. It is digital coherent monopulse radar, for nod-free tracking, automatically acquiring its targets at 2–6nm. Capabilities include the ability to acquire and track a missile fired by the platform it is tracking, *eg* an ASM dropped by an incoming aircraft.

RTN-30X radar, part of the Albatros system.
BY COURTESY OF *DEFENCE*

RTN-30X
RTN-30X is an automatic target-acquisition and monopulse tracking radar designed to work with weapon systems intercepting their targets at up to 8nm at low altitudes in intense clutter; among its anticlutter features are MTI and frequency agility. RTN-30X is particularly designed for the Albatros Mk 2 system.

RTN-16X
A private venture begun in 1969 to have maximum commonality with RTN-10X.

SPQ-2
One of the principal Italian Navy search radars, produced by SMA. An X-band radar which time-shares long and short pulses to achieve a combination of effective medium-range air and surface search and precise short-range navigational display. It sometimes functions as a target designation system; users include not only the Italian Navy, but also the Canadian Navy (*Iroquois* class frigates) and the Venezuelan Navy. Options include fast antenna rotation rate (40rpm), MTI with an L-band system using the same antenna for low-flyer detection, and provision for a transponder. Both stabilized and non-stabilized versions exist. The current model is SPQ-2F.

SPQ-3
A dual (X- and S-) band set for ground installation, illustrating just how complex a small search radar can be. The S-band antenna (1MW, 1.5-microsecond pulses at a PRF of 450 to 550) is a conventional parabolic section (1.5° × 7° beam, 4.5m × 1.8m, rotating at 4, 8 or 16rpm. The X-band antenna consists of two back-to-back units, either of which can vary its angle of elevation from −2 to +15 degrees (4.4m × 1.08m, 0.5° × 3.5° beam, rotating at 15, 30 or 60rpm) and transmitting signals at a peak power of 180kW (0.15 microseconds at a PRF of 2500–3000 or 1.5 microseconds at 450–550). Both X-band antennas employ slotted-waveguide feeders.

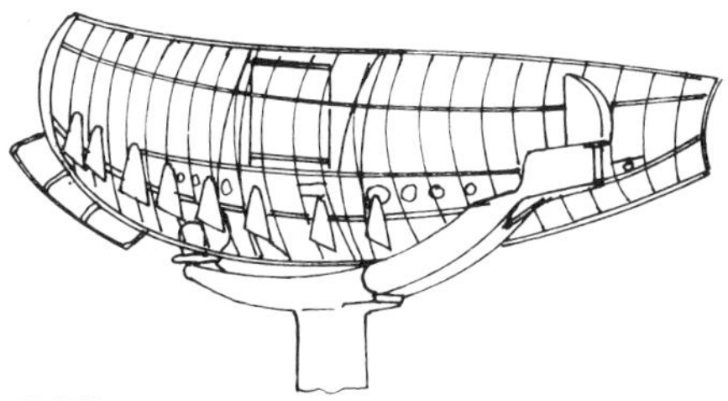

3RM7 antenna.

3RM series
The other principal Italian Navy series of surface search radars, also built by SMA. The 3RM series has some air target capability. All use slotted waveguides for horizontal beamwidths between 0.8° and 2° depending upon antenna size, a vertical beam pattern of 26° (shaped to 40), and a scanner rate of 25rpm. Two transmitters are available, with peak powers of 7 or 20kW (pulse lengths are 0.05-, 0.15-, 0.5- or 1.5-microsecond, with PRFs of, respectively, 6000, 3000, 1500 and 750).The 3RM20-H version is the US SPS-63, the NATO hydrofoil surface search radar; the Italian Navy uses the 3RM7-250 version in its Swordfish, with a different antenna and a second 250kW transmitter. Other versions are 3RM20 (MM/BPS-704), 3RM28-B (MM/SPN-703), MM/APS-705 and -707 for helicopters and MM/APQ-706 for the Marte system. SPN-703 is somwhat modified, with a 1.2° × 25° beam, and PRFs of (with the same pulse-lengths) 5200, 2600, 1300 and 650 respectively.

MM/SPQ-701, -702
Presumably developments of SPQ-2. No details are available, apart from X-band operation.

The Netherlands

DESIGNATIONS

Postwar Dutch search radars generally fall into four categories: LW (*Lucht*, or air, watch, in L-band); ZW (*Zee*, or sea, watch, in X-band, for surface search); DA (S-band target-indication and shorter range air search); and VI (height-finder; only one was produced). All are built by HSA, which applies its own SGR-series numbers. Dutch installation practice followed that of the Royal Navy, in that separate search and target indication sets were generally provided for each ship, but radar design practice differed markedly in that Dutch air search radars had very high resolution. In more recent ships, the DA radar has been eliminated, its target-acquisition function taken over by one element of the multiple-antenna fire control system (WM-20 series). For example, the Dutch destroyers of the 1950s and the frigates of the 1960s generally combined LW-02 with DA-01 (or, after recent refits, DA-05) which designated targets to a conventional fire control radar such as M45. In the new *Kortenaer* class, on the other hand, there is only an LW-08 long-range air search radar designating to a WM-25 two-antenna fire control system incorporating effective TWS performance.

Postwar HSA products also include a long series of fire control radars which, like the search sets, have enjoyed considerable commercial success. There is some confusion in designations, since the radars are often referred to by the numbers of the associated fire control systems, the M20 and M40 series being examples. The earliest models, which became operational in 1953, were mounted on the cabs of the M1, M2 and M3 directors.

RADARS

LW-01

The first generation of principal air search sets consisted of LW-01 and LW-02. The former was designed for cruiser-size ships and appeared aboard Dutch cruisers and the carrier *Karel Doorman*. It used a huge antenna, 11m × 4.88m (approximately 36.1ft × 16ft) and weighing about 3200lb, which formed a very narrow beam (about 1° wide, forming a cosecant-squared pattern in the vertical plane); it was more efficient than, for example, the US SPS-6 (for which the proportionate beamwidth at a similar antenna width would have been 1.65°). Peak power was 600kW, with relatively long pulses (2 or 5 microseconds, at PRFs of 500 or 250) and an antenna rotation rate of 0–8rpm. No performance data have been released. The HSA designation was SGR-114/06.

LW-02

The destroyer radar corresponding to LW-01. Exported to many navies, beginning with Sweden and including Australia, Colombia (Swedish-built destroyers) and West Germany. The Canadian SPS-501 combines its antenna (LW-03 variant) with a US SPS-12 radar transmitter. The 7.8m × 4.25m (25.6ft × 13.9ft) antenna weighs about 2100lb and produces a 2.2° wide beam forming a cosecant-squared pattern in the vertical. The transmitter matches that of its larger brother, although some sources claim a 500kW peak power; aircraft detection is claimed at 100nm and 59,000ft. Scan rate is 1–10rpm. HSA designation SGR-114/12.

LW-02 radar antenna (right) aboard HMAS Vendetta, *April 1977. The radome at left houses M22 fire control radar.* GIORGIO ARRA

LW-03

Similar to LW-02 but having a scan rate of 5 or 10rpm. These radars are now being replaced by LW-04 in the West German Navy.

LW-04

LW-04 substitutes an elliptical antenna for the characteristic nearly rectangular type of the earlier series. No SGR designation has been announced, and antenna dimensions are not available. Peak power is 1.1MW (LW-04/2 version), with 1.3-microsecond pulses (PRF not available), and beamwidth is about 2.2°. Range limits on a 2m² target are about 120nm at 55,000ft, with cover at

Evertsen *in May 1978 showing an LW-03 antenna with M44 Seacat fire control radar below. Note chaff launcher at base of radar tower.* PAUL BEAVER

60,000ft at about 90nm. LW-04 is fitted with a digital MTI and can operate on any one of six preset frequencies. It first entered service aboard the German destroyer *Hessen*, replacing an LW-02. No LW-05 has been announced, and LW-06 and LW-07 are land-based systems.

LW-08

The latest of the LW series, first appearing at sea aboard the Dutch frigate *Kortanaer,* which is also the first major Dutch warship to dispense entirely with a target-indicating system. It and LW-07 share a pulse-compression radar transmitter with alternative pulsewidths of 35 and 69 microseconds (comprising a 1-microsecond pulse for short-range resolution, and a 'chirp' for long-range performance) at PRFs of 1000 and 500, respectively; peak power is only 150kW, but the high *average* power (5.1kW, equivalent to a peak power of 5.1MW if applied to 1-microsecond pulses at a PRF of 1000) makes for a claimed detection range, on a $2m^2$ target, of 145nm at 85,000ft. As in earlier radars of this series, beamwidth is 2.2°. The scan rate is 7.5 or 15rpm and, as in LW-04, any of six preselected frequencies can be used; there is provision for pulse-to-pulse frequency jumping through the entire band. The 1-microsecond pulse provides a surface warning mode in addition to the pulse-compressed air search mode. Antenna dimensions are 7.9m × 4.85m (about 25ft × 15.9ft). The British Type 1022 is an LW-08 receiver/transmitter married to a Marconi antenna. LW-08 antenna weight is about 3300lb, with a corresponding below-decks weight of about 9200lb.

DA-01

The DA series is generally similar physically to the LW series, but operates in S-band. DA-01 bacame operational in 1954 aboard Dutch destroyers, and has been exported. The HSA designation is SGR-105. Peak power is 500kW (1.3- or 0.5-microsecond pulses at 500 or 1000 per second) and the 4.65m (15.3ft) wide antenna produces a 1.5° wide beam; it weighs a total of about 4400lb. Detection of a fighter (*ie* a $2m^2$ target) at 50nm and 35,000ft is claimed.

DA-02

A modified version of DA-01, using an antenna of wider beam (1.7°, for a range of 40nm at 30,000ft), first sold to Argen-

DA-01 radar antenna. AUTHOR

tina for the ex-US light cruiser *Neuve de Julio*. No DA-03 has been announced.

DA-04

The last of the initial series, a simplified version of DA-01 without antenna stabilization, for an antenna weight of only about 1200lb. Performance matches that of its prototype.

DA-04 (1), *M45 4.5in fire control* (2), *Kelvin-Hughes Type 925* (3) *and LW-03* (4) *radars aboard* Evertsen, *May 1978.*
PAUL BEAVER

DA-05

The first of a new generation; like LW-04, it employs an elliptical antenna, with a new and more powerful magnetron (1.2MW peak power with 1.3- or 2.6-microsecond pulses, with PRFs of, respectively, 1000 and 500). The 4.87m × 2.75m (about 16ft × 9ft) antenna weighs about 1600lb and produces a 1.5° beam. Like LW-04, DA-05 is fitted for digital MTI. Claimed performance is detection of a $2m^2$ target at 90nm and

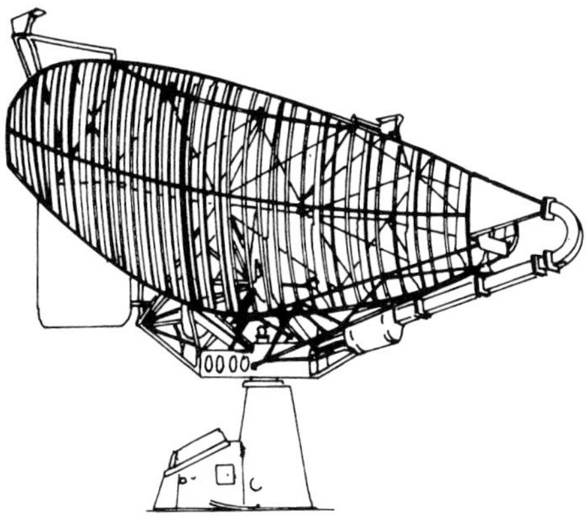

DA-05 antenna.

about 60,000ft. It is in use in the Belgian, Dutch and Spanish Navies. DA-05/M is a mobile land-based version, and both DA-06 and DA-07 are land-based types, DA-07 incorporating pulse-compression (as did LW-07). DA-05 is the primary air search radar of the Belgian *Wielingen* class.

DA-08

The latest of the DA series, an S-band set analogous to LW-08, with pulse compression (150kW peak power and 5.1kW average power, as in LW-08). Beamwidth (1.5°) matches that of earlier DA series sets, and range rises to about 115nm ($2m^2$ target). The basic design is derived from that of LW-08. DA-08 is to be the primary search radar of the German 'standard' frigates, in analogy to the use of DA-02 in the German *Köln* class; it has also been fitted to Geman *Hamburg* class missile destroyers as their primary search radars.

ZW-01, -03, -04, -05

Dutch surface search radars are conventional in appearance, particularly as compared to the early LW and DA types. The first of them, ZW-01, appeared in 1954 and is used extensively by the Dutch and West German navies. An improved version, SW-03, is in widespread service, and the first genera-

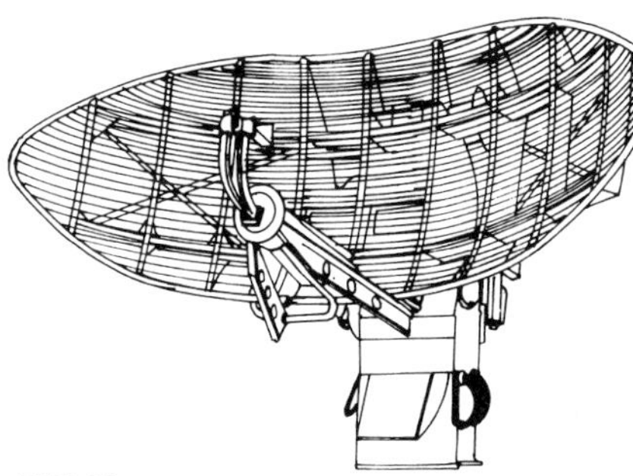

ZW-02 antenna.

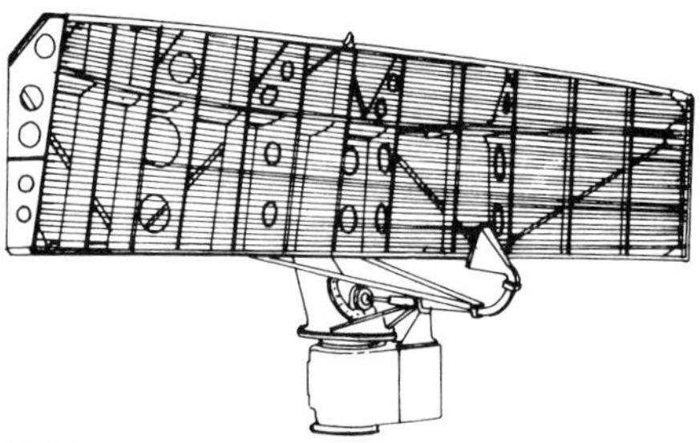

ZW-04 antenna.

tion of these radars is completed by ZW-04 (SGR-116/08). No details of these systems are available, nor has information on any ZW-05 been published.

ZW-06

The new generation of Dutch surface search radars is typified by ZW-06 with its more angular antenna (0.9° × 19° beam, 60kW peak power, scan rate 24rpm) with dimensions of 2.9m × 1.2m (about 9.5ft × 3.9ft). Very short pulses (0.06 microseconds) provide high resolution, and range on a 10m^2 target is about 14nm. At that range an airplane at 20,000ft can be detected. Features include surface coverage to the radar horizon, air coverage sufficient for helicopter control, provision for integration of helicopter transponder systems, sector scan facility and digital signal-processing. ECCM features include tunability, STC, a logarithmic receiver with pulse-length discrimination, and suppression of non-correlated pulses. ZW-06 is fitted to the Dutch *Kortenaer* class, to the Brazilian *Niteroi*, to the Spanish *F80*, and to the Indian *Leanders*.

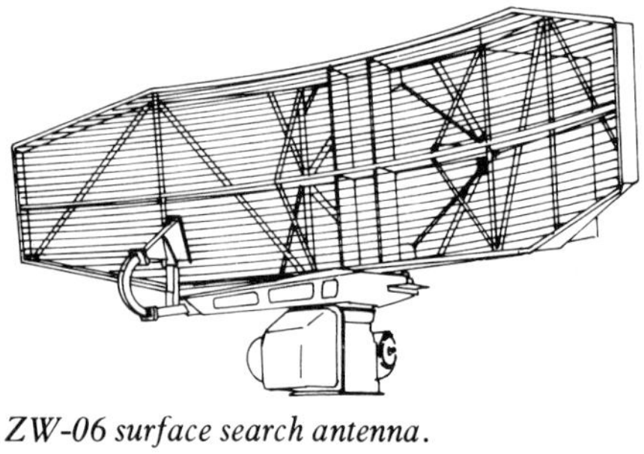

ZW-06 surface search antenna.

ZW-07

A submarine derivative of ZW-06, with peak power reduced to 40kW (pulse length increased to 0.22 microseconds) and a single-pulse transmission capability. It can also operate in a sector scan mode. Its smaller antenna reduces resolution to 2.4° in azimuth, and the longer pulse is required for single-pulse operation. A 10m^2 target can be detected reliably at about 16nm and 25,000ft.

ZW-08

The latest of the ZW series, with a stabilized antenna, 60kW peak power (0.06-microsecond pulses, 1° beamwidth). Claimed range performance on a 10m^2 target is 32nm, a figure bought to some extent by focussing the beam away from the vertical, so that where ZW-06 is officially described as a 'surface search and navigation and short-range air warning radar', ZW-08 is pure surface search, with a maximum altitude capability of less than 10,000ft. It forms the basis of a sea-skimming missile detection radar, SD-01. The latter also employs a stabilized antenna and has a powerful 200kW magnetron with digital MTI. Longer pulses (0.58 microseconds) are used to achieve quicker detection of a fast target, and bearing discrimination remains about the same with a 1° beamwidth. Range, measured against the much smaller cross-section of a small missile (about 0.1m^2) is about 26nm; presumably the digital MTI facility (absent from the ZW series) makes for superior performance against the extreme clutter of the sea surface.

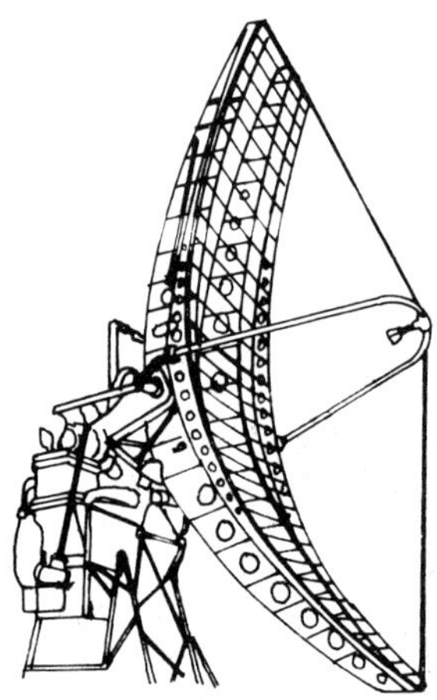

VI-01 height-finder antenna.

VI-01

The sole Dutch height-finder, HSA designation SGR-109/2, a conventional nodding-beam radar in which feed and antenna moved together, unlike the US Robinson-feed systems. It used the S-band transmitter of the contemporary DA-01 series, and had a 3.5° × 1.2° (vertical) beam. Vertical scan rates were 50 per minute for −2 to +8 degrees and 20 per minute for −2 to +50, the sector varying automatically with range. Meanwhile the antenna could rotate in bearing at 60° per second. Claimed performance for the very similar land version is a range of 150nm, with coverage up to about 120,000ft; up to six targets per minute, spread randomly over 360°, can be handled. The only surviving installation is aboard the Argentine carrier *25 de Mayo*, formerly the Dutch *Karel Doorman*. Two others were fitted to the cruisers *De Zeven Provincien* and *De Ruyter*. All of these antennas were designed with special attention paid to funnel gas and weather corrosion; stainless steel was used extensively in their construction.

SPS-01 radome, Tromp, *June 1977.*
GIORGIO ARRA

SPS-01

Although the LW, DA and ZW radars all showed considerable ingenuity in their designs, they were fundamentally conventional. That cannot be said of the other major Dutch search radar, SPS-01 (formerly known as MTTR, Multi-Target Tracking Radar). It combines search and target-tracking (*ie* the LW and DA functions) in a single multiple antenna, generally housed in a large radome.

SPS-01 was designed to meet the mutually contradictory requirements of search and multiple-target tracking. For search, it had to radiate over a very wide volume, with a relatively low data rate for long ranges and a higher one for short range; accuracy could be sacrificed for a high probability of target-detection. For tracking, on the other hand, precision was necessary, coupled with a concentration of radiated energy in the direction of the target, and a very high (or, better, continuous) data rate. Discussions with the Royal Netherlands Navy began in 1958, the goal being simultaneous high-precision tracking of multiple targets; SPS-01 is now credited with the capability to handle over a hundred aircraft tracks. Research began in 1959, and a first working model was available in February 1964. At that time the production of two systems began, the first beginning tests in 1967. They were completed in 1969, and two SP-01s have been installed aboard the two Dutch guided missile frigates (destroyers) of the *Tromp* class. The MTTR project was for a time a joint one with Great Britain, and a variant (Type 988, or 'Broomstick') was intended for the Type 82 missile destroyer (*Bristol*) and for the abortive British carrier project (CVA-01).

The radar employs six antennas, all mounted together on a stabilized base and rotating together at 20rpm: two paraboloids back-to-back, with feed horns moving to generate any of five alternative beams plus a sixth (fixed) low-angle beam; two back-to-back FRESCAN arrays at right angles to the parabolas; a multi-element antenna for high-angle search (omitted in production models); and a slotted waveguide for IFF mounted below one of the parabolas. In operation, the low-angle search beam is operated continuously to provide short-range warning with a high data rate (40 scans per minute). The other five beams are energized in sequence one by one during successive rotations, so that for each one the data rate is reduced to 8 per minute. The net search pattern is a cosecant-squared shape, but avoids the drawbacks of a more conventional fan-beam radar, *ie* low antenna gain and great susceptibility to clutter and jamming. Beam dimensions vary with elevation: in each case the paraboloid assures a horizontal width of 1.5° but the vertical beam size varies between 2° at an elevation of 2.5° to 30° at an elevation of 21°. Targets are detected automatically and their two-dimensional coordinates fed into the central computer, which instructs the two FRESCANS to find them in three dimensions for tracking.

Each of the FRESCANs consists of a series of slotted waveguides slanted to the horizontal, and fed by the usual sinuous waveguide for vertical scanning. The slant makes it possible for the system to perform a true vertical scan while the antenna rotates rapidly; in effect the FRESCAN can be used to scan both vertically and horizontally, and thus to obtain precision target data. Approximate beam dimensions are 1.5° × 1.5°, the latter depending on beam elevation. Back-to-back positioning provides a very high data rate of 40 scans per minute. In operation, the computer first ascertains that a target detected in two dimensions is not already being tracked. It then programs a frequency sweep which measures target elevation; on the next scan the frequency variation is programmed to give a horizontal scan for precision target-bearing location. Each target is scanned cross-wise at the rate of twenty measurements per minute. In this way a pure FRESCAN is used for true three-dimensional tracking, without phase shifters; however, its effective tracking rate is limited by the requirement for mechanical rotation.

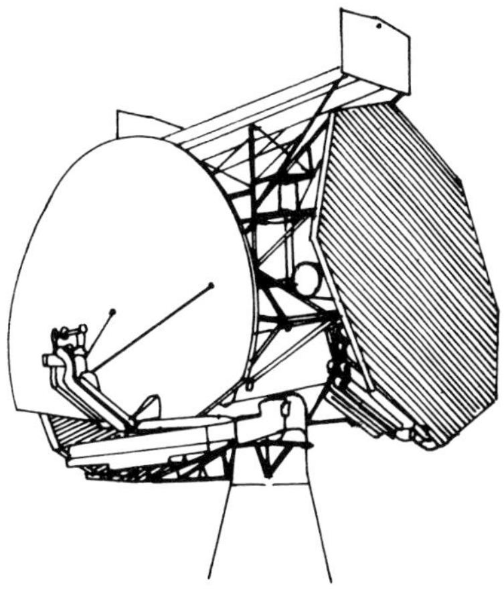

SPS-01 with radome removed.

The system as a whole operates in S-band, the frequency ranges of the paraboloids and the two FRESCANs differing slightly. A single power source produces 40-microsecond, 750kW pulses with staggered PRFs of 460 and 540, and energy is presumably time-shared among the antennas.

M1, M2, M4 radar systems

The M1 system (SGR-101) operated in the high S-band (about 8cm) and employed a 400kW magnetron. Precision was limited by its dimensions, and in some cases the beam was as much as 5° wide. M2 (SGR-107) employed X-band for higher precision (4.5° in a very small antenna) and M3 had a similar system. All of these radars employed small dish antennas with waveguide feeds across them, and with conically scanning feed horns. The M4 radars (SGR-108) differ in that they are cut mesh parabolas with central feeds supported by heavy pylons. Like SGR-107, SGR-108 is a relatively low-powered radar with a high PRF (1600) and short pulses (0.5 microseconds). Examples of the M4 fire control system include M4/1 for 40mm guns (1957), M4/2 (40mm guns, 1961), and M4/3 (40mm guns and Seacat point-defense missiles, for West Germany, 1961).

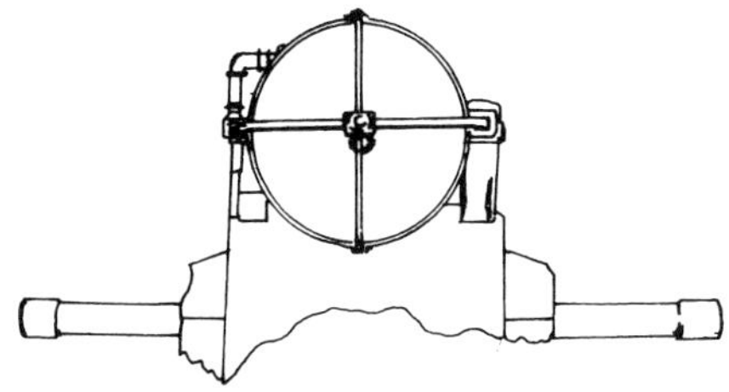

M1 (SGR-101) antenna.

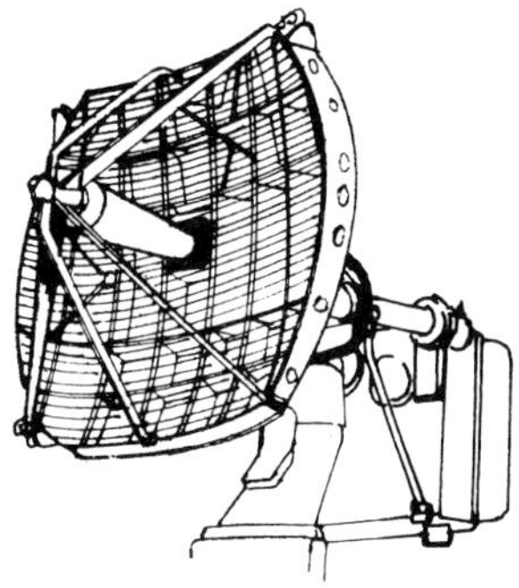

M4 (SGR-108) antenna.

M40 series systems

A second generation consisting of relatively conventional enclosed-dish radars associated with the M40 series fire control systems, particularly M44 for Seacat control and M45 for the control of medium- and light-caliber guns. The shape of these radomes suggests monopulse-feed, but no data have been released.

M20 series systems

An extremely successful series and one which can be recognized by its characteristic weatherproof radome, enclosing a pair of antennas fed from a common source and stabilized on a common platform, a cut paraboloid search antenna below and a tracking dish above. In one version (M26) only the search antenna is fitted, and it is mounted atop the stabilized platform, in a hemispherical radome. In all of these systems power is shared between the two antennas. Thus, during search, all power is fed to the search antenna, but once a target has been detected, some is fed into the tracker. HSA claims that this system has important advantages against small, rapidly closing targets such as cruise missiles: since the two antennas are mounted on a common platform, designation from one to the other does not suffer from the errors inherent in

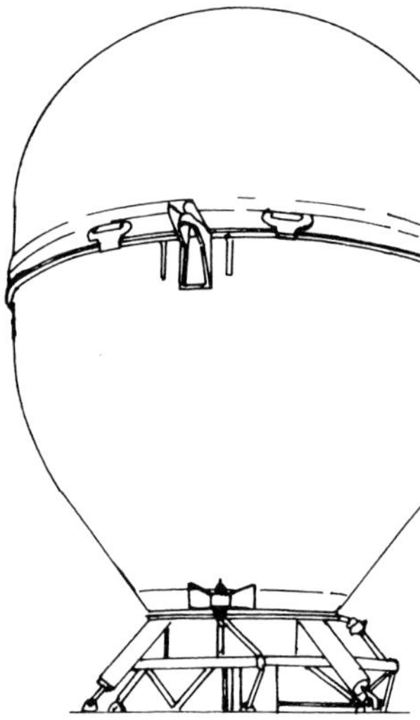

M20 radome.

separate-antenna systems. Moreover, in a small ship there is often only a single useful antenna location, which can be occupied by the M20 series radome. In many cases the search antenna can be used also to designate to a separate tracking antenna, or STIR. This is usually an HSA Cassegrain dish type, although in the US Mk 92 derivative of the M28 for the *Perry* class frigate it is a modification of the US SPG-60.

The first of the M20 series, the M20 itself, is designed for torpedo and gunfire control for small craft such as torpedo boats. It can control two light or medium guns against one air and one surface target simultaneously or two torpedoes against two surface targets simultaneously. One air and three surface targets can be tracked simultaneously. M22 is a simpler system for larger surface ships such as frigates, providing monopulse (dish) tracking against one air traget, and TWS (PPI auto-following) tracking against surface targets. Unlike M20, it is designed for use in conjunction with a long-range air search radar; it is intended to control two guns against one air and one surface target simultaneously. M24 can control up to three guns and provides for a link to an ASW fire control system. M25 incorporates CWI for the control of a surface-to-air missile; it can track simultaneously either one air target, one surface target and one shore target, or one air and two surface targets. Again, air targets are tracked by the pulse-doppler dish and surface ones by TWS use of the search antenna. M27 is very similar, with provision for torpedo fire control as well. It can control surface-to-surface semi-active homing missiles, but generally it controls one gun against one air or surface target, and missiles or torpedoes against surface targets. For surface-to-surface missile fire the tracker dish is slaved to the TWS target-track computer program and used for CWI. M26, on the other hand, is a simplified system for fire control (via TWS) and navigation, capable of controlling only a single gun.

M28, the system built under license in the United States as Mk 92, is designed to track one air and one surface target simultaneously, and to control one surface-to-air missile and two light or medium guns. As in the other systems, surface tracking is by TWS operation of the search antenna. It is described as capable of controlling a command-guided (line-of-sight) SAM, but in US service the Mk 92 controls the semi-active guided Standard. A separate Dutch-designed STIR antenna, 1.7m in diameter, permits control of two missiles against two targets simultaneously. M29 is a related system, with surface-to-surface rather than surface-to-air missile capability.

The Combined Antenna System (CAS) appears to be common to the entire M20 series of weapon control systems; characteristics have not been released, but they operate on X-band. Detection and tracking of targets at low level is emphasized, with such features as Digital MTI, variable polarization, and pulse doppler for clutter rejection. An enlarged search antenna, which entered production in 1973, increased effective range by about 20 per cent. Peak power is about 1MW; it is claimed that the effect of cumulative improvements is about a doubling of range: 20 per cent from the enlargement of the antenna; 20 per cent from the use of a tunnel diode amplifier; and 50 per cent from the power increase to 1MW by the use of a cross-field amplifier. A search range of about 25nm on a $1m^2$ target is claimed.

France

DESIGNATIONS

French postwar naval radar development followed a pattern markedly different from that of the Netherlands. The unique French designation system permits radar evolution to be traced in some detail: radars are coded functionally and, it appears, by frequency (first digit of the numerical designator). The functional designation is a four-letter code: 'DR' for radar ('DU' for sonar); 'B' for *bâtiment* (= 'surface ship') or 'U' for submarine; and then one of the letters 'A' (submarine attack radar), 'C' (fire control), 'I' (indication, *ie* target designation or height-finding), 'N' (navigation), 'R' (missile control) or 'V' (search). The band designators show the evolution of French radar transmitters – 10 (S-band) and 20 (P/L-band) at first, soon followed by 30 (X-band). No 40 series ever appeared (presumably it was K-band) but 50 appeared with the Masurca AA missile, which very much suggests C-band as in Terrier. The use of the 50 band in navigational and surface search sets bears out this hypothesis.

Thus, a typical French radar is DRBV-22, a 23cm surface-ship air search set, presumably the third P/L-band set of this kind. Unfortunately enough numbers remain blank (or have not been used) to make a detailed account of French development impossible. However, it is worthwhile to note some uniquely French approaches, which appeared primarily on the earliest sets and some of which still exist in considerable numbers. It should also be noted that Thomson-CSF exports many of its naval sets under commercial designations such as Castor or TRS-3200.

RADARS

DRBV-11

A two-beam target indication radar, one of the earliest French air search sets.

DRBV-13

The S-band, three-dimensional radome-enclosed radar of the corvette *Aconit*, also designated DRBI-13. It is officially described as a pulse-doppler radar capable of both search and tracking, but no details have ever been released.

DRBV-15 (Sea Tiger)

A current set which is being retrofitted to the T47 ASW destroyers and to the older *avisos*. It is a pulse-compression radar (60kW peak power, using either 5.5-microsecond or 12-microsecond pulses reduced to 0.5 microseconds; PRFs are, respectively, 1100 and 2200) using a roll-stabilized antenna with alternative rotation rates of 12 and 24rpm and generating a 50° cosecant-squared beam. Claimed performance includes detection of a missile (0.1m^2 target) at 20nm and of an airplane (2m^2) at 62nm.

DRBV-15 (Sea Tiger) air/surface surveillance radar. THOMSON-CSF, BY COURTESY OF *DEFENCE*

DRBV-20

The other early French air search set, a big P-band mattress looking like nothing so much as the standard image of a radar: a big parabolic section curved about its horizontal axis, with a row of three dipoles spaced along that axis; its limited dimensions and long wavelength guaranteed the kind of wide beam which required the presence of the target indicator. Long wavelength also guaranteed high power at the then-current state of radar technology. Performance figures have not been released, but the series continued with DRBV-20B and then with DRBV-20C, in which a large flat mattress was used to achieve a narrower beam and longer range, in rough analogy with the evolution from SRa through SPS-17 to SPS-17A. It is fitted to the carriers *Foch* and *Clemenceau* and to the missile cruiser *Colbert*.

DRBV-20 air surveillance (1), *DRBV-30 surface surveillance* (2) *and IFF Mk 10* (3) *antennas*, THOMPSON-CSF

DRBC-11 fire control (**1, 4**), *DRBI-10 3-D surveillance* (**2**) *and DRBV-22 air surveillance* (**3**) *antennas*. THOMPSON-CSF

DRBV-22, -23, -26, Jupiter

There was no immediate successor to DRBV-11, but the French soon shifted to the L-band for air search radars. No DRBV-21 has been announced, but DRBV-22 resembled (at least externally) the US SPS-6B, and may have evolved from that set. In DRBV-22C a new elliptical antenna was introduced, stabilized for roll and pitch; this antenna was combined with a new and much more powerful transmitter in DRBV-23. Development began in 1961, and maximum range is about 160nm.

Jupiter is an export version of DRBV-23, and its somewhat different characteristics suggest those of the French L-band series. Peak power is 2MW (2.5-microsecond pulses at a PRF of 450) for a detection range of over 110nm on a 2m² target. The 24.6ft × 9.9ft antenna generates a 2.5° × 50° beam, and alternative rotation rates are 7.5 and 15rpm. In 1972 work began on an improved Jupiter, designated DRBV-26, for the new generation of missile ships; a range of over 150nm is claimed, and such ECCM features as MTI and CFAR (Constant False Alarm Rate) are provided. No DRBV-24 or -25 has been announced.

DRBV-26 (**1**), *DRBV-51C* (**2**) *and DRBC-32C* (**3**) *antennas*. THOMPSON-CSF

DRBV-30, -31, DRBN-32

French surface search radar development began with the postwar DRBV-30. DRBV--31 was a license-built Decca navigational radar for destroyers and escorts. There is no DRBV-32, but a navigational radar for small warships such as minesweepers is designated DRBN-32.

DRBV-22A (**1**), *DRBV-50* (**2**) *and DRBC-32* (**3**) *radars as fitted to* D'Estrées, *1979*. STEFANO CIOGLIA

DRBV-50, -51, Triton

In 1961 development of a new low-flyer and surface search set, DRBV-50, was begun; it was mounted in the *Jeanne d'Arc*, and in the *Maillé Brézé* class ASW ships. The DRBV-51, for low-altitude air search and for Exocet target designation, is a far more recent development, derived from the Triton series of small- and medium-ship radars. Typical characteristics include a peak power of 250kW (0.7- and 0.15-microsecond pulses, with variable PRFs) and a range of over 16nm on a fighter (2m² target). In export versions, Triton is usually part of the Vega small-ship fire control system. There are several associated types. Triton II is most closely comparable to DRBV-51, and its characteristics have been listed. Triton-S is an S-band pulse-compression radar incorporating pulse-doppler processing; it has a range (fighter) of about 16nm. Similarly, Triton X operates in X-band, with 200kW (0.3-microsecond pulses), achieving a 1.2° beam, with a range of about 11nm on a fighter.

Triton surveillance and target indication radar. THOMSON-CSF, BY COURTESY OF *DEFENCE*

Saturne

A commercial S-band radar for air search, developed by Thomson-CSF, fitted in, for example, the Swedish *Visby*. It has a peak power of 1MW (0.5- and 2-microsecond pulses, PRF 4000 and 1000), with a maximum range of 62nm (2m² target) and a scan rate of 10 or 20rpm (data refer to Saturne II-30 version).

DRBI-10

Complementing the two-dimensional DRBV series is the DRBI series, the first of which was DRBI-10, an unconventional height-finder. It combines the Robinson-feed of US practice with a broad, roughly square antenna which produces a relatively narrow search beam. The feed moves up and down as the antenna rotates, to produce a wavy scanned area. The French Navy uses it as a three-dimensional radar: *Surcouf* class radar-picket destroyers, for example, had it in place of the DRBV-20 of their AA and pure-destroyer counterparts; both had DRBV-11. However, the AA cruiser *De Grasse* was completed with DRBV-11/22 *plus* a DRBI-10 aft, an arrangement close to US practice. DRBI-10 was apparently evolved from the land-based Picador system; antenna dimensions are 3.4m (11.2ft) on a side, and a peak power of 1–2MW is said to confer a range of 100–140nm on a fighter.

DRBI-23

The only other announced DRBI-series radar. The large radome-enclosed

DRBI-23 radome, aboard the missile frigate Duquesne, *1972.* GIORGIO ARRA

antenna aboard the missile frigates *Suffren* and *Duquesne,* it employs a big Cassegrain reflector, consisting of a parabolic mirror onto which microwave horns feed and which itself illuminates a flat-plate reflector. The latter in turn rotates the polarization of the reflected radiation, so that it can pass through the small mirror – in effect, therefore, the entire feed system casts little or no effective microwave 'shadow' on the incoming or outgoing beam. Published references to 'monopulse' height-finding suggest physical parallels to SPS-2.

DRBJ-11

Thomson-CSF finally moved to modern electronic scanning methods in this new set, which is also referred to as SYA and which is intended for the C70 AA air-defense destroyers. The single photograph released suggests a spherical surface covered with phase-shifters which are themselves connected directly to the radiating elements.

DRUA-30, -32, -33, Calypso

In 1956 DRUA-30, the first French submarine surface search radar, was produced, followed by DRUA-32 and DRUA-33 (1964). The latter led to the Calypso series of export and domestic submarine attack radars; the Calypso III version was developed in conjunction with the German IKL submarine design firm. Calypso II is described as a search and navigational radar operating in X-band, with 70KW peak power (alternative 0.15- and 0.5-microsecond pulses). Operating modes are: continuous transmission, sector surveillance, short-time transmission (0.1 or 1 second only) and silent operation with a dummy load to permit maintenance testing. A large airplane (10m²) can be detected at 16nm. The improved Calypso III has greater power, for a range of about 18nm on a typical ASW airplane at about 8000ft. In one version of Calypso, a separate antenna is mounted on the attack periscope for a quick range measurement prior to attack.

Pollux (Vega 2 system) antenna.

DRBC-11, -30, -31, -33, Castor, Pollux

The earliest generation of French naval radars included two X-band antiaircraft fire control radars, DRBC-30 and DRBC-31, and the S-band (long-range fire control) DRBC-11. None was exported, and for none have details been published. DRBC-32 was designed specifically to operate with the new 100mm automatic gun (1961); it has been built in -32A through -32E models and remains in production.

Thomson-CSF produces a range of small-ship fire control radars, Castor and Pollux; the former has an alternative DRBC-33 designation in its Castor II version, suggesting French Navy use. Castor is an X-band monopulse radar tracker with a peak power of 80kW (4-microsecond pulses, acquisition and tracking of a 0.1m² target at 8nm). Pollux is more powerful, with the Triton X radar transmitter (200kW, 0.3-microsecond pulses, PRF 1500) capable of acquiring and tracking a small aircraft at 16nm.

DRBR-51

The Masurca guidance radar. Masurca exists in both beam-riding (command to line-of-sight) and semiactive versions, and DRBR-51 is thus roughly comparable to the US SPG-55. Each radar group consists of a pair of antennas, a large-diameter target-tracker (which can also measure the deviation of the missile from line-of-sight) and a small diameter

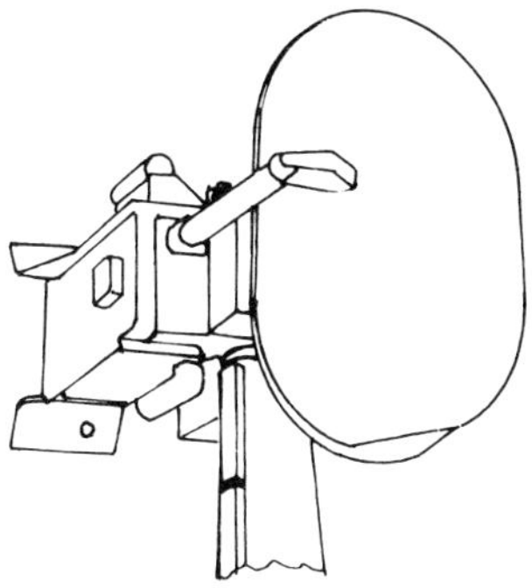

DRBC-32E antenna.

unit for command transmission; a third antenna is used to generate the initial gathering beam. The main dish is fitted for CW injection for semiactive operation. Reportedly missiles are tracked at 5cm (C-band) while command signals are transmitted at 7cm; the tracking channels are designated 'blue' and 'yellow'.

Appendices

I. Some Technical Notes

Thus far we have avoided quantitative discussion and above all mathematics. Nonetheless, some readers may find such analysis useful. In this section we discuss the questions of radar range and interference of radar signals, the latter of great importance in antenna design and in understanding fading.

Maximum radar range is often estimated by means of a 'radar equation'. Imagine first a point source of radar signals, radiating a signal strength S equally in all directions. At a range R the signal is spread over a sphere of area $4\pi R^2$; a target area A will, therefore, pick up a fraction $(A/4\pi R^2)$ of the signal. In fact, of course, a radar sends out a very directional signal, and the extent of its directionality is usually indicated by a gain, G. Then the signal S_1 picked up by a target area A at a distance R is

$$S_1 = G(A/4\pi R^2)S$$

The target area ('radar cross-section') A now reflects the signal and sends it back to the radar. In effect it acts like a radio transmitter. Usually the target area A is by no means the physical area of the target; it is defined to take into account the directionality and efficiency with which the target reflects radar signals. For example, a corner reflector is far more effective than a smooth sphere.

The radar antenna, too, has an area, which we can call A_1; the signal it receives, S_2, is

$$S_2 = (A_1/4\pi R^2)S_1$$
$$= (A_1/4\pi R^2)G(A/4\pi R^2)S$$

which shows that the received signal falls off very rapidly indeed with range, as the fourth power. What actually determines the range at which some object can be detected is the ratio of the received signal S_2 to the noise in which it is embedded, which for a particular radar is a complex question of receiver design and the form of the radar signal.

The radar antenna functions of reception and transmission are closely related; the antenna patterns for both are identical (as long as one antenna serves both). In fact the effective antenna *area* for reception, in square wavelengths, is proportional to the *gain* involved in transmission; if L is the wavelength,

$$G = 4\pi(A/L^2)$$

so that

$$S_2 = \frac{1}{(4\pi)^3} G^2 L^2 (\frac{A}{R^4})S$$

For an airplane, a typical effective area might be one square meter. For a typical P-band (1.5m) long-range air search set, G would be less than 100. Such a set might well be effective at 200km, at which distance the equation would predict a return signal about

0.000000000000000000007

of that sent in the first place, *ie* for every megawatt sent out, seven billionths of a millionth of a watt return. Such a figure gives some idea of the demands placed upon a radar antenna: at one instant to transmit a short pulse of megawatt power, at very nearly the next to detect an echo so much fainter.

In fact matters improve as the radar sends out pulse after pulse and adds up the results. For example, a typical radar PRF might be a thousand per second. A typical air search radar might rotate at a speed of 6rpm, *ie* one scan every ten seconds; and a typical beam might be 3.6° wide. From the radar's point of view, there is little difference between targets located at various points *within* the beamwidth. The radar sweeps over 36° – ten beamwidths – every second. It 'sees', then, a hundred pulses in each direction that it can distinguish.

For each pulse there is some probability that the radar will or will not be able to pick it out of the noise – noise from outside as well as the noise the radar itself cannot avoid producing within its circuits. As the pulses multiply, the odds in favor of detection improve. Radar range theory is thus fundamentally a matter of statistics, and the design of a radar receiving system – even the 'design' of the radar signal itself – can affect the odds decisively. For example, one is accustomed to think of a radar signal as a sharp pulse, usually drawn as a little rectangle. It returns as a bump not so very much larger than many other randomly positioned bumps (noise). If the radar pulse is lengthened and made more complex, it can be distinguished more easily from the bits of noise – it no longer looks to the radar like just one more random fluctuation.

The radar range equation just presented cannot take such effects into account. Rather, a radar engineer will estimate the noise with which his system will have to contend, and then use a sophisticated form of the equation to estimate the signal, so that he can design equipment to contend with that S/N ratio. However, the equation does provide some estimate of relative ranges on various targets, given a radar's performance against a given target. For example, ten airplanes will have roughly ten times the effective target area of one; they will return a signal ten times as large, at a given range. That is, the radar signal returned will, by the radar equation, be the same as long as an increased range (reflected in the fourth power of R) is balanced by the greater signal. An increase of 70 per cent in R gives a factor of 10 in R^4, enough to balance off a tenfold increase in effective target area. Then a radar which can detect one airplane at 100nm can detect a formation of ten at about 170. Similarly, a bomber with roughly twice the radar cross-section of a fighter can be detected at about 20 per cent greater range: the fourth root of 2 is about 1.2.

The equation shows the value *to a target* of reducing its effective area. The propeller-driven aircraft of World War II had complex shapes which might be visualized as masses of corner reflectors. On the other hand, modern jets are quite streamlined, and therefore present far smaller target areas. For example, a radar handbook of 1945 estimated the radar cross-section of a fighter as about $20m^2$; modern figures for a fighter might be as little as $0.5m^2$.

The radar equation also gives some idea of the value of anti-radar paint. If the wavelength of a radar is known fairly precisely, it is possible to paint a radar target so as to trap most of the radar energy which strikes it; the paint is of little value at other wavelengths, however. A typical paint job might result in a reduction of effective radar cross-section to a tenth of its previous value; in that case a radar formerly effective to 90nm might not detect its target beyond about 50. The reduction in range is not by a factor of ten, but only by a factor of 1.7, because of the fourth-power of range. Anti-radar paints eventually fell from favor because they could not cope with the variety of radar systems typically used, *eg* both P- and L-band air search in the US fleet.

There are radar-absorbing materials other than paints, and the aim has always been to broaden the bands for which they are applicable. Presumably the recently-announced 'Stealth' technology combines such materials with careful shaping (for example to avoid corner reflectors) and with ECM techniques which further reduce the effectiveness of search radars. Such technology is applicable both to airborne vehicles (including missiles) and to ships; moreover, a ship, which is massive, is more likely to be able to accommodate the weight penalty inherent in special non-structural materials and in additional countermeasures. In general, radar countermeasures do not make a ship or missile completely invisible. However, they may make it far less visible than, say, a chaff cloud, so that an approaching missile can be decoyed away.

The radar equation also gives a more concrete feeling for ECM, in particular for radar intercept and jamming devices. Both operate *one way, ie* involve an antenna substituted for the radar target. Thus a radar intercept device with gain G^1 would obtain a signal S_3 at range R:

$$S_3 = (1/4\pi)^2\,(L^2G^1G/R^2)S$$

Because S_3 falls off only as the second and not as the fourth power of R, the intercept device can pick up signals far beyond the effective ranges of the original radar. For example, many air search radars are calibrated in terms of $1m^2$ targets. In theory a radar which could detect such a target at 100km could itself be detected by a similar radar, using the same signal to noise criterion, at several *million* kilometers. In fact intercept antennas tend to be quite small, since such detection ranges are worthless. To achieve detection at 150km, the intercept system would need a gain less than a ten-billionth that of the original radar. In fact, the air search radar would provide scant illumination below its radar horizon, *ie* at the surface of the sea 100 or 150km away; but the very small demand on the intercept set suggests just how little illumination it would need to pick something up.

In general the intercept antenna will not point directly at the search antenna. Alternatively, it may be an antenna capable of pick-up (at very low gain) from nearly any direction. In either case the equation means that the chances of radar intercept well outside radar range are excellent. For example, some of the radar intercept sets used by the US Navy employ small spinning antennas. Perhaps such an antenna picks up 1 per cent or less of the stream of pulses sent its way; and its gain may be less than 1 per cent of that of the radar it intercepts. Moreover, the intercept set may not be able to employ complex signal processing simply because its designers may be unaware of the precise details of enemy radar signals; and it may have to search over a wide spectrum to take into account enemy use of tunable radars. These considerations may lower its efficiency by further factor – but even so, that margin of ten billion at 50 per cent more range gives considerable comfort.

These figures actually rather overstate the case for the passive receiver. However, a 1945 naval report suggests that a good rule of thumb for airborne intercept receivers might be counter-detection at about twice radar range. That was before the probabilistic character of radar was really well understood, and so it is more concrete than should be the case. Other counter-detection ranges are given elsewhere in this book.

A jammer, too, can benefit from the one-way range equation. Typically air- and surface-borne jammers have been unable to afford the kind of power produced by large radars, and, more importantly, they often put out a continuous stream of power, rather than the pulses of a radar. Thus a megawatt radar producing 1000 one-microsecond pulses per second puts out an *average* of 1kW. At a range of 100km the radar described above receives back a stream of micro-microwatt pulses of narrowly defined frequency. The band on which a jammer works may be as much as ten thousand times as wide, especially if the radar uses frequency-agility as a countermeasure. Then a kilowatt jammer covering the whole radar band may *in effect* deliver only a fraction of 1 per cent of its power – 0.1W or less – to the radar at the right frequency. Once more, the jammer gain is unlikely to match that of the search radar; 1 per cent might be optimistic. At 100km the jammer delivers to the radar a power of about a milli-microwatt, a billionth of a watt, still perhaps enough to swamp out the radar signal and completely wash out range information. However, if the radar is directional enough, it picks up the jamming only when it is pointed at the jammer. Two or more ships can triangulate the jammer. Moreover, matters are close enough that signal coding or a little more power can defeat the jammer – hence the rise of specialized deceptive devices and also of jammers more closely correlated with their targets. The same one-way equation also applies to transponders and, indeed, to beacons of all types, and explains why IFF equipment is a relatively small burden; its signals go one way only.

The reader who wishes to apply the radar equation to other radar systems will find gain (and several other quantities) expressed most often not as numbers but as numbers of *decibels* (db). These are merely logarithms, used because of the great range of numbers involved. The gain in decibels is ten times the logarithm (base 10) of the gain as a pure number. For example, 30db corresponds to 1000, 3db to 2, and therefore 33db to 2000. In the accompanying tables, all quantities have been transformed into the corresponding numbers.

The radar equation does not take into account two important factors: losses in the air due to absorption of radar waves by the astmosphere, and the interference phenomenon wich causes fading and range enhancement in many naval radars. Nor does it take into account the subtle weather effects already discussed.

Several times we have referred to the way radar signals and their reflections add or cancel out, which more technically would be called destructive 'interference'. A slightly more detailed account provides some useful insights.

We can think of radar signals as composed of elementary waves of pure frequency; it is those waves we use for calculation. The theory of interference is, essentially, that waves add when their peaks coincide, and cancel when a peak and a trough coincide. Very crudely, waves can be represented as sine waves; a wave of length L presents, at a point X from its origin, an amplitude

$$\sin(2\pi X/L)$$

where the time behavior of the wave can be disregarded; we will be interested in the behavior of a system of waves at a fixed moment of time. The quantity $2\pi/L$ is the *phase* of the wave. Since the sine repeats every 2π, two waves whose phases are separated by some multiple of 2π – *ie* which have traveled distances separated by whole wavelengths – will add, in other words they will be in phase.

The simplest case of interference is *reflection*. Most long-wave antennas use reflectors a quarter-wavelength back from the dipole radiation source. The phenomenon of reflection in itself flips the phase of a wave through 180° (π). Then the quarter-wavelengths going and coming back bring the reflected wave into phase with the wave emitted by the dipole.

The case of microwave antennas is different; there is no attempt to achieve phase coherence between the horn feed and the reflector; indeed, in many cases the effort is to remove the horn from the stream of outgoing reflected signals, on the theory that it will merely block them. One reason is that, whereas manufacturing tolerances of tenths of 1 per cent of a meter are easy to achieve with large structures such as radar antennas, similar tolerances on a base of centimeters are far more difficult – and far harder to maintain, in practice, in a big open structure subject to weathering and to the wind. On the other hand, modern FRESCAN and slotted planer arrays do require very high standards of manufacturing – indeed, their performance is entirely dependent on the distance from slot to slot.

A phased array is a far more subtle use of the phase idea. In effect each elementary radiator acquires a phase delay; imagine a pair of radiators spaced *d* apart, one with no delay and one with delay T. We can imagine both radiators putting out signals; the effect of the delay T is to vary the *angle* of the beam their combined signals form. As the diagram shows, the two signals match in phase when

$$2\pi(d/L)s \text{ in } A + T$$

forms a multiple of 2π, A being the angle to dead ahead. As T is varied, the angle A varies.

FESCAN is a particularly elegant variation on this idea. In a FRESCAN antenna the elementary radiators are slots cut in a long waveguide folded to increase the delay from slot to slot. Once more we can imagine two radiators *d* apart; but now the delay from one to the next is a matter of the waveguide length D and the effective wavelength *in* the waveguide, L^1:

$$2\pi(d/L) \sin A + 2\pi(D/L^1)$$

must be a multiple of 2π. As L (and, with it, L^1) changes, the delay changes and with it the angle A. The greater the delay length D, the greater the impact of small changes in L.

The case of fading is very similar. Radar signals can follow different paths from antenna to target; the question is, given the lengths of those paths, whether the signals arrive at the target in or out of phase.

II. Examples of Radar Operation during World War II

taken from Action Reports of the cruiser SAN FRANCISCO

USS *San Francisco* fought in most of the major actions of the Pacific War, including the night engagements around Guadalcanal in the fall of 1942. After each action her commanding officer produced a report, including lessons learned; and these reports included comments on such *matériel* as radar and radios. The excerpts reproduced here provide some indication of how shipboard radar performed, and of what its problems were.

Night Action 11–12 October 1942

FC radars gave excellent performance except for a period of about 10 minutes each when they were inoperative. They were used for initial range and bearing on five different targets and in each instance the data furnished was very accurate. One target (probably a CL) was fired upon using radar control exclusively. Radar indicator showed straddles in range, but fall of shot could not be determined.

SC radar was not used for search after 1800 upon order from task group commander. FC radars of the four cruisers were used for search, augmented by the SG of the *Helena* and *Boise*. Search sector assigned *San Francisco* FC was 315 to 040 True. It should be noted that radar contacts made were initially in the neighborhood of 300 True or slightly outside the *San Francisco* sector.

SC was kept manned, warmed up, and ready for instant use and was used after commence firing. The reason SC radar was not used was because of highly secret intelligence received that Japanese had radar receivers covering the frequency band employed by the SC. It was assumed that Japanese units might possibly obtain radar bearings from the emissions of our SC radar and the valuable element of surprise might thereby be lost.

. . . This ship was never able to obtain a regular TBS extension to Flag Plot, and ship's force improvised one when ship became flagship in early September. This improvised equipment gave good performance insofar as transmissions to other ships were concerned, but transmissions from our own flag plot could be barely heard on our own navigation bridge. This TBS extension (flag plot) went out of action during the firing and stayed out until repairs could be effected many hours later. The regular installations in radio central and on navigation bridge were unaffected. The need for a regular TBS extension is most pronounced and must be undertaken at the earliest opportunity.

The TBS receiver should be parallel by an RBK receiver so that the failure of the TBS receiver will not leave the officer in tactical command completely deaf to the TBS transmissions of his force. This ship previously had an RBK receiver and it was removed by order of CinCPac. Subsequent efforts to obtain an RBK receiver in anticipation of just such a situation were unavailing.

It might be well to duplicate the TBS installation in ships designed as flagships with another complete TBS installation. The value of TBS in a night encounter cannot be overemphasized and the failure of TBS communication will leave an officer in tactical command almost incapable of exercising tactical command over his units.

This ship had obtained and installed an inter-office communication system ('Talk-Back') between Radio Central, Flag Plot, Nav Bridge, Conning Tower, and Control Forward. When TBS extension in Flag Plot went out, the 'Talk-Back' proved most valuable. Messages to and from flag plot were relayed over this system. TBS transmissions were made from Radio Central on instructions over the 'Talk-Back'. This non-standard piece of equipment proved of great value. A Hallicrafter 'sky buddy' on the aircraft frequency was of great value.

This ship obtained a TBM-7 transmitter during availability at Mare Island in May, but remote control panels to stations such as flag plot, conning tower, nav bridge, plot, radio central, and Radio II, could not be obtained and installed. This transmitter was located in Radio III in accordance with existing installation plans. The ship's force obtained cable and, using a spare TBS handset, improvised an extension to radio central; while the performance of the transmitter itself was excellent, the improvised extension causes transmitter to squeal during voice transmission and planes have reported some difficulty in reception . . .

The installation of remote control panels for this transmitter in the places enumerated is important. It would have been distinctly advantageous for the Admiral or Captain or Staff Duty Officer to have talked directly to the pilot of our plane without having to resort to the delays and possibility of misunderstanding by relaying messages over interior communication circuits.

This transmitter is our only voice transmitter and therefore of more value to the ship and Flag than any other transmitter in the ship. The importance of voice in a night action and for communication with aircraft (particularly during night aircraft operations) cannot be overemphasized. Radio III is a poor location because [it is] below decks, [with] poor ventilation, the overheating which occurs during long periods of condition Zed, and the necessity for breaking watertight integrity to obtain access. This transmitter should be moved to Radio II. Little used low frequency transmitters should be put in Radio III in its stead.

Bridge radio was not used, and under existing circumstances would have been of little value.

Battery locker in Radio II was demolished and batteries broken by shaking up from own gunfire. Acid was sprayed on Radio II personnel but prompt first-aid prevented any burning.

Two FC radars rendered excellent performance and were

of inestimable value. FD radars had they been installed would have been invaluable to the AA battery for night firing at surface targets. These should be installed as soon as practicable. The SC radar was not used for search but performed well when cut in at commence firing. The performance of the SC radar has never been equalled to that of the SC-1 radar of other cruisers in company. SC should be replaced by an SC-1 or later model. In the field of radar the need for an SG with indicator in flag plot, nav bridge, and also in control forward was most keenly felt. In the early stages of the action the flag and commanding officers experienced the greatest concern over the location of our own destroyers. Fire was checked at a crucial time because it was believed that destroyer targets might be own instead of enemy. Had the flag and commanding officers been able to look into the screen of an SG and see our disposition laid out before them, there could have been no question whatsoever of the hostile nature of the destroyers on starboard beam. The frequency band employed by the SG is outside of the band employed by known Japanese receivers, and can be used at night without fear of enemy taking bearings on its emanation, at least until intelligence indicates otherwise. The installation of SG radar must be undertaken at the earliest possible moment.

Commenting on this action, Admiral Halsey, then Commander South Pacific Area and South Pacific Forces, wrote that

The value of radar in obtaining surprise was amply demonstrated in the early stages of this action when enemy destroyers were sunk without having fired a shot . . . The lack of FD and SG radars on the *San Francisco* seriously impairs her fighting efficiency. The installation of these radars . . . during her present Navy Yard overhaul is highly recommended.

Night Action 12–13 November 1942

There was no evidence that the Japanese employed fire control Radar. It is strongly recommended that future actions when possible be fought at ranges which preclude the use of searchlight by the enemy which he is very skillful in using. Usually one enemy ship will illuminate a ship while other ships concentrate fire on it . . .

Strongly recommend duplicate TBS installation of mainmast of large ships as well as the one on the foremast. We have come to depend so completely on this equipment that we are literally lost when we no longer have it.

Radio material personnel did excellent work in rerigging antennae, overhauling and returning to service various equipment, and during the afternoon of the 13th all radio equipment was back in service except that TBS transmitter could not be used due to the carrying away of large sections of transmission line up foremast . . .

Bombardment of Wake Island, 5 October 1943

The SG radar performed in its usual satisfactory manner though it was not needed for fire control purposes during the bombardment.

The SC-2 radar, red antenna, was handicapped by severe interference, caused by BK, BL emissions. Fighter direction was thus made impracticable although many reported bogies were identified by the SC-2 and so reported. Close liaison was maintained between our fighter director and fighter director on the *Minneapolis* and *New Orleans*; due to intermittent failures of the VHF it was frequently necessary for messages to be relayed to and from the CAP by someone other than the fighter director actually controlling – thus reasonably effective fighter direction was accomplished. *Independence* VF shot down one of three 'Zekes' 30 miles south of base and one lone 'Betty' 18 miles southeast of base.

Both FC radars and one FD radar were put out of commission due to shock of main battery fire. The main power transfer switch in the after transmitter room was shaken from the bulkhead on the first salvo. Subsequently both FC and FD regulator rectifiers were jarred to such an extent that wiring was broken, mounting legs sheared and tubes smashed. The power transfer switch is supposedly shock mounted. Additional steps must and will be taken to relocate and remount these instruments. Similar trouble has been experienced in the past but remedial action taken has not been sufficient.

The FD radar antennas are mounted on top of the AA directors in such a manner that the control officer's view of aircraft is obstructed except at low position angles. Further, these antennae are mounted on bracing installed inside the director (Mark 33). The director was not designed for the additional structure inside it and personnel inside the director are too crowded for efficient operation.

As a result of experience gained in this action it is recommended that all SC series radar with red antennae be converted to higher frequency bands. The red band SC radar definitely does trigger Mark III IFF. Since the antenna radiation pattern has strong minor side and back lobes, the interference is received over a wide sector and sometimes through out the entire 360 degrees. This interference during the Wake Island raid made it very difficult to detect small echoes and to determine bearings. It was serious enough to render Fighter Direction impracticable.

It is also recommended that some policy be formulated for BL frequencies. The BL receiver had so many spurious signals that it was almost impossible to identify aircraft. This condition is made even more critical by the non-directional BL antennae. With as many as one hundred or more planes in the air the present installation in this ship is of little use. The *San Francisco* was dependent on reports over the fighter director and warning circuits for most of the information on air targets.

The advantage of discarding the SC red band in favor of the higher bands would far outweigh the additional interference between SC radars. If ships could have taken BL frequencies staggered over the 30 megacycle band of the Mark III IFF systems, it would be much more efficient.

The TBS transmitting and receiving equipment performed well, however the circuit was badly jammed due to all of the ships in the task force being on the same frequency, and, in cases, transmitters were badly adjusted, causing complete blocking of the circuit, resulting in garbled transmissions and repeated requests for acknowledgements and repetitions.

It is suggested that each task group be allocated a high

frequency channel with a common channel assigned task group and force commanders. This solution would require the installation of additional TBS or similar high frequency equipment. All such equipment should be installed, thoroughly tested and adjusted before leaving port.

Capture of Marshall Islands, 30 January–5 February 1944

A number of casualties to equipment occurred . . . On D–1 day the starter box and motor generator unit of the SC-2 radar were put out of commission by shock from own gun fire. On D Day the search radar antenna ceased rotating for a period of time for an undetermined cause, presumably associated with shock . . . There were no casualties in the four fire control radars (two Mark 3 and two Mark 4) nor was there any failure detected in the IFF equipment.

This operation, as those previous, stressed the fact that the output in the form of useful information from CIC is dependent to a considerable degree upon the amount of the information available. Activities furnishing plans, orders, charts, etc to forces afloat might well consider the demands of CIC in distribution of operational data. Intelligent functioning of CIC in an operation can only take place when there is complete knowledge of the operation plans. When only two copies of the plan are made available to a ship the multiple demands of the CO, Executive Officer, Navigator and Gunnery Officer make it difficult for CIC to have available the material which it urgently needs.

Being a relatively new activity, knowledge of the functions performed by CIC is not as broad as it might be. It was brought out during the operation that at one time while the turrets were in local control their rangefinders were useless at short range and they were, therefore, without range data. As CIC regularly maintains a DRT track such data is generally available instantaneously; had the turrets made use of the information available, some advantage made have accrued . . .

It is recommended that improvement and repair of the radar equipment be provided for at an early date. The failures mentioned above are due, in a considerable degree, to improper mounting . . . The ship is also under the handicap of having a single SG radar and that one improperly installed produces a 'blind' sector astern of over 20 degrees; air search efficiency would be greatly inscreased with installation of the improved SK search equipment to replace the present SC-2. Our IFF equipment (BL) is non-directional which limits ability to identify . . .

The radio tower (210 feet) on Carlson Island, Kwajalein Atoll, furnishes an excellent landmark for both visual and radar ranges and bearings. In moving about Kwajalein lagoon at night to take bombardment stations, primary reliance was on radar ranges and bearings on this tower, using FC and FD radar.

Mindoro Landing, January 1945

The *San Francisco* has recently had a new CIC installed which is believed to be the most up-to-date cruiser CIC now in the fleet. It served both the ship and the flag continuously during the operation.

Although not productive of important direct results, the services performed by CIC were of a comprehensive nature. The ship had the 'Charlie' (long range surface guard) from December 30 to January 7; the 'Able' guard (long range air search) on January 1 – during the temporary absence of the *Massachusetts* – and from January 7 to January 20; the 'Baker' guard (medium range air search) from January 20 to the end of the operation. Throughout the operation this ship had the 'King' RCM assignment (40–80 mcs band) and was prepared to jam on the 150–160 mcs with the TDY equipment.

Surface plots kept in CIC enable ship control stations and flag readily to follow movements of friendly forces – no inconsiderable task, considering the number of units involved. CIC was able on one occasion to identify and report to Task Group Commander a unit of DE's whose movements were puzzling to several ships on the task group.

CIC's task requiring the greatest amount of skill and effort and presumably productive of most important results is that of contributing to the air picture while operating as either a long range or short range radar guard ship. Despite the excellence and number of air plots maintained by the carriers and the advantage of SM radar, there still is an important function to be performed by heavy vessels. Search radars have different fade zones and varying efficiency on different bearings which account for the fact that some radars will have echoes on planes where others do not. Appreciating this fact, CIC at all times kept up an air plot which was composed of all bogey and friendly tracks as received from own radar, Task Group IDF (inter-fighter director circuit) and Inter Task Group circuit (2096 kcs). With this information in hand CIC officers were enabled on numerous occasions to identify as friendly contacts reported as bogies thus lightening the load of the Task Group FDO charged with the interceptions of suspicious contacts. Moreover, it placed this vessel in a position to take over fighter direction if ordered.

This vessel was given no assignment of visual fighter direction (control of Jacks) although we were prepared to take over at any time. The position of this ship in the formation made the *San Francisco* a last choice for such assignment. The circuit controlling the Jacks was constantly monitored and CIC kept the standby visual FDO informed of the air picture.

The operation afforded this vessel an opportunity to become familiar with and test out its RCM radar gear. Operators have improved in ability to detect enemy radar activity and numerous reports were made while in enemy territory. On one occasion – during the presence of a night snooper – the strength of the enemy radar signals coresponded almost directly with the distance of the plane from the ship as shown on SK radar. At the time when RCM watch (located in Aux CIC) reported 'signals disappeared' the plane was lost on the radar screen at about 100 miles.

CIC was used extensively as an intelligence center for other activities during this operation. A copy of the Operation Plan was made available in CIC and consulted frequently by Bridge, Gunnery, Aviators, etc. A strategic chart is kept in CIC as well as navigation charts of operating areas and officers concerned are able to identify places referred to in the operation plans. This was particularly helpful to aviators standing by for air-sea rescue missions. The reference points and locations of lifeguard submarines were kept continuously plotted on the strategic plot. On one occasion when a request for a rescue plane was made over the voice circuit to the TG Commander, the location of the downed plane with respect to the reference point, lifeguard submarine and ship's position was made available to Bridge within about three minutes . . .

Getting underway from an unlighted and very crowded anchorage such as Ulithi before dawn may frequently be a strategic necessity. It is, however, hazardous and may become costly . . . radar navigation in a crowded anchorage such as Ulithi is well nigh impossible because of the saturation of the scope with so many pips so close at hand. Small islands and marker beacons are lost in the maze or hidden in the radar shadow of a large ship.

Long range surface radar guard is normally given to a heavy ship, within the task group formation. During the past operation the *San Francisco* had this guard while stationed in the center of the formation *seven thousand* yards from the screen. This reduced the effective range of the maximum search ahead by at least 25%. It is recommended that the long range surface search be given to a destroyer in the van especially so designated (or as might be the case, the CLAA in the screen) or at least to a heavy ship in the van along the line of advance.

TBS-2 was normally used for administrative traffic. Occasionally, however, important operations traffic was sent on this circuit. Circuit discipline was bad, and distance often so great that unusual delays occurred. In two instances delays of more than two hours occurred in the delivery of OP traffic. It is not believed, therefore, that TBS-2 is a suitable circuit for a TF common and it is recommended that a more suitable circuit be used whenever the condition of radio silence will permit.

Although the discipline on TBS-1 has improved slightly in recent weeks, conditions were still bad. It is believed that too much administrative traffic – which could be sent by other means – was sent on TBS-1.

It is further recommended that a restricted channel be selected for encryption of information for which wide distribution is desired, thus eliminating the practice of labeling unencrypted visual relays SECRET or CONFIDENTIAL.

Iwo Jima Operation, February 1945

Every opportunity is taken to track planes by radar. Frequent drills have increased the ability of the Mk 28 radars to pick up targets by designation from CIC. In CIC one man manning the 41JS phones and one manning the 42JS phones directly to the forward and after AA directors, pointers, and trainers give each director designation. Experience has shown that when target bearing changes rapidly a three degree lead is given the directors. This method has proven satisfactory. Once the radar is on target is quite easy to stay on unless target bearing line crosses ships and then sometimes the target is lost momentarily because of the additional pip on the screen. This is more apparent when surface target is about the same range as air target. With training the above condition has been remedied to a great extent. Frequent checks are made to insure that the Mk 28 radars are properly aligned.

During the air attack on the formation on the evening of 21 February, CIC coached the Mk 28 directors on one plane approaching ship on a free bearing. It was noted that at two different times when the enemy plane was approaching our formation some kind of deceptive device was dropped from the plane. From the comparatively short time that the echoes appeared on the screen, it is believed that this was not window but rather some device with a tuned dipole or single-element reflectors. The echoes were solid, narrow, and steady, coming in strong, for only about two minutes – then fading completely. No echoes from these reflectors or dipoles appeared on either the Mk 28 or SG scopes. This deception presented no difficulty as the stationary position showed clearly that the echoes was not planes and their quick disappearance eliminated any difficulty in discriminating between plane and deceptive echoes.

Shackle Cipher. While recognizing that we should not present pertinent information of ship movements in plain language to the enemy, we should not be led astray with the false security offered by the shackle cipher. The present practice of shackling every course and speed change – regardless of the proximity of the enemy and the length of time the ships concerned will be at that course and speed – indicates lack of appreciation of reality. The continuous manoeuvers of a carrier task group engaged in launching and landing aircraft require ever changing courses and speeds of relatively short duration. There is no value in trying to deceive the enemy under such conditions – even if we assume he catches out every transmission. The practice of shackling such courses and speeds is not only valueless; it is in fact dangerous. In rapid task group maneuvers it is imperative that each ship know with certainty the exact course and speed signaled – yet we make this obscure by encipherment and so require a continual check and double check on each transmission to avoid error.

This pernicious habit of shackling everything reaches its nadir when two US ships already identified to each other encipher their respective bearings and ranges from each other – or when Zigzag Plan Six is shackled.

But to approach the question of the use of the shackle cipher from another point of view, are we not being entirely inconsistent when we assumed the enemy is *not* listening to our TBS and broadcast the morrow's operations in detail, giving times of strikes, launching points, targets, etc? Certainly if TBS is sufficiently secure for such a secret transmission, then it is secure enough for plain language transmissions of most tactical signals.

Task Group VHF. The new task group VHF recently installed and described by the task group communication officer as an 'informal' circuit is an aid and in inter-task group communications and certainly removes a load from the already overcrowded TBS-1 circuit. However, it must be clearly recognized that this circuit cannot and must not be allowed to become a tactical circuit. The OOD and Signal Officer are at present already strained to catch TBS transmissions concerning the ship (it is not considered safe to delegate this responsibility for tactical signals to a radioman) and it is physically impossible for the OOD and Signal Officer to listen to any other circuit at the same time.

Circuit Discipline. Generally speaking, circuit discipline on voice circuits, particularly TBS-1, has improved, a condition for which all of us are thankful. There still remain, however, certain bad habits which hamper communication seriously. The incessant 'How do you hear me . . . over?' on certain circuits almost precludes their use for other messages. Again, particularly on TBS-1, certain operators speak entirely too fast while transmitting important traffic. Task group commanders should insure that tactical signals are transmitted slowly and clearly. And again, it is believed that the now universal practice of requiring *two* acknowledgements for all tactical signals is unnecessary and leads to unjustified confusion, and to additional traffic on an already over-burdened circuit. A dozen acknowledgements will not insure that the thirteenth ship has received the message.

Okinawa Operation, March–April 1945

Radar fire control was used in all night AA actions and radar ranges used when target could be seen visually. Constant drilling has decreased the time in shifting from search radar to Mk 28. CIC did a good job in keeping the ship's position plotted accurately using radar, so that very little time was lost when asked to deliver fire either by Shore Fire Control Party or plane observers.

Up to L Day there were a limited number of destroyers in the radar picket stations. When this ship was not land locked, we beleive we gave considerable assistance to the FD ship by reporting bogie plots over the IFD circuit, 2096 kcs. Several low flying enemy planes were never picked up on the search gear but were first seen visually; they are believed to have taken off from fields on Okinawa . . . it should be noted that at no time was this ship able to track bogies over land masses, and from the tracks plotted from IFD reports no other ship had much success either.

Starting with L Day scattered heavy aerial attacks were attempted by the Japanese. The usual time of day varied, but the heavier strikes occurred when the day CAP could be used. The *Eldorado* was the area control ship, and enough radar pickets were on station to give adequate warning. A most excellent job was done by the pickets in controlling CAP and in breaking up raids. At times when a great number of enemy planes were in the area, the IFD channel was overcrowded and some confusion resulted. On the whole however great credit is due those in charge of air defense in the objective area who, through the IFD net, put out a brand of radar telling which enabled all ships to be fully cognizant of the air situation even though many ship's screens were blocked out by land echoes. Without this information the CIC on this vessel – because of limitation of SK performance due to land echoes – would have been unable to keep the CO and Flag advised of information requisite to maneuver the ship or task unit.

During air attacks where enemy planes were attacking this vessel or were close in no difficulty was encountered in coaching the fire control radars on target when we had them on SK screen. The ship's doctrine of separate designators at the SK scope sending ranges and bearings directly to the pointer, trainer, and range reader of each director over a separate voice hook-up (41JS and 41-100JS forward and 42JS and 42-100JS aft) seemed to work satisfactorily. It may well be that the optimum arrangement under continuous air attack is to have the condition watches handle the first wave of an attack (this is what always occurs during sudden, close-in attacks).

Throughout the operation the SK and BL were excellent up to their capabilities under fairly constant use. The need of installation of an SP radar became apparent during the operation. In order to defend the ship it was necessary to keep the SK constantly on the 20 mile scale in order to coach the Mk 28 fire control radars on bogies within that range. We were unable to keep a continuous picture of the raids developing outside of that range except through the medium of plots received via the IFD circuit. If we had been called on to handle interceptions, there would have been serious conflict in the performance of both duties. Another difficulty would have arisen if we had been called on to handle interceptions, namely that the VHF positions in CIC were taken up by gunfire frequencies and the IFD channel. Furthermore the SK radar did not give sufficiently accurate altitude determination due to the many minor fade zones on either side of the major zones – this was noted at altitudes except between 200 and 2000 feet and 7000 and 15,000 feet . . .

There were instances of interference on at least two Naval Gunfire Support circuits by the enemy. The ship received at least two call-ups by an unknown party and attempts were made each time either to gain information or to interrupt transmissions. On channel 3905 kcs there was noted much enemy traffic of some kind, but it is not known whether this was an intent to jam the circuit. All transmissions made by the ship on this circuit were uninterrupted by this interference. From this it is thought that there was no intentional jamming of the circuit by the enemy since they probably used it as an operational channel of their own.

This ship was assigned the intercept guard from 300 to 1000 mcs while operating with TF 54 . . . no interceptions were made . . . While operating with fire support units, the intercept equipment was used to check enemy radars reported by the pickets. Many were confirmed, but none were new enemy frequencies . . .

No electronic jamming was done by this ship. The TDY was set to jam on the current airborne frequencies (140–190 mcs). A Japanese Mk CHI radar at 72 mcs caused a great deal of interference on TBS primary during the first three weeks of the operation.

On numerous occasions the Japanese employed window while making attacks on our shipping. They never used a sufficient quantity to make its presence troublesome. Also, it was easy to detect. They also make use of a reflector which would remain in the air from thirty to forty-five minutes. This would give a large single pip on the SK and would have a course and speed comparable with that of the wind. Several plots are needed to distinguish this from an orbiting plane. The echo received was comparable to that of a large plane . . .

There was no noticeable improvement in the use of the Shackle Cipher. An example of improper use was the following intercepted TBS transmission: 'Instructions for tomorrow (shackle-date) canceled'. It is obvious that the shackling of the date was unnecessary since 'tomorrow' was sufficient; and furthermore, such usage was a compromise of the shackle cipher. It is believed that greater care and discretion in the use of this cipher is still needed.

The introduction of two TBS channels (75.7 mcs Primary and 72.1 mcs Secondary) for Gunfire Support Ships aided greatly in reducing congestion on the Primary Tactical Channel Strict supervision by Task Force commanders on the TBS Primary Tactical Circuit, with a view to eliminating all administrative traffic on this net, will be a further improvement. It is suggested that as means to this end the task force commander monitor the TBS primary and submit a discrepancy report to the offending commands.

It was an excellent plan to assign each task force a Gunfire common circuit. Good administrative control was exercised, and the definite assignment to ships of a Naval Liaison Officer and a Shore Fire Control Party, with frequencies to be used, provided effective communications. Occasionally there was some difficulty in establishing contact, when the assigned NLO did not answer up or a different one answered, leaving some doubt as to whether friend or foe. At first the NLO and SFCP did not seem familiar with the shackle card or did not have the effective one. These difficulties, however, were not of serious consequence, and the gunfire support communications were in general entirely satisfactory.

III. Glossary

ADT Automatic Detection and Tracking. The automatic decision as to whether a given return is a target, and the association of such detections to form target tracks.
AEW Airborne Early Warning
Aided Tracking A means of tracking a target so that tracking will persist despite jamming, assuming constant target course and speed.
ATD Automatic Target Dection. An automated decision as to whether a given return is a target.
Bandwidth The range of frequencies over which a radar operates; for example, a radar receiving signals between 199 and 201mc/s has a bandwidth of 2mc/s. In the case of a filter, *band-pass* indicates the range of frequencies the filter will accept.
Barrage-Jamming The simultaneous jamming of a number of adjacent channels of frequencies.
Beamwidth The sector (in angle) over which an antenna radiates. Usually it is measured between the angles at which radiated power is halved.
Blip Enhancement An ECM technique which augments the return from small target to make it appear to be a much larger one; alternatively, an ECCM measure to enhance a real target in the presence of jamming or clutter.
Blip–Scan Ratio A measure of the number of scans needed to produce a recognizable blip, actually its inverse, so that the higher the ratio, the fewer scans per target are required.
Burn-Through ECCM method: power is added until the radar can overcome the noise level imposed by the jammer.
CFAR Constant False Alarm Rate. A method of automatic target detection in which the radar accepts only targets showing sufficiently above the level of noise surrounding them, so that it keeps its false alarm rate constant.
Chaff Strips of aluminum foil dropped to form false radar targets, as a countermeasure. Its wartime code name was 'Window'.
Chirp A signal varying, usually either up or down, in frequency over time. Such variation is often described as an FM (frequency-modulation) ramp or slide.
Clutter Unwanted signals, echoes or images on a radar display which interfere with the desired signal.
Coasting An ECCM technique in which the system continues to project forward target motion at the last known rate, in the presence of intense jamming.
Coaxial A cable consisting of a central conductor and a conducting tube, separated by insulation, useful at the lower end of the radar frequency range.
Coherence The maintenance of a stable relationship from pulse to pulse over multiple pulses. Examples include phase coherence, *ie* the internal structure of successive pulses is kept constant so that their echoes can be compared in detail. Pulse-coding requires a high degree of coherence in most cases.
Conical Scan The process of swinging the radar beam about an axis a few degrees off the center of the beam so that the beam describes a cone in space, the apex of which is at the antenna. It is used for tracking, the variation of target signal with scan angle indicating target motion relative to the radar.
Cosecant-squared A beam shape which maintains uniform coverage over a wide variety of ranges for aircraft flying at a limiting altitude.
CW Continuous Wave, *ie* non-pulsing, operation. In addition, Morse Code is often described as CW radio transmission; in it, the CW radio wave is interrupted to form dots and dashes.
Data Rate The rate at which the radar renews its store of data. In a rotating radar, it is the inverse of the scanning rotation rate, so that a 5rpm radar has a data rate of once every 12 seconds, which is often corrupted to a data rate of '12 seconds'.
Dicke Fix An ECCM technique against swept-frequency jamming (wideband limiting). Widely used in current US naval radars.
Dipole An antenna consisting of a single wire, generally fed at its center, and generally half a wavelenth long. It forms the basis of many antenna systems, radiating in all directions about the dipole's lengthwise axis.
Doppler Effect The shift of frequency with the speed of a source. For example, signals reflected from a moving object are shifted in frequency corresponding to its speed. The shift in frequency is proportional to speed.
Ducting A mode of radar propagation in which signals are trapped as they are bent by atmospheric refraction; they partially follow the earth's curvature far beyond normal radar range.
Duplexer A TR box.
Duty Cycle The cycle of operation of the radar, often referred to the percentage of the time during which it transmits.
ECCM Electronic Counter-Countermeasures, *ie* the protection of radar (or radio) against ECM.
ECM Electronic Countermeasures, *ie* attempts to interfere directly with radar or radio operation.
Eclipsing Loss In a high-PRF radar, the loss in signal return due to the action of the duplexer, *ie* the receiver is turned off while a pulse returns.

EMCON Emission Control – totally shutting down electronic emissions so as to deny information to an enemy.
ESM Electronic Support Measures (sometimes referred to as passive ECM). A means of intercepting enemy emissions so as to gain intelligence. Intercept receivers are employed for ESM.
Fade Loss of radar signal; in discussions here, generally due to interference between different reflection paths. Angles at which signals fade are also called nulls.
FM Ramp/Slide Frequency-modulation ramp/slide, meaning a systematic variation of frequency up or down. A linear variation in frequency (*ie* a variation proportional to time) is often imposed on the pulses of a doppler radar to permit some ranging, as frequency can then be used to differentiate successive pulses, avoiding the STA problem.
Frequency Agility The ability of a radar to change its operating frequency. An ECCM feature. Frequency can generally be changed on a pulse-to-pulse basis.
Frequency Diversity The provision of a wide range of operating frequencies, any one of which may be used. It refers to the tunability of the radar, not to its ability to change frequencies while operating. Generally it is a counter to finely tuned enemy ECM devices.
Frequency Hopping The sequential use of different discrete frequencies (ECCM).
FRESCAN Frequency Scanning, a system in which changes in frequency (*eg* from pulse to pulse or within a pulse) move the beam.
Gain The rating of radar output power compared to the output of a simple non-directional dipole. It is usually expressed in decibels (db), which are logarithmic, but can be expressed as a simple number.
Gate-Stealing An ECM technique for self-protection to overcome automatic tracking. Signals larger than the echoes from the tracking victim are returned, gradually attracting the attention of the tracker, and moving the tracking gate away from the victim. After a specified walk-off the jammer is turned off, and the tracking radar must search once more for its target.
Gating A technique of limiting radar observation to a desired interval of range or bearing, partly to overcome noise.
IF Intermediate Frequency, the fixed frequency to which the carrier wave is converted in the radar after reception, for subsequent processing. Operation at several alternative IF frequencies and bandwidths can be an ECCM measure.
IFF Identification Friend or Foe, a system (usually called secondary radar) permitting positive identification of friendly objects under surveillance.
Interrogator In an IFF system, the unit which transmits signals asking targets to identify themselves. It is answered by a transponder.
Jitter The variation in PRF as an ECCM feature.
Kilocycle Thousands of cycles (kc or kc/s) per second.
Klystron A type of radar power tube, useful because of its stability from pulse to pulse.
Lobe-Switching (Lobing) Switching a radar beam back and forth between two or more positions (lobes). Comparison between the returns in the two directions permits the beam to be pointed more accurately at a target than might be possible given its inherent beamwidth.
Lockbreaking ECM against a tracking radar.
Look-Through When jamming, a technique whereby the jamming emission is turned off irregularly for extremely short intervals to permit the victim signal to be monitored.
LORO Lobe On Receive Only, a form of lobing in which a constant signal is sent, but the receiver varies its beam form for tracking. The constancy of the transmitted signal makes ECM more difficult.
LPI Low Probability of Intercept, referring to a radar or radio system specifically designed to avoid detection by ESM systems. Radars of this type are also referred to as 'quiet'.
Magnetron A type of radar power generator, the first available at high power in microwave frequencies, but characterized by its instability, *ie* its inability to maintain coherence between successive pulses.
Matched Filter A device for separating a waveform of known shape from the surrounding noise.
Megacycle Millions of cycles per second (mc or mc/s)
Microsecond A millionth of a second.
Microwave Wavelengths of 20cm or less, characterized by the use of magetrons or klystrons, and generally requiring waveguides.
Monopulse A tracking mechanism employing two or more feeds and a single reflector, so that multiple beams with a small angular displacement between them are produced. Differences in their returns indicate the extent to which the target is off the radar axis, and can be used in tracking.
MTI Moving Target Indicator, a means of eliminating all stationary targets from the radar screen.
Multipath Transmission and reception simultaneously via a number of different physical paths, due to reflection or refraction; the signals thus sent interfere at the receiver, either constructively (*ie* they add) or destructively. Multipath transmission causes fading in radars over water. In radio signals, it may limit the extent to which a signal can be compressed, as a very short signal may arrive garbled by it.
Narrow-Band Limiting The use of a filter which passes only a narrow frequency band and thus reduces the effect of extended targets such as chaff and clutter, followed by a limiter which clips all signals and noise to a common level. It is useful in a pulse-compression radar which can then extract the signal from the surrounding noise.
Nodding Beam A method of height-finding in which the radar beam itself 'nods', *ie* moves up and down

through possible elevation angles.
Null A direction from an antenna in which the return is equal to, or nearly equal to, zero, generally due to multipath interference.
Nutating Feed An antenna feed used to produce a conical scan without rotating the polarity of the signal. It is similar to the nodding motion of a top.
Peak Power The maximum power level of a radar pulse, generally measured in kilowatts (thousands of watts, kW) or in megawatts (millions of watts, MW).
Pencil Beam A narrow conical beam, generally with equal horizontal and vertical beamwidths.
Phase The position at any instant which a periodic wave occupies in its cycle, often measured in degrees (with a 360° period). Two waves 180° apart cancel completely when they are added; at 360° they add in phase. Thus waves out of phase by odd multiples of 180° cancel, and at even multiples add. Phase interference is the basis of the fading phenomenon.
Polarization The direction of the electric field in a radio or radar wave; it can be vertical, horizontal, or a mixture (sometimes varying) of the two. Polarity is often used to distinguish between separate signals sent by a single antenna.
PRF Pulse Repetition Frequency, the rate at which pulses repeat, generally given in terms of pulses per second (pps). PRF defines the maximum unambiguous range of a conventional radar.
Pulse-Compression The use of long coded pulses to gain high average energy and good resolution without excessive peak power.
Pulse-Doppler A radar technique for moving target direction, using the doppler shift of returning echoes.
Pulse Length The time duration of a single radar pulse, generally measured in microseconds. It is also referred to as *pulsewidth*, or as pulse *duration*.
Range Gate Pulloff (RGPO) An anti-tracking ECM technique. See gate-stealing.
Resolution The ability to distinguish two nearby targets. Range resolution is the minimum distance between two distinguishable targets at the same bearing.
Repeater A deception device (ECM) which intercepts incoming pulses, modifies them (*eg* by delaying and amplifying) and then retransmits them to present erroneous information.
Search Receiver A calibrated ESM receiver which can be tuned over an unusually wide frequency range to detect and measure enemy signals.
Sector Scan Radar scanning over a limited sector of the total possible scan range, *eg* over 30° of a possible 360°.
Side Lobe A portion of the radar beam other than that from the main lobe, usually much smaller.
Spot-Jamming The jamming of a particular channel or frequency. The ECCM countermeasure is some form of channel or frequency switching.
STA Second Time Around, referring to pulses arriving back at the radar after the next pulse has been sent out, and therefore causing an ambiguity in the range of the signals they represent. In some very high PRF radars, such as pulse-doppler radars, successive signals can be modified (*eg* by changing their frequency or their phase) so that STA signals can be distinguished and the range limits associated with a high PRF overcome.
Stacked Beam A radar transmitting multiple beams covering different angular ranges, *eg* in height, so that comparisons of the returns on those beams allow height to be measured. Sometimes referred to as monopulse height-finding.
Sweep-Jamming Jamming by sweeping space with pulses of the same frequency as those received; continuous change in jammer frequency at an optimum rate to jam all or a portion of the jammer operating frequency range.
Sweep-Through Jammer transmitter that sweeps through a band and jams each frequency briefly, producing a sound like that of an aircraft engine.
TR Box The device which permits a radar to use the same antenna for transmission and reception, turning off the reception channel during pulse transmission.
Transponder In IFF and other secondary radars, the device which receives the interrogator signal and automatically replies.
TVM Track-Via-Missile, a missile control system whereby all information received by the missile antenna(s) is transmitted to a mother radar system for data-processing, and then used to generate missile control commands. The missile is, then, no more than a remote antenna, auto-pilot, and data link transmitter-/receiver.
TWS Track-While-Scan, a capability to track one target while searching for others.
TWT Travelling Wave Tube, an electron tube in which a beam of electrons interacts continuously with a guided electromagnetic wave to produce amplification at microwave frequencies.
Velocity Gate Pulloff (VGPO) An antitracking ECM technique, analogous to RGPO, used against a radar tracking in target velocity.
Waveguide A hollow pipe, usually of rectangular cross-section, used to conduct radar signals.
Wideband Limiter An ECCM technique applied to the intermediate frequency to overcome swept-frequency jamming which otherwise can affect IF operation. The IF bandwidth is limited to exclude such noise.
Yagi A type of directional antenna consisting of a driven half-wave dipole, a reflecting dipole, and one or more 'directors', the dipoles being in the direction of transmission.

IV. Source Material

American Radio Relay League: *ARRL Antenna Book* (Newington, Conn, 1974)

Bagnasco, E: *Le Armi Delle Navi Italiane Nella Seconda Guerra Mondiale* (Parma, 1978)

Barton, D K: 'Real World Radar Technology', IEEE International Radar Conference (1975)

Bittencourt, R P: *Electronic Warfare Technology* (Thesis: US Naval Postgraduate School, July 1976)

Boyd, J A *et al: Electronic Countermeasures* (Handbook prepared by the Institute of Science and Technology of the University of Michigan for the US Army, 1961, and later declassified)

Brandt, L: *Für Geschichte der Radartechnik in Deutschland und Grossbrittanien* (XV Convegno delle Communicazioni, Genoa, 1967)

Brookner, E: *Radar Technology* (Dedham, Mass, 1977)

Corse, C D: *Introduction to Shipboard Weapons* (Annapolis, 1975)

Croney, J: 'Ship-Borne Radar Antennas – A Survey', *IEEE Conference Publication 105: Radar - Present and Future* (London, October 1973)

Eustace, H F (ed): *The International Countermeasures Handbook* (6th Edn, 1980–81: Palo Alto, 1980)

Gebhardt, L A: *Evolution of Naval Radio-Electronics and Contributions of the Naval Research Laboratory* (NRL Report, 8300, Washington, 1979)

Guerlac, H E: *Radar in World War II* (US Office of Scientific Research and Development, Washington, 1945)

Hezlett, Sir A: *Electronics and Sea Power* (New York, 1976; published in Britain as *The Electron and Sea Power,* 1975)

Hill, R T: 'Trends in European Radar Tchnology; *Electro '76,* IEEE Professional Program (May 1976)

Howeth, L S: *History of Communications-Electronics in the United States Navy* (Washington, GPO, 1963)

Johnston, S L: *Radar Electronic Counter-Countermeasures* (Dedham, Mass, 1979)

Lenton, H T: *Warships of the British and Commonwealth Navies,* (1966 and 1969 Edns: London)

Mitchell, A: 'The Development of Radar In the Royal Navy 1935–45', *Warship* (1980).

——————: 'The Development of Radar in the Royal Navy, 1945–1960', *Warship* (1981)

Raven, A, and Roberts, J: *British Battleships of World War Two* (London, 1976)

——————: *British Cruisers of World War Two* (London, 1980)

Raytheon Co: *Electronic Progress* (4 issues devoted to detection and tracking systems, 1974–75)

Reference Data for Radio Engineers 5th Edn: H W Sams Co, New York, 1968)

Ridenour, L N (ed): *Radar System Engineering* (New York, 1947 and 1956)

Rohwer, J: *The Critical Convoy Battles of March 1943* (Annapolis, 1977)

Schlesinger, R J: *Principles of Electronic Warfare* (New York, 1961)

Sherwin, S A, and Miller, R T: 'The Impact of Electronics on Warship Design', *Transaction of the Society of Naval Architects and Marine Engineers* (1962)

Skolnick, M I: *Radar Handbook* (New York, 1970)

——————: 'The Role of the Antenna In Radar', *NRL Memorandum Report 2574* (Washington, 1973)

Taylor, D, and Westcott, C H: *Principles of Radar* (Cambridge, 1948)

Trenkle, F: *Die Deutschen Funkmessverfahren bis 1945* (Stuttgart, 1979)

US Army Air Force, Operations Analysis, AC/AS-3: *A Short Survey of Japanese Radar,* Vol III (Shipboard Radar) (Washington, November 1945)

US Bureau of Naval Personnel: *Naval Search and Fighter-Director Radar* (NAVPERS 10837, January 1949)

——————: *Supplementary Readings in Fundamental Naval Electronics* (NAVPERS 10809, March 1949)

——————: *Naval Electronics,* Vol III (October 1952)

——————: *Principles of Naval Ordnance and Gunnery* (NAVPERS 10783, 1959)

——————: *Shipboard Electronic Equipments* (NAVPERS 10794, 1963 and 1969 Edns)

——————: *Radarman 322* (NAVPERS 10146-F, *c* 1964)

US Bureau of Ordnance: *Armament Summary* (various dates, for fire control and radar system identification)

——————: *History of the Bureau of Ordnance in World War Two* (MS, Vol V, on fire control systems)

US Bureau of Ships: *Electronic Equipment System Design and Application* (A course offered in the Bureau of Ships during the spring of 1963: course notes)

US Combined Intelligence Objectives Sub-Committee: *CIOS Targets: Electronic Equipment Aboard German Naval Units* (Washington, 1946)

US Department of Defense: *Military Standardization Handbook: United States Radar Equipment* (MIL-HDBK-162B), Vol I (Unclassified volume; December 1973)

US Fleet, Headquarters of the C-in-C: *Capabilities and Limitations of Shipborne Radar (RADONE-A)* (July 1945)

————: *Description of Amphibious Force Flagship* (June, 1944)

————: *Battle Experience* (Washington, 1942–46)

————: *US Radar* (FTP-217, 1943)

US National Defense Research Committee: *US Radar Survey* (Washington, 1945)

————: *MARS* (Military Airborne Radar Systems) (Washington, 1946)

————: *Radio Countermeasures* (Washington, 1946)

————: *Radar: Summary Report and HARP Project* (Washington, 1946)

US Naval Academy, Department of Weapons and Systems Engineering: *Shipboard Weapon Systems* (Annapolis, *c* 1976)

US Naval General Board: *Hearings* (1948 and 1950)

US Naval Research Laboratory: 'Interim Report on Development of Model SPS-3 (XDK) Shipborne Hemispheric Search Radar' (August 1947)

————: 'Evaluation of the AN/SPS-3 (XB-1) Radar at the Chesapeake Bay Annex' (September 1956)

————: 'Preliminary Design of the AN/SPS-2(XDQ) Antenna' (February 1959)

————: 'A Study of 220- and 425-Megacycle Shipboard Radar for Long-Range Detection' (November 1956)

US Naval Technical Mission to Japan: *Electronics Target Reports* (Washington, 1946)

US Navy, Office of the Chief of Naval Operations: *CIC* magazine (published irregularly 1944–52, now declassified)

————: *Combat Readiness* magazine (quarterly, 1953–69; some issues now declassified)

————: *Presentation to WSEG on Fleet Air Defense as Applied to the Carrier Task Force* (Washington, August 1950)

US Navy, Operational Development Force: *Conduct a Fleet Operational Investigation of the SPS-39/39A Series III Field Changes* (29 January 1964)

————: *Evaluation of the AN/SPS-26 (XN-1) Radar Equipment* (26 April 1958)

US Navy, Operational Test and Evaluation Force: *AN/SPS-40 Air Search Radar Preliminary Information Pamphlet* (Norfolk, 1962)

————: *Concurrent Evaluation of AN/SPS-49(XN-1) and AN/SPS-50(XN-1)* (Norfolk, 1965 and 1966)

US Pacific Fleet: 'Report of a Board Convened by CINCPAC to Make Recommendations on Radar and Countermeasures on Experience in the War which Resulted in the Defeat of Japan'. (Horne Board, 3 December 1945)

————: 'Report . . . on Radio Equipment . . .' (Horne Bord, 29 November 1945)

Van Brunt, L B: *Applied ECM* (Dunn Loring, Va, 1980)

Wheeler, G J: *Radar Fundamentals* (Englewood Cliffs, 1967)

Much of the material used was taken from official sources, particularly those at the Naval Historical Center at the Washington Navy Yard and those of the US National Archives in Washington. In addition, considerable use was made of defense journals, whose technical articles were too numerous to list: *Aviation Week and Space Technology*, *Defence* (Eton), the *International Defense Review*, *Maritime Defence* (London), the *Proceedings* of the US Naval Institute, and *Ships of the World* (Tokyo). Articles included detailed descriptions of particular items of shipboard electronics as well as considerable reference photography. Under these heads should also be listed *Jane's Weapon Systems* and *Jane's Fighting Ships*, and *Combat Fleets*. However, the primary source for modern Western European radars was company-produced descriptive brochures, from which operating data have generally been taken.

Much of the US radar data was taken from unclassified electronics nomenclature cards produced at Fort Monmouth and held by, among others, NAVSEA Code 06G4D. I am grateful to Jim Glander of that office.

Index

This index covers radar and associated systems only. Page references in **bold** type indicate main catalog entries; those in *italics* denote illustrations.